INTRODUCTION TO VLSI SYSTEMS

CARVER MEAD
*Professor of Computer Science, Electrical Engineering,
and Applied Physics, California Institute of Technology*

LYNN CONWAY
*Research Fellow, and
Manager, VLSI System Design Area
Palo Alto Research Center, Xerox Corporation*

ADDISON-WESLEY PUBLISHING COMPANY

*Reading, Massachusetts • Menlo Park, California
London • Amsterdam • Don Mills, Ontario • Sydney*

This book is in the
Addison-Wesley Series in Computer Science

Consulting Editor
Michael A. Harrison

Library of Congress Cataloging in Publication Data

Mead, Carver A
  Introduction to VLSI systems.

  1.   Integrated circuits—Large scale integration.
2. Microcomputers.   3. Digital electronics.
4. Computer architecture.   I. Conway, Lynn A.,
joint author.   II. Title.
TK7874.M37   621.3819′535   78-74688
ISBN 0-201-04358-0

Second printing, October 1980

ISBN 0-201-04358-0
ABCDEFGHIJK-HA-89876543210

TO W. R. SUTHERLAND

PREFACE

As a result of improvements in fabrication technology, Large Scale Integrated (LSI) electronic circuitry has become so dense that a single silicon LSI chip may contain tens of thousands of transistors. Many LSI chips, such as microprocessors, now consist of multiple complex subsystems, and thus are really *integrated systems* rather than integrated circuits.

What we have seen so far is only the beginning. Achievable circuit density now doubles with each passing year or two. Physical principles indicate that transistors can be scaled down to less than 1/100th of their present area and still function as the sort of switching elements with which we can build digital systems. By the late 1980s it will be possible to fabricate chips containing millions of transistors. The devices and interconnections in such very large scale integrated (VLSI) systems will have linear dimensions smaller than the wavelength of visible light. New high-resolution lithographic techniques have already been demonstrated that will enable fabrication of such circuitry.

VLSI electronics presents a challenge, not only to those involved in the development of fabrication technology, but also to computer scientists and computer architects. The ways in which digital systems are structured, the procedures used to design them, the trade-offs between hardware and software, and the design of computational algorithms will all be greatly affected by the coming changes in integrated electronics. We believe this will be a major area of activity in computer science on through the 1980s.

Until recently the design of integrated circuitry has been the province of circuit and logic designers working within semiconductor firms. Computer architects have traditionally composed systems from standard integrated circuits designed and manufactured by these firms but have seldom participated in the specification and design of these circuits. Electrical Engineering and Computer Science (EE/CS) curricula reflect this tradition, with courses in device physics and integrated circuit design aimed at a different group of students than those interested in digital system architecture and computer science.

This text is written to fill a current gap in the literature and to introduce all EE/CS students to integrated system architecture and design. Combined with individual study in related research areas and participation in large system design projects, this text provides the basis for a graduate course-sequence in integrated systems. However, it is primarily intended for use in intensive undergraduate courses on the subject. The material can also be used to augment courses on computer architecture. We assume the reader's background contains the equivalent of introductory courses in computer science, electronic circuits, and digital design.

There have been major obstacles in the way of those seeking an overall understanding of integrated systems. Integrated electronics has developed in a heatedly competitive and often secretive business environment. There has been a proliferation of different device technologies, circuit design families, logic design techniques, maskmaking and wafer fabrication techniques, etc. Many of these technologies have sprung up from the grass roots of "Silicon Valley" in the San Francisco Bay Area of California, and thus many of the "experts" are located in that one region. Most workers in the industry have concentrated on narrow specialties. Separate integrated electronics cultures have independently evolved within many companies, and thus the terminology and practices of the specialties vary from company to company.

As a result of this background, texts on integrated electronics have tended to give detailed accounts of some very narrow horizontal segment of the overall subject, such as device physics or circuit design, and are often tied in subtle ways to some specific context, thus limiting their general applicability.

We have chosen instead to provide just enough essential information about devices, circuits, fabrication technology, logic design techniques, and system architecture to enable the reader to fully span the entire range of abstractions from the underlying physics to complete VLSI digital computer systems. A rather small set of key concepts is sufficient. Only by learning the essence of each topic, and by carrying along the least amount of mental baggage at each step, will the student emerge with a good overall understanding of the subject. This understanding can then be mapped into the reader's own space of application, technology, and technical culture.

The high rate of change of integrated electronics presents another obstacle: information often becomes obsolete very rapidly. The major force for obsolescence is the ongoing improvement in fabrication technology, leading to smaller and smaller devices as time passes and thus to a constant change in device characteristics. We attack this obstacle by stressing the effects of the scaling-down of device dimensions. Many of the coming changes in system architectural parameters are thus anticipated. The reader will learn what is common to systems composed of 6 μm, 2 μm, and 0.5 μm devices, and what is not.

While the material in this text is presented in a particular order, it need not be read in that order. Each chapter presents material from a distinct level in the hierarchy of disciplines involved in integrated systems. The material falls into four major groupings: Chapters 1 and 2 provide the basics of devices, circuits, and fabrication; Chapters 3 and 4 give the basics of system design and implementation; Chapters 5 and 6 present an example of LSI system design; topics of current research interest are discussed in Chapters 7, 8, and 9. We recommend that readers start in the chapter where they are most knowledgeable, and read until information is required from an adjacent area described in some other chapter. By using this algorithm and consulting the suggested references where necessary, readers can gradually work through the primary material of all chapters. Although much of the material in this text is previously unpublished, it nevertheless contains only basic concepts. However, these concepts cover quite a wide range of disciplines and are easily visualized only after the overall context becomes clear.

In any given technology, form follows function in a particular way. The most efficient first step towards understanding the architectural possibilities of a technology is the study of carefully selected existing designs. However, system architecture and design, like any art, can only be learned by doing. Carrying a small design from conception through to successful completion provides the confidence necessary to undertake larger designs. The space of possibilities unfolds only as the medium is worked. This book provides a set of selected design examples and also describes procedures for implementing one's own designs. Because of the density, speed, and topological properties of nMOS, and the easy access to nMOS wafer fabrication, that technology is used for our examples. The architectural skill of mapping function into form, when once acquired, can then be extended to other technologies.

The general availability of courses in VLSI system design at major universities marks the beginning of a new era in electronics. The rate of systems innovation using this remarkable technology need no longer be limited by the perceptions of a handful of semiconductor companies and large computer manufacturers. New metaphors for computation, new design methodologies, and an abundance of new application areas are already arising within the universities, within many system firms, and within a multitude of new small enterprises. There may never have been a greater opportunity for free enterprise than that presented by these circumstances.

An atmosphere of excitement and anticipation pervades this field. A growing community of workers from many backgrounds, computer scientists, electrical engineers, mathematicians, and physicists are collaborating on a common problem area that has not yet become classical. The territory is vast, and largely unexplored. The rewards are great for those who simply press forward.

Pasadena, California   C.M.
Belmont, California   L.C.
July 1979

BACKGROUND

This text has its origins in a series of courses in integrated circuit design given by Carver Mead at Caltech, beginning in 1970. Starting in 1971, students in these courses designed and debugged their own integrated circuits. The students undertook increasingly complex system designs, using only rather simple implementation aids. The structured design methodology presented in this text has evolved within this milieu. These early courses greatly benefited from interactions with friends in industry, particularly Robert Noyce, Gordon Moore, Frederico Faggin, Dov Frohman-Bentchkowsky, Ted Jenkins, and Joel Sorem.

A separate Computer Science activity was created at Caltech in 1976, with integrated systems as a focus. An early, informal association was formed with systems architects in industry, in particular with the then newly formed LSI Systems Area, led by Lynn Conway, at Xerox Palo Alto Research Center (PARC). The increased interaction of Caltech students and faculty with industrial researchers stimulated the research on both sides.

Work on this text began in August 1977. Collaborators from a number of universities and industrial firms joined in the enterprise. Prior to commercial publication, several limited printings were distributed to a selected group of universities as notes for courses on integrated systems. The first three chapters were used as course notes during the fall of 1977, in courses given by Carver Mead at Caltech and by Carlo Séquin at U.C. Berkeley. The first five chapters were used during the spring of 1978 in courses given by Ivan Sutherland and Amr Mohsen at Caltech, by Robert Sproull at Carnegie-Mellon University, by Dov Frohman-Bentchkowsky at Hebrew University, Jerusalem, and by Fred Rosenberger at Washington University, St. Louis. The third and final prepublication printing of all nine chapters, in the fall of 1978, was used in the courses at Caltech and U.C. Berkeley, and in new courses by Kent Smith at the University of Utah, and by Lynn Conway, while visiting at M.I.T.

The 1978 M.I.T. course provided a final test, prior to publication of this text, to confirm the transportability of the project-oriented form of the course (in which as much emphasis is placed on creative architectural activity as on formal analysis), and to also confirm the technical and economic feasibility of the remote-entry of student LSI designs to a central facility for fast-turnaround implementation. In the future, M.I.T. will offer this as course 6.371, with Jonathan Allen teaching it in the fall of 1979.

The following information concerning the M.I.T. experience may be useful to those planning similar activities. The course began in mid-September and was attended by 30 students (mostly graduate EE/CS students). Most of the formal lecture material necessary for undertaking projects (covering selected portions of Chapters 1 through 6 of this text) was completed by early November. The students then defined and began work on their LSI design projects. A design cut-off date of December 5 was set, and most designs were completed by that time. The projects included a LISP Microprocessor, a Graphics Memory Subsystem for mirroring and rotating bit-map data, a Writeable PLA project, the Data Path for a Bit-Slice Microprocessor, an LRU Virtual-Memory Paging subsystem, a Bus-Interfaceable Real-Time Clock-Calendar, a Multifunction Smart Memory, several digital signal-processing projects, several subsystems for data-base operations, and many other innovative designs.

Students described their layouts in a simple subset of CIF2.0, using a standard text-editor running on a DEC-20. The only hardware added to the DEC-20 system to support the course were several CRT terminals, two HP four-color pen plotters, and a connection to the local ARPANET host machine. The only software developed were programs for parsing the CIF subset, for instantiating data for plotting, and for driving the plotters. A small library of useful cells, namely input pads with lightning arrestors, output pads with cascaded drivers, and a set of PLA cells, were made available in CIF form. Some students developed their own symbolic layout languages and translaters to CIF, in order to make layout encoding less tedious. By using a structured design methodology the students were able to complete substantial LSI projects in a short period of time, using only primitive design tools. Each project contained on the order of several hundred to several thousand transistors. The logistics of interacting with the large group of student designers to organize the multiproject chip (updating the rules of the game, selecting projects for inclusion, negotiating space allocations, answering individual questions, etc.) were expedited by using the message system on the DEC-20. The corresponding interactions with the remote implementation facility (PARC) were handled via electronic mail.

On December 6, 1978, the individual CIF2.0 design files were transmitted from M.I.T. via the ARPANET to Xerox PARC. This was an additional test by these ARPA contractors of the use of such packet-switching networks for transmitting LSI design files and organizing multiproject chip sets. At PARC all the student designs were merged into a multiproject chip design file, from which masks were generated using Micro Mask, Inc.'s electron-beam maskmaking facility. Wafers were fabricated at Hewlett-Packard's Deer Creek Laboratory, which cooperated in the feasibility test. The wafers were returned to M.I.T. and electrically characterized by tests of the project-set electrical test patterns. They were then diced and the resulting chips packaged. Packaged chips, with wire-bonding customized for each project, were made available to all students by January 18, 1979. Many of the projects have since undergone thorough functional testing by the students. A number of these function completely correctly. Most exhibited only minor bugs, typically at the logic level of abstraction, of a sort reminiscent of one's first efforts at constructing large programs in a new language.

As a common VLSI system design culture spreads, as higher-level design aids are developed and shared within this culture, and as standard-interface commercially-accessible implementation facilities are established, we will undoubtedly see far more ambitious courses, projects, and research activities undertaken by students and faculty within the universities. And thus the period of exploration begins.

ACKNOWLEDGEMENTS

We wish to express our gratitude to the many individuals who have contributed their ideas, their time, and their energy toward the creation of this textbook. In particular, we wish to thank the following:

For contributions to the text: Chuck Seitz, California Institute of Technology, for contributing Chapter 7, *System Timing*, to the text; Martin Rem, Eindhoven University of Technology, and Sally Browning, Caltech, for their contributions to Chapter 8; David Johannsen, Caltech, for his major contributions to Chapters 5 and 6; H. T. Kung and Charles Leiserson, Carnegie-Mellon University, for permission to reprint their original, copyrighted work, *Algorithms for VLSI Processor Arrays*, as Section 3 of Chapter 8; Robert Sproull, CMU, and Richard Lyon, Xerox PARC, for the CIF section; Carlo Séquin, U.C. Berkeley, for his contributions to Chapter 1, his detailed review of the text, and his many suggestions for improvements; John Best, Chuck Seitz, Jim Kajiya, and Tom McGill, Caltech, and Johan de Kleer, Xerox PARC, for their stimulating discussions and suggestions for Chapter 9; Robert Sproull and Wayne Wilner, for their assistance in the preparation of Chapter 8; Hank Smith, M.I.T., for his contributions to the section on high-resolution lithography; Douglas Fairbairn, Xerox PARC, and James Rowson, Caltech, for the ICARUS section; Dale Green and Liz Bond, Xerox EOS, for technical assistance in the preparation of the text, figures, and color plates; Barbara Baird of Xerox PARC, for overseeing the printing and distribution of the prepublication versions of the text.

For valuable discussions and comments: Robert Sproull; Fred Rosenberger and Charles Molnar, Washington University; Richard Lyon, Doug Fairbairn, Wayne Wilner, and Leo Guibas, Xerox PARC; Chuck Seitz and Ivan Sutherland,

Caltech; Craig Mudge, DEC; Bill Heller, IBM; Gerry Parker, Dick Pashley, and John Wipfli, Intel; Harry Peterson, BNR Ltd.; Al Perlis, Yale University; Wesley Clark.

For participation in the OM projects: Dave Johannsen, Mike Tolle, Chris Carroll, Rod Masumoto, Ivan Sutherland, Chuck Seitz, Danny Cohen, and Leslie Froisland.

For helping to establish the multiproject chip capability: Richard Lyon, Doug Fairbairn, and Alan Bell, Xerox PARC; Bob Hon, CMU; Carlo Séquin, U.C. Berkeley; Ted Jenkins, Intel; Jim Rowson, Ron Ayres, and Steve Trimberger, Caltech.

For contributions to the M.I.T. 1978 VLSI design course: Jonathan Allen, Paul Penfield, Jr., Paul E. Gray, Fernando Corbató, Bill Henke, Glen Miranker, Joy Thompson, Dimitri Antoniadis, Stephen Senturia, Gerald Sussman, Jack Holloway, and Tom Knight, M.I.T.; Robert Noyce and Gordon Moore, Intel Corporation; Merrill Brooksby, and Patricia Castro, Hewlett-Packard; Richard Lyon and Alan Bell, Xerox PARC; and, especially, all the students who participated in course 6.978.

For long-standing support of integrated systems research at Caltech: The Office of Naval Research; Robert Noyce and Gordon Moore, Intel Corporation.

We are very grateful to Xerox Corporation for providing us access to the *Alto-Ethernet-Dover* research-prototype office systems at Xerox PARC and elsewhere within Xerox. Throughout the development of the ideas for this textbook, these personal, distributed computer systems enabled a large, scattered group of collaborators to function as a closely knit research community. These systems also enabled rapid creation and distribution of the prepublication printings of the text, thus helping us obtain valuable feedback from those teaching the early courses in the universities.

We are especially grateful to W. R. Sutherland, Manager, Systems Science Laboratory of Xerox PARC, for providing us with inspiration, guidance, and support; to Ivan Sutherland, Fletcher Jones Professor of Computer Science, Caltech, for his role in starting the Computer Science activities at Caltech, and for establishing links between the computer science community and the emerging activities in integrated systems; to Robert E. Kahn, Chief Scientist and Director for Information Processing Techniques, Defense Advanced Research Projects Agency, for encouraging and supporting integrated systems research in the universities; and to George E. Pake, Vice-President, Xerox Corporate Research, for creating the research environment that made this book possible.

CONTENTS

DATA AND CONTROL FLOW IN SYSTEMATIC STRUCTURES   60

IMPLEMENTING INTEGRATED SYSTEM DESIGNS: FROM CIRCUIT TOPOLOGY TO PATTERNING GEOMETRY TO WAFER FABRICATION   91

OVERVIEW OF AN LSI COMPUTER SYSTEM, AND THE DESIGN OF THE OM2 DATA PATH CHIP   145

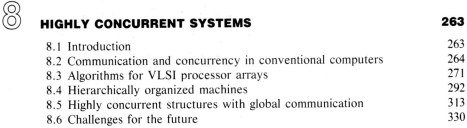

1

MOS DEVICES
AND CIRCUITS

We begin with a discussion of the basic properties of the n-channel, metal-oxide-semiconductor (MOS), field-effect transistor (FET). We then describe and analyze a number of circuits composed of interconnected MOS field-effect transistors. The circuits described are typical of those we will commonly use in the design of integrated systems. The analysis, though highly condensed, is conceptually correct and provides a basis for the solution of most system problems typically encountered.

Integrated systems in MOS technology contain three levels of conducting material separated by intervening layers of insulating material. Proceeding from top to bottom, the levels are termed *metal, polysilicon,* and *diffusion*, respectively. Patterns for paths on the three levels, and the locations of contact cuts through the insulating material to connect certain points between levels, are transferred into the levels during the fabrication process from *masks* similar to photographic negatives. (Details of the fabrication process will be discussed in Chapter 2.)

In the absence of contact cuts through the insulating material, paths on the metal level may cross over paths on either the polysilicon level or the diffusion level with no significant functional effect. However, wherever a path on the polysilicon level crosses a path on the diffusion level, a transistor is created. Such a transistor has the characteristics of a simple switch, with a voltage on the polysilicon-level path controlling the flow of current in the diffusion-level path. Circuits composed of such transistors, interconnected by patterned paths on the three levels, form our basic building blocks. With these basic circuits, we will design integrated systems, to be fabricated on the surface of monolithic crystalline chips of silicon.

1.1 THE MOS TRANSISTOR

An MOS transistor will be produced on the integrated system chip wherever a polysilicon path crosses a diffusion path, as shown in Fig. 1.1. The electrical sym-

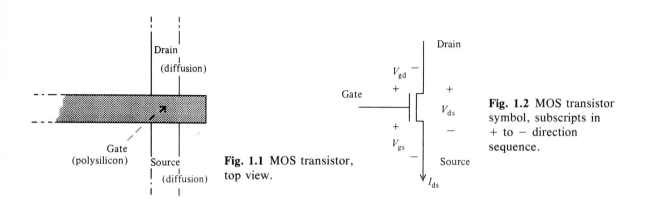

Fig. 1.1 MOS transistor, top view.

Fig. 1.2 MOS transistor symbol, subscripts in + to − direction sequence.

bol used to represent the MOS transistor in our circuit diagrams is shown in Fig. 1.2, along with symbols and polarities of certain voltages of interest. Note that the source and drain terminals of the device are physically symmetrical. For the n-channel MOSFET's the terminal labels are assigned such that drain-to-source voltage V_{ds} is normally positive. A more detailed view of the rectangular region called the gate, where the polysilicon (poly) crosses the diffusion, is given in Fig. 1.3. During fabrication the diffusion paths are formed after the poly paths are formed (as explained more fully in Chapter 2). The poly gate, and the thin layer of oxide beneath it, mask the region under the gate during diffusion. Therefore, no diffusion path forms under the gate, and there is no direct connection on the diffusion level between the source and drain terminals of the transistor. Metal, poly, and diffusion paths all conduct electricity well enough to be considered "wires" until further notice.

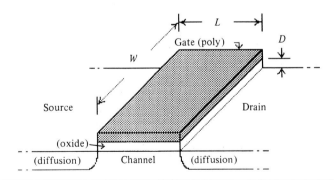

Fig. 1.3 MOSFET gate dimensions.

In the absence of any charge on the gate, the drain-to-source path through the transistor is like an open switch. The gate, separated from the substrate by the layer of thin oxide, forms a capacitor. If sufficient positive charge is placed on the gate so that gate-to-source voltage V_{gs} exceeds a *threshold voltage* V_{th}, electrons will be attracted to the region under the gate to form a conducting path between drain and source. Most of the transistors we will use in our systems have threshold voltages greater than zero. These are called *enhancement mode* MOSFET's and their threshold voltage typically is ≈ 0.2 VDD, where VDD is the positive supply voltage for the particular technology.

The basic operation performed by the MOS transistor is to use charge on its gate to control the movement of negative charge between source and drain through the channel under the gate. The current from source to drain equals the charge induced in the channel divided by the *transit time* or average time required for an electron to move from source to drain. The transit time itself is the distance the electron has to move divided by its average velocity. In semiconductors under normal conditions, the velocity is proportional to the electric field driving the electrons. The relationship between drain-to-source current I_{ds}, drain-to-source voltage V_{ds}, and gate-to-source voltage V_{gs} is sketched in Fig. 1.4. For small V_{ds}, the transit time τ is given by Eq. 1–1:

$$\tau = \frac{L}{\text{velocity}} = \frac{L}{\mu E} = \frac{L^2}{\mu V_{ds}} .\qquad (1\text{–}1)$$

The proportionality constant μ is called the *mobility* of the charge carriers, in this case electrons, under the influence of an electric field in the conducting material of the channel region. It is a velocity per unit electric field (cm^2/volt-sec.). We shall see that the transit time is the fundamental time unit of the entire integrated system.

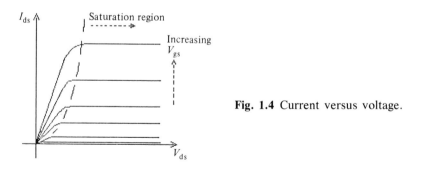

Fig. 1.4 Current versus voltage.

The amount of negative charge in transit Q is just the gate capacitance times the voltage on the gate in excess of the threshold voltage. The capacitance of two parallel conductors of area A, separated by insulating material of thickness D, equals $\epsilon A/D$. The proportionality constant ϵ is called the permittivity of the insulating material and has a simple interpretation. It is the capacitance of parallel conductors of area $A = 1$ cm^2, separated by a thickness $D = 1$ cm of the insulator material, and is in the units farad/cm. Therefore, the gate capacitance equals

$\epsilon WL/D$. Thus the charge in transit is given by Eq. (1–2), and the current is given by Eq. (1–3).

$$Q = - C_g(V_{gs} - V_{th}) = - \frac{\epsilon WL}{D}(V_{gs} - V_{th}) \qquad (1\text{--}2)$$

$$I_{ds} = - I_{sd} = - \frac{\text{charge in transit}}{\text{transit time}} = \frac{\mu \epsilon W}{LD}(V_{gs} - V_{th})(V_{ds}) \qquad (1\text{--}3)$$

Note that for small V_{ds}, the drain current is proportional to the source-drain voltage and also to the gate voltage above threshold. Any device with a current through it proportional to the voltage across it may be viewed as a resistor, and in the case of an MOS device with *low* drain-to-source voltage, the resistance is controlled by the gate voltage as given in Eq. (1–4).

$$\frac{V_{ds}}{I_{ds}} = R = \frac{L^2}{\mu C_g(V_{gs} - V_{th})} \qquad (1\text{--}4)$$

In Eqs. (1–2) and (1–4), C_g is the gate-to-channel capacitance of the turned-on transistor. In the simple case where this transistor is driving the gate of an identical transistor, the time response of the system will be an exponential with a time constant RC_g, given in Eq. (1–5). This time constant is similar in form to the expression for transit time τ given in Eq. (1–1).

$$RC_g = \frac{L^2}{\mu(V_{gs} - V_{th})} \qquad (1\text{--}5)$$

Although the above equations are greatly simplified, they provide sufficient information to make many design decisions that we will face, and they also give us insight to the scaling of devices to smaller sizes. In particular, the transit time τ can be viewed as the basic time unit of any system we shall build in the integrated technology. In almost all situations, the fastest operation that we can perform is to transfer a signal from the gate of one MOS transistor onto the gate of another. The transit time is the minimum time in which a charge placed on the gate of one transistor results in the transfer of a similar charge through that transistor's channel onto the gate of a subsequent transistor. For example, a transfer of charge from one transistor onto two identical transistors requires a minimum of two transit times. Thus, the transit time of the basic transistor in an integrated system can be viewed as the unit of time in which all other times in the system are scaled. Although it is a somewhat optimistic approximation, we will use τ as the primary time metric in calculating the delay through elementary inverting-logic stages. More accurate predictions of circuit behavior can be produced using any one of a number of available circuit simulation programs.[1,2]

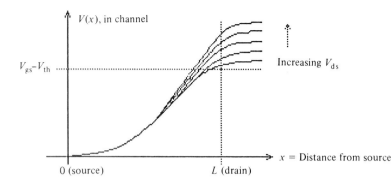

Fig. 1.5 Voltage profile across channel.

As V_{ds} is increased, not all the drain-to-source voltage is available for reducing the transit time. Drain voltage in excess of one threshold below the gate voltage creates a short region of high electric field, adjacent to the drain, that the carriers cross very quickly. The electric field in the major portion of the channel from the source up to this region is proportional to $V_{gs} - V_{th}$, as shown in Fig. 1.5. For $V_{ds} > (V_{gs} - V_{th})$, the drain current becomes independent of V_{ds}. Further increases in V_{ds} neither increase I_{ds} nor decrease the transit time. This range of V_{ds} values is known as *saturation*. In saturation,

$$I_{ds} = \frac{\mu\epsilon W}{2LD} (V_{gs} - V_{th})^2. \tag{1-6}$$

With the exception of the factor of 2 in the denominator, Eq. (1–6) is similar to Eq. (1–3), with the V_{ds} factor in (1–3) replaced by its maximum effective value, $V_{gs} - V_{th}$. The factor of 2 in Eq. (1–6) arises from the nonuniformity of the electric field in the channel region when in saturation.[3] (Richman, 1973)

1.2   THE BASIC INVERTER

The first logic circuit we will describe is the basic digital inverter. Analysis of this circuit is then extended to analysis of basic NAND and NOR logic gates. The inverter's logic function is to produce an output that is the complement of its input. When describing the logic function of circuits in integrated systems, we assign the value logic-1 to voltages equaling or exceeding some defined logic threshold voltage, and logic-0 to voltages less than this threshold voltage.

Were there an efficient way to implement resistors in the MOS technology, we could build a basic digital inverter circuit using the configuration of Fig. 1.6. Here, if the inverter input voltage V_{in} is less than the transistor threshold voltage V_{th}, then the transistor is switched off and V_{out} is "pulled up" to the positive supply voltage VDD. In this case the output is the complement of the input. If V_{in} is greater than V_{th}, the transistor is switched on and current flows from the VDD supply through the resistor R to GND. If R were sufficiently large, V_{out} could be

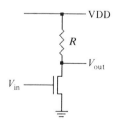

Fig. 1.6 An inverter.

"pulled down" well below V_{th}, thus again complementing the input. However, the resistance per unit length of minimum-width lines of various available conducting elements is far less than the effective resistance of the switched on MOSFET. Implementing a sufficiently large inverter pull-up using resistive lines would require a very large area compared to that occupied by the transistor itself.

To circumvent this problem a *depletion mode* MOSFET is used as a pull-up for the basic inverter circuit, symbolized and configured as shown in Fig. 1.7. In contrast to the usual enhancement mode transistor, the depletion mode transistor has a threshold voltage, V_{dep}, that is less than zero. During fabrication, one of the masks is used to select any desired subset of transistors in the integrated system for processing as depletion mode transistors. For a depletion mode transistor to turn off, a voltage is required on its gate relative to its source that is more negative than V_{dep}. But the depletion mode pull-up transistor's gate is connected to its source, and thus it is always turned on. Hence, when the enhancement mode transistor is turned off (for example, by connecting zero voltage to its gate) the output of the inverter will be equal to VDD. We will find that for reasonable ratios of the gate geometries of the two transistors, input voltages above a defined logic threshold voltage, V_{inv}, will produce output voltages below that logic threshold voltage, and vice versa.

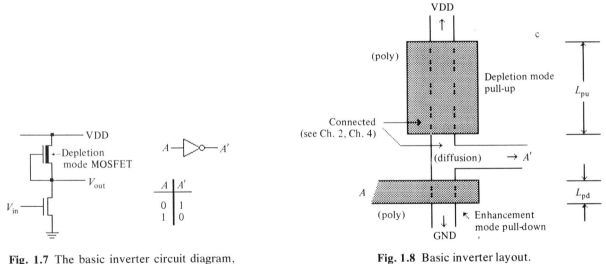

Fig. 1.7 The basic inverter circuit diagram, logic symbol, and logic function.

Fig. 1.8 Basic inverter layout.

The top view of the layout of an inverter on the silicon surface is sketched in Fig. 1.8. It consists of two polysilicon regions overhanging a path in the diffusion level that runs between VDD and GND. This arrangement forms the two MOS transistors of the inverter. The inverter input A is connected to the poly that forms the gate of the lower of the two transistors. The pull-up is formed by connecting the gate of the upper transistor to its source. (The layout geometry and fabrication details of such connections are described in Chapter 2.) The output of the inverter is shown emerging on the diffusion level, from between the drain of the pull-down and the source of the pull-up. The pull-up is a depletion mode transistor, and it is usually several times longer than the pull-down in order to achieve the proper inverter logic threshold.

Figures 1.9 and 1.10 show the characteristics of a typical pair of MOS transistors used to implement an inverter. The relative locations of the saturation regions of the pull-up and pull-down differ in these characteristics, due to the difference in the threshold voltages of the transistors.

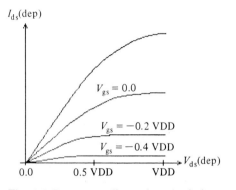

Fig. 1.9 Inverter pull-up characteristics.

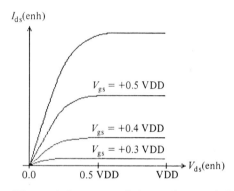

Fig. 1.10 Inverter pull-down characteristics.

We can use a graphical construct to determine the actual transfer characteristic, V_{out} versus V_{in}, of the inverter circuit. From Fig. 1.7 we see that $V_{ds}(enh)$ of the enhancement mode transistor equals VDD minus $V_{ds}(dep)$ of the depletion mode transistor. Also, $V_{ds}(enh)$ equals V_{out}. In a steady state and with no current drawn from the output, the I_{ds} of the two transistors is equal. Since the pull-up has its gate connected to its source, only one of its characteristic curves is relevant, namely the one for $V_{gs}(dep) = 0$. Taking these facts into account, we begin the graphical solution (Fig. 1.11) by superimposing plots of $I_{ds}(enh)$ versus $V_{ds}(enh)$ and the one plot of $I_{ds}(dep)$ versus $[VDD - V_{ds}(dep)]$. Since the currents in both transistors must be equal, the intersections of these sets of curves yield $V_{ds}(enh)$ = V_{out}, versus $V_{gs}(enh) = V_{in}$. The resulting transfer characteristic is plotted in Fig. 1.12.

While studying Figs. 1.11 and 1.12, consider the effect of starting with $V_{in} = 0$ and then gradually increasing V_{in} towards VDD. While the input voltage is below the threshold of the pull-down transistor, no current flows in that transistor, the output voltage is constant at VDD, and the drain-to-source voltage across the

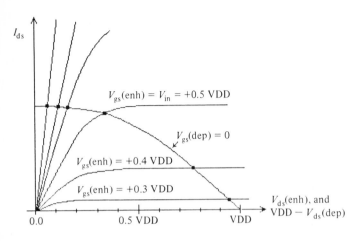

Fig. 1.11 I_{ds}(enh) versus V_{ds}(enh), and I_{ds}(dep) versus [VDD − V_{ds}(dep)].

pull-up transistor is equal to zero. When V_{in} is first increased above the enhancement mode threshold, current begins to flow in the pull-down transistor. The output voltage decreases slowly as the input voltage is first increased above V_{th}. Subsequent increases in the input voltage rapidly lower the pull-down's drain-to-source voltage, until the point is reached where the pull-down leaves its saturation region and becomes resistive. Then as V_{in} continues to increase, the output voltage asymptotically approaches zero. The input voltage at which V_{in} equals V_{out} is known as the *logic threshold voltage* V_{inv}. Figure 1.12 also shows the effect of changes in the transistor length-to-width ratios on the transfer characteristics and on the logic threshold voltage. The resistive impedance of the MOS transistor is proportional to the *length-to-width ratio Z* of its gate region. Using the subscripts pu (for the pull-up transistor) and pd (for the pull-down transistor) we find that if $Z_{pu} = L_{pu}/W_{pu}$ is increased relative to $Z_{pd} = L_{pd}/W_{pd}$, then V_{inv} decreases, and vice versa. The gain G, or negative slope of the transfer characteristic near V_{inv}, increases as Z_{pu}/Z_{pd} increases. The gain must be substantially greater than unity for digital circuits to function properly.

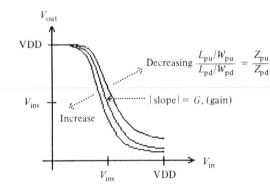

Fig. 1.12 V_{out} versus V_{in} for the basic inverter.

1.2.1   Inverter Logic Threshold Voltage

The most fundamental property of the basic inverter circuit is its *logic threshold voltage*, V_{inv}. The logic threshold here is *not* the same as V_{th} of the enhancement mode transistor; it is that voltage on the input of the enhancement mode transistor that causes an equal output voltage. If V_{in} is increased above this logic threshold, V_{out} falls below it, and if V_{in} is decreased below V_{inv}, V_{out} rises above it. The following simple analysis assumes that both pull-up and pull-down are in saturation, so that Eq. (1–6) applies. Usually the pull-up is not quite in saturation, but the following is still nearly correct: V_{inv} is approximately that input voltage that would cause saturation current through the pull-down transistor to be equal to saturation current through the pull-up transistor. Referring to Eq. (1–6), we find the condition for equality of the two currents given in Eq. (1–7). Currents are equal when

$$\frac{W_{pd}}{L_{pd}} (V_{inv} - V_{th})^2 = \frac{W_{pu}}{L_{pu}} (-V_{dep})^2 ; \tag{1-7}$$

thus,

$$V_{inv} = V_{th} - \frac{V_{dep}}{\sqrt{Z_{pu}/Z_{pd}}} . \tag{1-7a}$$

Here we note that the current through the depletion mode transistor is dependent only on its geometry and threshold voltage V_{dep}, since its $V_{gs} = 0$. Note that V_{inv} is dependent on the thresholds of both the enhancement and depletion mode transistors and also on the square root of the ratio of the $Z = L/W$ of the enhancement mode transistor to $Z = L/W$ of the depletion mode transistor.

   Various possible choices of values for these threshold voltages can be traded off against the areas and current driving capability of transistors in the system's inverters. To maximize $(V_{gs} - V_{th})$ and increase the pull-downs' current driving capability for a given area, V_{th} should be as low as possible. However, if V_{th} is too low, inverter outputs won't be driveable below V_{th}, and inverters won't be able to turn off transistors used as simple switches. The original choice of $V_{th} \approx 0.2$ VDD is a reasonable compromise here.

   Similarly, to maximize the current driving capability of pull-ups of given area, we might set the system's V_{dep} as far negative as possible. However, Eq. (1–7a) shows that for chosen V_{inv} and V_{th}, decreasing V_{dep} requires an increase in L_{pu}/W_{pu}, typically leading to an increase in pull-up area. The compromise made in this case is often as follows. The negative threshold of depletion mode transistors is set during fabrication such that with gate tied to source, they turn on approximately as strongly as would an enhancement mode transistor with VDD connected to its gate and its source grounded. In other words, depletion mode transistors and enhancement mode transistors of equal gate dimensions would have equal drain-to-source currents under those conditions. Applying Eq. (1–7) in those conditions, we find that

$$(-V_{dep})^2 \approx (VDD - V_{th})^2.$$

Therefore, $-V_{dep} \approx$ (VDD $- V_{th}$), and $V_{dep} \approx -0.8$ VDD. Adjustments in the details of the choice of V_{dep} are often made in the interest of optimization of processes for particular products. Perhaps the most common choice is that of V_{dep} ~ -0.6 VDD (leading to smaller pull-up areas than would $V_{dep} \sim -0.8$ VDD). Substituting this choice of V_{dep} into Eq. (1–7a), we find that

$$V_{inv} \approx 0.2 \text{ VDD} + \frac{0.6 \text{ VDD}}{\sqrt{Z_{pu}/Z_{pd}}}. \tag{1–8}$$

In general it is desirable that the margins around the inverter threshold be approximately equal, i.e., that the inverter threshold, V_{inv}, lie approximately midway between VDD and ground. We see from Eq. (1–8) that this criterion is met by a ratio of pull-up Z to pull-down Z of approximately 4:1.

1.3 INVERTER DELAY

A minimum requirement for an inverter is that it drive another identical to itself. Let us analyze the delay through a string of inverters of identical dimensions. This is the simplest case in which we can estimate performance. Inverters connected in this way are shown in Fig. 1.13(a). We define the *inverter ratio k* as the ratio of Z of the pull-ups to Z of the pull-downs. We will sometimes use the alternative "resistor-with-gate" pull-up symbol, as in Fig. 1.13(a), to clarify its functional purpose.

Let us assume that prior to $t = 0$, the voltage at the input of the first inverter is zero, hence the voltage output of the second inverter will be low. At time $t = 0$, let us place a voltage equal to VDD on the input of the first inverter and visualize the sequence of events that follows. The output of the first inverter, which leads to the gate of the second inverter, will initially be at VDD. Within approximately one transit time, the pull-down transistor of the first inverter will remove from this node an amount of charge equal to VDD times the gate capacitance of the pull-down of the second inverter. The pull-up transistor of the second inverter is now faced with the task of supplying a similar charge to the gate of the third inverter, to

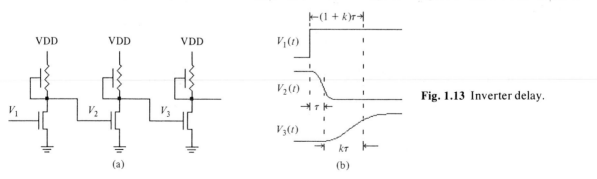

Fig. 1.13 Inverter delay.

(a)   (b)

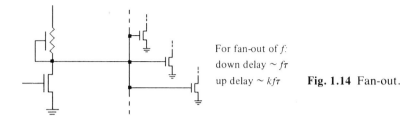

For fan-out of f:

down delay $\sim f\tau$

up delay $\sim kf\tau$ **Fig. 1.14** Fan-out.

raise it to VDD. Since it can supply at most only $1/k$ of the current that can be supplied by the pull-down transistor, the delay in the second inverter stage is approximately k times that of the first.

It is thus convenient to speak of the *inverter pair delay* that includes the delay for one "low-going" transition and one "high-going" transition. Inverter pair delay is approximately $(1 + k)$ times the transit time, as shown in Fig. 1.13(b). The fact that the rising transition is slower than the falling transition, by approximately the geometry ratios of the inverter transistors, is an inherent characteristic of any ratio-type logic. It is not true of all logic families. For example, in families such as complementary MOS (CMOS), where there are both pMOS and nMOS devices on the same silicon chip and both types operate strictly as pull-down enhancement mode devices, any delay asymmetry is a function of the difference in mobilities of the p- and n-type charge carriers rather than of the transistor geometrical ratios.

Figure 1.14 shows an inverter driving the inputs of several other inverters. In this case, for a "fan-out" factor f, it is clear that in either the pull-up or pull-down direction, the active device must supply f times as much charge as it did in the case of driving a single input. In this case, the delay in both the "up-going" and "down-going" directions is increased by approximately the factor f. In the case of the down-going transition, the delay is approximately f times the transit time of the pull-down transistor, and in the case of the up-going transition, the delay is approximately the inverter ratio k times the fan-out factor times the pull-down transit time.

In the discussions of transit time given earlier, it was assumed that both the depletion mode pull-up device and the enhancement mode pull-down device were operating in the resistive region. It was also assumed that all capacitances were constant, and not a function of voltage. These conditions are not strictly met in the technology we are discussing. Delay calculations given in this text are based on a "switching model" where individual stages spend a small fraction of their time in the midrange of voltages around V_{inv}. This assumption introduces a small error of the order of $1/G$. Because of these and other second-order effects, the switching times actually observed vary somewhat from those derived.

1.4 PARASITIC EFFECTS

In integrated systems, capacitances of circuit nodes are due not only to the capacitance of gates connected to the nodes but also to capacitances to ground of signal paths connected to the nodes and to other stray capacitances. These stray capacitances, sometimes called parasitic capacitances, are not negligible. While gate capacitances are typically an order of magnitude greater per unit area than capacitances of the signal paths, the signal paths are often much larger in area than the associated gate regions. Therefore, a substantial fraction of the delay encountered may be accounted for by stray capacitance rather than by the inherent properties of the active transistors. In the simplest case where the capacitance of a node is increased by the presence of parasitic area attached to the node, the delays can be accounted for by simply increasing the transit time by the ratio of the total capacitance to that of the gate of the transistor being driven. Time is required to supply charge not only to the gate itself but also to the parasitic capacitance.

There is one type of parasitic capacitance, however, which is not accounted for so simply. All MOS transistors have a parasitic capacitance between the drain edge of the gate and the drain node. This effect is shown schematically in Fig. 1.15. In an inverter string, this capacitance will be charged in one direction for one polarity of input and in the opposite direction for the opposite polarity input. Thus, on a gross scale its effect on the system is twice that of an equivalent parasitic capacitance to ground. Therefore, gate-to-drain capacitances should be approximately doubled, and added to the gate capacitance C_g and the stray capacitances, to account for the total capacitance of the node and thus for the effective delay time of the inverter. The effective inverter pair delay then is equal to $\tau(1 + k)\, C_{total}/C_g$.

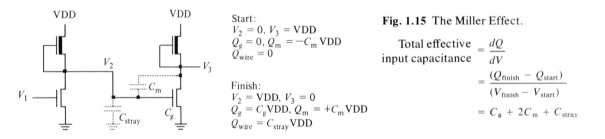

Start:
$V_2 = 0$, $V_3 = \mathrm{VDD}$
$Q_g = 0$, $Q_m = -C_m\, \mathrm{VDD}$
$Q_{wire} = 0$

Finish:
$V_2 = \mathrm{VDD}$, $V_3 = 0$
$Q_g = C_g\, \mathrm{VDD}$, $Q_m = +C_m\, \mathrm{VDD}$
$Q_{wire} = C_{stray}\, \mathrm{VDD}$

Fig. 1.15 The Miller Effect.

$$\frac{\text{Total effective}}{\text{input capacitance}} = \frac{dQ}{dV}$$

$$= \frac{(Q_{finish} - Q_{start})}{(V_{finish} - V_{start})}$$

$$= C_g + 2C_m + C_{stray}$$

1.5   DRIVING LARGE CAPACITIVE LOADS

As we have seen, the delay per inverter stage is multiplied by a fan-out factor. The overall performance of a system may be seriously degraded if it contains any large fan-outs, where one circuit within the system is required to drive a large capacitive load. As we shall see, this situation often occurs in the case of control drivers required to drive a large number of inputs to memory cells or logic function blocks. A similar and more serious problem is driving wires that go off the silicon chip to other chips or input/output devices. In such cases the ratio of the capacitance that must be driven to the inherent capacitance of a gate circuit on the chip is often many orders of magnitude, causing a serious delay and a degradation of system performance.

Consider how we may drive a capacitive load C_L in the minimum possible time given that we are starting with a signal on the gate of an MOS transistor of capacitance C_g.[4] Define the ratio of the load capacitance to the gate capacitance, C_L/C_g, as Y. It seems intuitively clear that the optimum way to drive a large capacitance is to use our elementary inverter to drive a larger inverter and that larger inverter to drive a still larger inverter until at some point the larger inverter is able to drive the load capacitance directly. Using an argument similar to the fan-out argument, it is clear that for one inverter to drive another inverter, where the second is larger in size by a factor of f, results in a delay f times the inherent inverter delay, τ. If N such stages are used, each larger than the previous by a factor f, then the total delay of the inverter chain is $Nf\tau$, where f^N equals Y. Note that if we use a large factor f, we can get by with few stages, but each stage will have a long delay. If we use a smaller factor f, we can shorten the delay of each stage, but we are required to use more stages. What value of N minimizes the overall delay for a given Y? We compute this value as follows: Since $f^N = Y$, $\ln(Y) = N\ln(f)$, and the delay of one stage equals $f\tau$; thus the total delay is

$$Nf\tau = \ln(Y)[f/\ln(f)]\tau. \qquad (1\text{-}9)$$

Notice that the delay is always proportional to $\ln(Y)$, a result of the exponential growth in successive stages of the driver. The multiplicative factor, $f/\ln(f)$, is plotted as a function of f in Fig. 1.16, normalized to its minimum value (e). Total delay is minimized when each stage is larger than the previous one by a factor of e, the

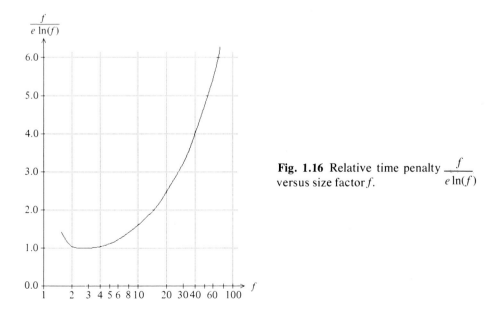

Fig. 1.16 Relative time penalty $\dfrac{f}{e\ln(f)}$ versus size factor f.

base of natural logarithms. Minimum total delay, t_{min}, is the elementary inverter delay τ times e times the natural logarithm of the ratio of the load capacitance to the elementary inverter capacitance:

$$t_{min} \approx \tau e \ln \left(\frac{C_L}{C_g} \right). \qquad (1\text{--}10)$$

Minimum delay through the driver is seldom the only design criterion. The relative time penalty introduced by the choice of other values of f can be read directly from Fig. 1.16.

1.6   SPACE VERSUS TIME

From the results of the sections on inverter delay, parasitic effects, and driving large capacitances, we see that areas and distances on the silicon surface trade off against delay times. For an inverter to drive another inverter some distance away, it must charge not only the gate capacitance of the succeeding inverter but also the capacitance to ground of the signal path connecting the two. Increasing the distance between the two inverters will therefore increase the inverter pair delay. This effect can be counterbalanced by increasing the area of the first inverter, so as to reduce the ratio of the load capacitance to the gate capacitance of the first inverter. But the delay of some previous driving stage is then increased. There is no way to get around the fact that transporting a signal from one node to another some distance away requires either charging or discharging capacitance and therefore takes time. Note that this is not a velocity of light limitation, as is often the case outside the chip. The times are typically several orders of magnitude longer than those required for light to traverse the distances involved. To minimize both the time and space required to implement system functions, we will tend to use the smallest possible circuits and locate them in ways that tend to minimize the interconnection distances.

The results of a previous section can be used here to illustrate another interesting space-versus-time effect. Suppose that the minimum-sized transistors of an integrated system have a transit time τ and gate capacitance C_g. A minimum-sized transistor within the system produces a signal that is then passed through successively larger inverting logic stages and eventually drives a large capacitance C_L with minimum total delay equal to t_{min}. With the passage of time, fabrication technology improves. We replace the system with another in which all circuit dimensions, including those vertical to the surface, are scaled down in size by dividing by a factor α, and the values of VDD and V_{th} are also scaled down by dividing by α. The motivation for this scaling is clear: the new system may contain α^2 as many circuits. As described in a later section, we will find that the transit times of the smallest circuits will now be $\tau' = \tau/\alpha$, and their gate capacitance will be $C_g' = C_g/\alpha$. The new ratio of load to minimum gate capacitance is $Y' = \alpha Y$.

Referring to Eq. (1–10), we find the new minimum total delay, t'_{min}, to drive C_L scales as follows:

$$t'_{min} = t_{min} \left(\frac{1}{\alpha} \right) \left(1 + \frac{\ln \alpha}{\ln Y} \right).$$

Therefore, as the inverters scale down and τ gets smaller, more inverting logic stages are required to obtain the minimum "off-chip" delay. Thus the relative delay to the outside world becomes larger. However, the absolute delay becomes smaller.

1.7 BASIC NAND AND NOR LOGIC CIRCUITS

NAND and NOR logic circuits may be constructed in integrated systems as simple expansions of the basic inverter circuit. The analysis of the behavior of these circuits, including their logic threshold voltages, transistor geometry ratios and time delays, is also a direct extension of the analysis of the basic inverter.

The circuit layout diagram of a two-input NAND gate is shown in Fig. 1.17. The layout is that of a basic inverter with an additional enhancement mode transistor in series with the pull-down transistor. NAND gates with more inputs may be constructed by simply adding more transistors in series with the pull-down path. The electrical circuit diagram, truth table, and logic symbol for the two-input NAND gate are shown in Fig. 1.18. If either of the inputs A or B is a logic-0, the

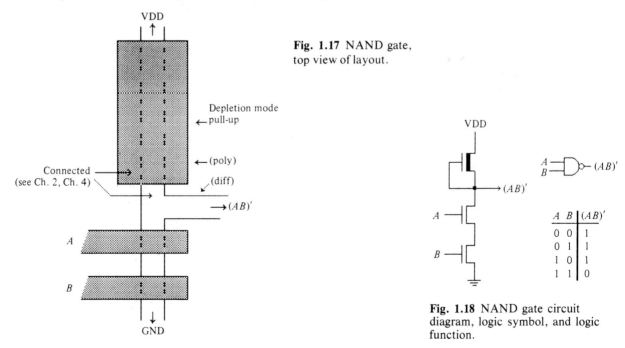

Fig. 1.17 NAND gate, top view of layout.

Fig. 1.18 NAND gate circuit diagram, logic symbol, and logic function.

A	B	$(AB)'$
0	0	1
0	1	1
1	0	1
1	1	0

pull-down path is open and the output will be high and therefore a logic-1. For the output to be driven low, to logic-0, both inputs must be high, at logic-1. The logic threshold voltage of this NAND gate is calculated in a manner similar to that of the basic inverter, except that Eq. (1–8) is rewritten with the length of the pull-downs replaced with the sum of the lengths of the two pull-downs (assuming their widths are equal) as follows:

$$V_{thNAND} \approx \frac{VDD}{\sqrt{\dfrac{L_{pu}/W_{pu}}{(L_{pd_a} + L_{pd_b})/W_{pd}}}} .$$

This equation indicates that as pull-downs are added in series to form NAND gate inputs, the pull-up length must be enlarged to hold the logic threshold voltage constant.

The logic threshold voltage of an n-input NAND gate, assuming all the pull-downs have equal geometries, is

$$V_{thNAND} \approx \frac{VDD}{\sqrt{\dfrac{L_{pu}/W_{pu}}{nL_{pd}/W_{pd}}}} .$$

As inputs are added and pull-up length is increased, the delay time of the NAND gate is also correspondingly increased, for both rising and falling transitions:

$$\tau_{NAND} \approx n \ \tau_{inv} .$$

The circuit layout diagram of a two-input NOR gate is shown in Fig. 1.19. The layout is that of a basic inverter with an additional enhancement mode transistor in parallel with the pull-down transistor. Additional inputs may be constructed by simply placing more transistors in parallel with the pull-down path. The circuit diagram, truth table, and logic symbol for the two-input NOR gate are shown in Fig. 1.20. If either of the inputs A or B is a logic-1, the pull-down path to ground is closed and the output will be low and therefore a logic-0. For the output to be driven high, to logic-1, both inputs must be low, at logic-0. If one of its inputs is kept at logic-0, and the other swings between logic-0 and logic-1, the logic threshold voltage of the NOR gate is the same as that of a basic inverter of equal pull-up to pull-down ratio. If this ratio were 4:1 to provide equal margins, then $V_{thNOR} \approx VDD/2$ with only one input active. However, if both pull-downs had equal geometries, and if both inputs were to move together between logic-0 and logic-1, V_{thNOR} would be reduced to $\approx VDD/(8)^{1/2}$. The logic threshold voltage of an

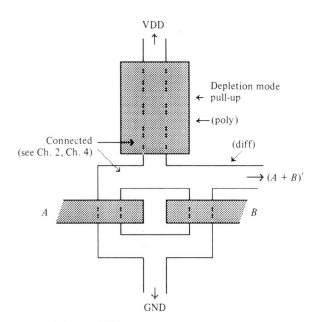

Fig. 1.19 NOR gate, top view of layout.

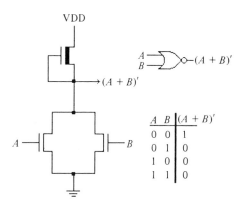

Fig. 1.20 NOR gate circuit diagram, logic symbol, and logic function.

n-input NOR circuit decreases as a function of the number of active inputs (inputs moving together from logic-0 to logic-1). The delay time of the NOR gate with one input active is the same as that of an inverter of equal transistor geometries, except for added stray capacitance. Its delay time for falling transitions is decreased as more of its inputs are active.

1.8   SUPER BUFFERS

As we have noted, ratio-type logic suffers from an asymmetry in its ability to drive capacitive loads. This asymmetry results from the fact that the pull-up transistor has of necessity less driving capability than the pull-down transistor. There are, however, methods for avoiding this asymmetry. Shown in Figs. 1.21 and 1.22 are circuits for both inverting and noninverting drivers that are approximately symmetrical in their capability of sourcing (sinking) charge into a capacitive load. Drivers of this type are called *super buffers*.

Both types of super buffer are built using a depletion mode pull-up transistor and an enhancement mode pull-down transistor, with a ratio of Z's of approximately 4:1 as in the basic inverter. However, the gate of the pull-up transistor, rather than being tied to its source, is tied to a signal that is the complement of that driving the pull-down transistor.

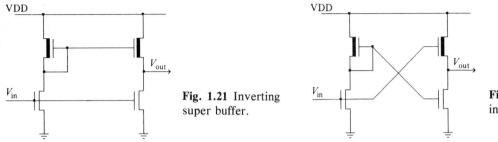

Fig. 1.21 Inverting super buffer.

Fig. 1.22 Noninverting super buffer.

When the pull-down transistor gate is at a high voltage, the pull-up transistor gate will be approximately at ground, and the current through the super buffer will be similar to that through a standard inverter of the same size. However, when the gate of the pull-down transistor is put to zero, the gate of the pull-up transistor will go rapidly to VDD since it is the only load on the output of the previous inverter, and the depletion mode transistor will be turned on at approximately twice the drive it would experience if its gate were tied to its source. Since the current from a device in saturation goes approximately as the square of the gate voltage, the current-sourcing capability of a super buffer is approximately four times that of a standard inverter. Hence, the current-sourcing capability of its pull-ups is approximately equal to the current-sinking capability of its pull-downs, and waveforms from super buffers driving capacitive loads are nearly symmetrical.

The effective delay time, τ, of super buffers is thus reduced to approximately the same value for high-going and low-going waveforms. Needless to say, when large capacitive loads are to be driven, super buffers are universally used. The arguments used in the last section to determine how many stages are used to drive a large capacitive load from a small source apply directly to super buffers. For that reason we have not explicitly indicated an inverter-ratio k in that section.

1.9   A CLOSER LOOK AT THE ELECTRICAL PARAMETERS

Up to this point we have talked in very simple terms about the properties of the MOS transistors. They have a capacitance associated with their gate input and a transit time for electrons to move from the source to the drain. We have given simple expressions for the drain-to-source current. For very low V_{ds}, the MOS transistor's drain-to-source path acts as a resistor whose conductance is directly proportional to the gate voltage above threshold, as given in Eq. (1–3). For values of V_{ds} larger than $V_{gs} - V_{th}$, the device acts as a current source, with a current proportional to $(V_{gs} - V_{th})^2$, as given in Eq. (1–6). As V_{ds} passes through the intermediate range between these two extremes, there is a smooth transition between the two types of behavior,[3] as given in the following equation:

$$I_{ds} = \frac{Q}{\tau} = \frac{\mu C_g}{L^2}\left[(V_{gs} - V_{th})\,V_{ds} - \frac{V_{ds}^2}{2}\right]. \qquad (1\text{–}11)$$

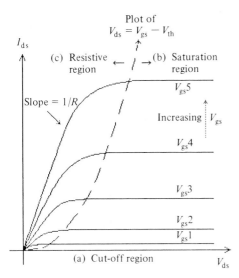

Plot of
$V_{ds} = V_{gs} - V_{th}$

I_{ds}

(c) Resistive
region

(b) Saturation
region

$V_{gs}5$

Slope = $1/R$

Increasing V_{gs}

$V_{gs}4$

$V_{gs}3$

$V_{gs}2$
$V_{gs}1$

(a) Cut-off region

V_{ds}

Fig. 1.23 Summary of MOS transistor characteristics.
(a) Cut-off region: $V_{gs} < V_{th}$, $I_{ds} = 0$.
(b) Saturation region: $V_{gs} \geq V_{th}$. V_{ds} sufficiently high so $V_{gd} < V_{th}$, that is, $V_{ds} \geq (V_{gs} - V_{th})$. MOSFET acts as current source, with I_{ds} proportional to $(V_{gs} - V_{th})^2$.
(c) Resistive region: $V_{gs} \geq V_{th}$. V_{ds} sufficiently low so $V_{gd} \geq V_{th}$, that is, $V_{ds} < (V_{gs} - V_{th})$. MOSFET acts as resistor, with resistance inversely proportional to $(V_{gs} - V_{th})$.

Figure 1.23 plots I_{ds} versus V_{ds}, summarizing the various regions of MOS transistor operation.

There is another electrical characteristic we may occasionally have to take into account. The threshold voltage of an MOS transistor is not a constant; it varies slightly as a function of the voltage between the source terminal of the transistor and the silicon substrate. This is called the *body effect* and it is illustrated in Fig. 1.24. The silicon substrate is usually connected to our system's circuit ground during packaging. However, a fixed bias voltage is sometimes applied between circuit ground and substrate as shown in Fig. 1.24, and this bias must be taken into account in estimating the body effect. If the source-to-bulk (substrate) voltage, V_{sb}, equals zero, then V_{th} is at its minimum value of approximately 0.2 VDD. As V_{sb} is increased, V_{th} increases slightly.

For enhancement mode transistors fabricated using typical processes, V_{th} reaches a maximum value of about 0.3 VDD when V_{sb} is increased to $\approx$ VDD. The

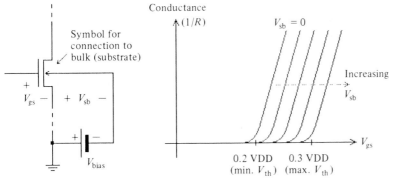

Symbol for
connection to
bulk (substrate)

$+$
V_{gs} $-$ $+$ V_{sb} $-$

$+$ $-$

V_{bias}

Conductance
$(1/R)$

$V_{sb} = 0$

Increasing
V_{sb}

V_{gs}

0.2 VDD   0.3 VDD
(min. V_{th}) (max. V_{th})

Fig. 1.24 The body effect.

value of the depletion mode transistor threshold, V_{dep}, is similarly affected, ranging from about -0.8 VDD to -0.7 VDD as V_{sb} is raised from zero to VDD volts. As shown in Fig. 1.24, it is possible to insert a fixed bias voltage between the circuit ground and the substrate, rather than just connect them. Such a *substrate bias* provides an electrical mechanism for setting the threshold to an appropriate value.

1.10 DEPLETION MODE PULL-UPS VERSUS ENHANCEMENT MODE PULL-UPS

With its gate tied to VDD, an enhancement mode transistor will be on for all $V_{ds} > V_{th}$, and thus can be used for a pull-up device in inverting logic circuits. Early MOS processes used pull-up devices of exactly this type.

In this section we will make a comparison of the rising transients of the two types of pull-up circuits. As noted earlier, rising transients in ratio-type logic are usually slower than falling transients, and thus rising transients generally have greater impact on system performance. In the simplest cases, this asymmetry in the transients results from the current-sourcing capability of the pull-up transistor being less than that of its pull-down counterpart. The simple intuitive time arguments given earlier are quite adequate for making estimates of system performance in most cases. However, there are situations in which the transient time may be much longer than a naive estimate would indicate. The rising transient of the enhancement mode pull-up is one of these.

A depletion mode pull-up transistor feeding a capacitive load is shown schematically in Fig. 1.25. Since $V_{gs} \geq V_{th}$ and $V_{gd} \geq V_{th}$, the pull-up transistor is in the resistive region. The final stages of the rising transient are given by the following exponential:

$$V(t) = \text{VDD}[1 - e^{-t/(RC_L)}] .$$

For an inverter-ratio k, pull-down transit time τ, and gate capacitance C_g, the time-constant of the rising transient is given by

$$RC_L = k\tau \frac{C_L}{C_g} .$$

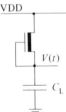

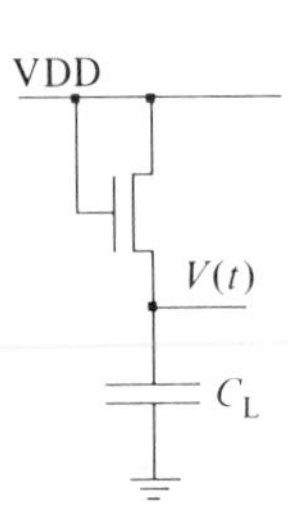

Fig. 1.25 Depletion mode MOSFET pulling up a capacitive load.

A somewhat more complicated situation is presented by an enhancement mode transistor sourcing charge into a capacitive load. This situation is shown schematically in Fig. 1.26. Note that since $V_{gd} = 0$, the transistor is in saturation whenever $V_{gs} > V_{th}$. The problem with sourcing charge from the enhancement mode transistor is that as the voltage at the output node gets closer and closer to one threshold below VDD, the amount of current provided by the enhancement mode transistor decreases rapidly.

The dependence of the enhancement mode pull-up current, I_{ds}, on output voltage, V, is given in Eq. (1–12):

$$Q = -\frac{\epsilon WL}{D} [(\text{VDD} - V_{th}) - V] ;$$

Fig. 1.26 Enhancement mode MOSFET pulling up a capacitive load.

$$\tau = \frac{2L^2}{\mu[(\text{VDD} - V_{\text{th}}) - V]} \;;$$

$$I_{\text{ds}} = -\frac{Q}{\tau} = \frac{\mu \epsilon W}{2LD} \, [(\text{VDD} - V_{\text{th}}) - V]^2. \qquad (1\text{--}12)$$

The fact that the pull-up current decreases as the output voltage nears its maximum value causes the rising transient from such a circuit to be of qualitatively different form than that of a depletion mode pull-up. Equating $I_{\text{ds}} = C_{\text{L}} dV/dt$ with the expression in Eq. (1–12), and then solving for $V(t)$, we find the rising voltage transient, for large t:

$$V(t) = \text{VDD} - V_{\text{th}'} - C_{\text{L}} \, \frac{LD}{\mu \epsilon W t} \, . \qquad (1\text{--}13)$$

Note that in this configuration, the threshold voltage $V_{\text{th}'}$ of the pull-up is near its maximum value as $V(t)$ rises towards VDD, due to the body effect.

A comparison of the rising transients of the preceding two circuits, assuming the same load capacitance and the same pull-up source current at zero output voltage, is shown in Fig. 1.27. The rising transient for the depletion mode pull-up transistor is crisp and converges rapidly towards VDD. However, the rising transient for the enhancement mode pull-up transistor, while starting rapidly, lags far behind, and within the expected time response of the system it never even comes close to one threshold below VDD. Even for very large t, $V(t) < \text{VDD} - V_{\text{th}'}$.

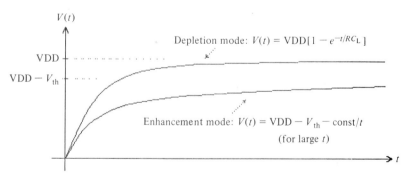

Fig. 1.27 Comparisons of rising transients for the two types of pull-ups.

The practical effect of this property of enhancement mode transistors is that circuits designed to work from the output of such a circuit should be designed with an inverter threshold V_{inv} considerably lower than that of circuits designed to work with the output of a depletion mode pull-up circuit. In order to obtain equal inverter margins without sacrificing performance, we will normally use depletion mode pull-ups.

1.11 DELAYS IN ANOTHER FORM OF LOGIC CIRCUITRY

Enhancement mode transistors, when used in small numbers and driving small capacitive loads, may often be used as switches in circuits of simple topology to provide logic-signal steering functions of much greater complexity than could be easily achieved in ratio-type inverting logic. These circuits are reminiscent of relay-switching logic, and transistors used in this way are referred to as "pass transistors" or "transmission gates." Examples of circuits using this type of design are given in Chapter 3. A particularly interesting example is the Manchester carry chain,[5,6] used for propagating carry signals in parallel adders. In each stage of the adder, a carry-propagate signal is derived from the two input variables to the adder, and if it is desired to propagate the carry, this propagate signal is applied to the gate of an enhancement mode pass transistor. The source of the transistor is "carry-in" to the present stage, and the drain of the transistor is "carry-out" to the next stage. In this way, a carry can be propagated from less significant to more significant stages of the adder without inserting a full inverter delay between stages. The circuit is shown schematically in Fig. 1.28.

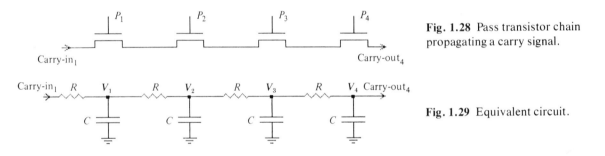

Fig. 1.28 Pass transistor chain propagating a carry signal.

Fig. 1.29 Equivalent circuit.

The delay through such a circuit does not involve inverter delays but is of an entirely different sort. A voltage along the chain divides into V_{ds} across each pass transistor. Thus V_{ds} is usually low, and the pass transistors operate primarily in the resistive region. We can think of each transistor as (1) a series resistance in the carry path, (2) a capacitance to ground formed by the gate-to-channel capacitance of each transistor, and (3) the strays associated with the source, drain, and connections with the following stage. An abstraction of the electrical representation is shown in Fig. 1.29. The minimum value of R is the turned-on resistance of each enhancement mode pass transistor, while the minimum value of C is the capacitance from gate to channel of the pass transistor. Strays will increase both values, especially that of C. The response at the node labeled V_2 with respect to time is given in Eq. (1–14). In the limit as the number of sections in the network becomes large, Eq. (1–14) reduces to the differential form shown in Eq. (1–15), where R and C are now the resistance and capacitance per unit length, respectively.

$$C \, dV_2/dt = [(V_1 - V_2) - (V_2 - V_3)]/R \qquad (1\text{--}14)$$

$$RC \, dV/dt = d^2V/dx^2 \qquad (1\text{--}15)$$

Equation (1–15) is the well-known diffusion equation, and while its solutions are complex, in general the time required for a transient to propagate a distance x in such a system is proportional to x^2. One can see qualitatively that this might be so. Doubling the number of sections in such a network doubles both the resistance and the capacitance and therefore causes the time required for the system to respond to increase by a factor of approximately four. The response of a system of n stages to a step function input is shown in Fig. 1.30.

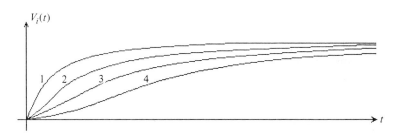

Fig. 1.30 Response to step function input.

If we add one more pass transistor to such a chain of n pass transistors, the added delay through the chain is small for small n but very large for large n. Therefore, it is highly desirable to group the pass transistors used for steering logic, multiplexing logic, and carry-chain-type logic into short sections and interpose inverting logic between these sections. This approach applied to the carry chain is shown in Fig. 1.31, where an inverter is inserted after every n pass transistors. The delay through a section of n pass transistors is proportional to RCn^2. Thus the total delay through one section is approximately RCn^2 plus the delay through the inverter τ_{inv}:

$$\approx RCn^2 + \tau_{inv}.$$

The average delay per pass transistor stage is

$$\approx RCn + \tau_{inv}/n. \tag{1–16}$$

To minimize the delay per stage, we choose n such that the delay through n pass transistors equals the inverter delay:

$$RCn^2 \approx \tau_{inv}.$$

Since logic done by steering signals with pass transistors does not require static power dissipation, a generalization of this result may be formulated. It pays to put as much logic into steering-type circuits as possible until there are enough pass transistors to delay the signal by approximately one inverting logic delay. At this point, the level of the signal can be restored by an inverting logic stage.

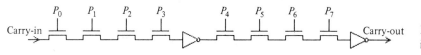

Fig. 1.31 Minimizing delay by interposing inverters.

The pass transistor has another important advantage over an inverting logic stage. When used to control or steer a logic signal, the pass transistor has only one input, one control, and one output connection. A NAND or NOR logic gate implementing the same function, in addition to containing two more transistors and thus occupying more area, also requires VDD and GND connections. As a result, the topology of interconnection of pass transistor circuits is far simpler than that of inverting-logic circuits. The topological simplicity of pass transistor control gates is an important factor in the system-design concepts developed in later chapters.

1.12 PULL-UP/PULL-DOWN RATIOS FOR INVERTING LOGIC COUPLED BY PASS TRANSISTORS

Earlier we found that when an inverting logic stage directly drives another such stage, a pull-up to pull-down ratio $Z_{pu}/Z_{pd} = (L_{pu}/W_{pu})/(L_{pd}/W_{pd})$ of 4:1 yields equal inverter margins and also provides an output sufficiently less than V_{th} for an input equal to VDD. Rather than coupling inverting logic stages directly, we often couple them with pass transistors for the reasons developed in the preceding section, thus affecting the required pull-up to pull-down ratio.

Figure 1.32 shows two inverters connected through a pass transistor. If the output of the first inverter nears VDD, the input of the second inverter can rise at most to $(VDD - V_{thp})$, where V_{thp} is the threshold of the pass transistor. Why does this effect occur? Consider the following: The output of the first inverter is at or above $(VDD - V_{thp})$, the pass transistor gate is at zero volts, and the input gate of the second inverter is also at zero volts. The pass transistor's gate voltage is now driven quickly to VDD, turning on the pass transistor. As current flows through the pass transistor, from drain to source, the input-gate voltage of the second inverter rises and the gate-to-source voltage of the pass transistor falls. When the gate voltage of the second inverter has risen to $(VDD - V_{thp})$, the pass transistor's gate-to-source voltage has fallen to its threshold value, and the pass transistor will switch off.

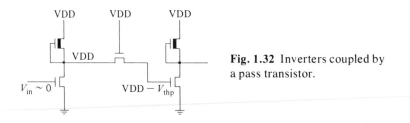

Fig. 1.32 Inverters coupled by a pass transistor.

If the second inverter is to have its output driven as low with an input of $(VDD - V_{thp})$ as would a standard inverter with an input of VDD, then the second inverter must have a pull-up/pull-down ratio larger than 4:1. The larger ratio is calculated as follows: With inputs near VDD, the pull-ups of inverters are in saturation, and the pull-downs are in the resistive region. Figure 1.33 shows equivalent circuits for two inverters. In part (a), VDD is input to one inverter, and in part (b) (VDD − V_{thp}), is input to the other inverter. For the output voltages of the two inverters to be equal under these conditions, $I_1 R_1$ must equal $I_2 R_2$. Referring to Eqs. (1–4) and (1–6), we find

$$\frac{Z_{pu1}}{Z_{pd1}}(\text{VDD} - V_{th}) = \frac{Z_{pu2}}{Z_{pd2}}(\text{VDD} - V_{th} - V_{thp}).$$

Since V_{th} of the pull-downs is approximately 0.2 VDD, and V_{thp} of the pass transistor is approximately 0.3 VDD due to the body effect, then $Z_{pu2}/Z_{pd2} \approx 2 Z_{pu1}/Z_{pd1}$. Thus a ratio of $(L_{pu}/W_{pu})/(L_{pd}/W_{pd}) = 8$ is usually used for inverting logic stages placed as level restorers between sections of pass transistor logic.

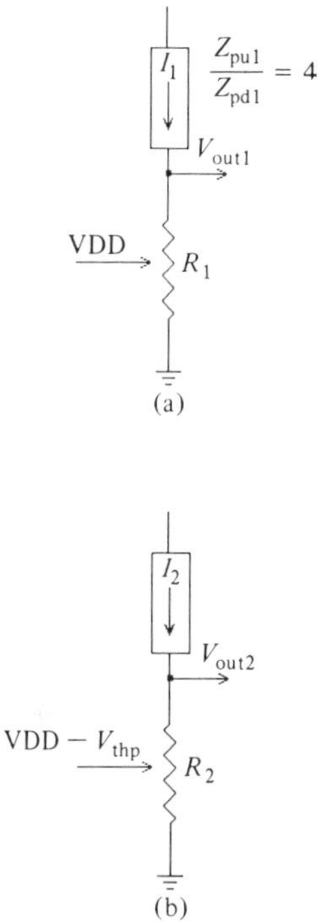

1.13 TRANSIT TIMES AND CLOCK PERIODS

In Chapter 3 we will develop a system design methodology in which we will be able to construct and estimate the performance of arbitrarily complex digital systems, using only the basic circuit forms presented in the preceding sections. The basic *system* building block in the design methodology is a register-to-register transfer through combinational logic, implemented with pass transistors and inverting logic stages. Using the basic ideas already presented, we may anticipate the results of that chapter in order to estimate the maximum clocking frequency of such systems.

The design methodology uses a two-phase nonoverlapping clock scheme. During the first clock phase, data passes from one register through combinational logic stages and pass transistors to a second register. During the second clock phase, data passes from the second register through still more logic and pass transistors to a third (or possibly back to the first) register. The data storage registers are implemented by using charge stored on the input gates of inverting logic stages, the charge being isolated by pass transistors controlled by clock signals (as described in Chapter 3).

Since pass transistors are used to connect inverting logic stages, inverter ratios of $k \approx 8$ are required. If the combinational logic between registers is implemented using only pass transistors, and if the delays through the pass transistors have been carefully matched to those of the inverting logic stages, the total delay will be twice that of the simple $k = 8$ inverter. In the absence of strays, the $k = 8$ inverters have a maximum delay (in the case of the output rising toward VDD) of 8τ, and hence a minimum of 16τ must be allowed for the inverter plus logic delay. However, in most designs the stray capacitance is at least equal to that inherent in the circuit. Thus the minimum time required for one such operation is $\approx 30\tau$. Control lines to the combinational logic and pass transistors each typically

$$\frac{Z_{pu1}}{Z_{pd1}} = 4$$

(a)

(b)

Fig. 1.33 For

$$V_{out2} = V_{out1},$$
$$Z_{pu2}/Z_{pd2} = 8.$$

drive the gates of 10 to 30 transistors. Even when using a super buffer driver, the delay introduced by this fanout is at least the minimum driving time for a capacitive load. With $Y = 30$, this time is $\approx 9\tau$. To this we must add an 8τ inverter delay for operation of the drivers.

Thus the total time for one clock phase is $\approx 50\tau$. Since two clock phases are required per cycle, a minimum clocking period of $\approx 100\tau$ is required for a system designed in this way. In 1978, $\tau \approx 0.3$ nanoseconds (ns), and clocking periods of 30 ns to 50 ns are achievable in carefully structured integrated systems where successive stages are in close physical proximity. If it is necessary to communicate data over long distances, longer periods are required.

1.14   PROPERTIES OF CROSS-COUPLED CIRCUITS

In many applications of control sequencing and data storage, memory cells and registers are built using two inverters driving each other, as shown in Fig. 1.34. This circuit can be set either in the state where V_1 is high and V_2 is low or in the state where V_1 is low and V_2 is high. In either case, the condition is stable and will not change to the other condition unless it is forced there through some external means. The detailed methods of setting such cross-coupled circuits into one state or another will be discussed later. However, it is important at the present time to understand the time evolution of signals impressed on cross-coupled circuits, since they exhibit properties different from circuits not having a feedback path from their output to an input.

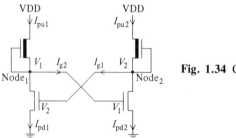

Fig. 1.34 Cross-coupled inverters.

We have seen that there exists a voltage at which the output of an inverter is approximately equal to its input voltage. If a cross-coupled circuit is inadvertently placed in a situation where its input voltage is equal to this value, then an unstable equilibrium condition is created where voltages V_1 and V_2 are equal. Since the net current flowing onto either gate is now zero, there is no forcing function driving the system to any voltage other than this equilibrium one, and the circuit can stay in this condition for an indefinite period. However, if either voltage changes, even very slightly, the circuit will leave this unstable equilibrium. For example, if the voltage V_1 is increased from its unstable equilibrium value V_{inv} by a slight amount, this will in time cause a lowering of voltage V_2, as net current flows from gate 1. This lowering of V_2 will at some later time cause V_1 to increase further. As time goes on, the circuit will feed back on itself until it rests in a stable equilibrium state.

The possibility of such unstable equilibria in cross-coupled circuits has important system implications,[7] as we will later see. For this reason, we will make a fairly detailed analysis of this circuit's behavior near the metastable state. While it is not essential that the reader follow all the details of the analysis, the final result should be studied carefully. The time constant of the final result depends in detail on the regions of operation of the transistors near the metastable state, as given in the following analysis. However, the exponential form of the result follows simply from the fact that the forcing function pushing the voltage away from the metastable point is proportional to the voltage's distance away from that point. This general behavior is characteristic of bistable storage elements in any technology. However, more complex waveforms are observed in logic families having more than one time constant per stage.

The time evolution of this process can be traced as follows. At the unstable equilibrium, the current in the pull-ups equals that in the pull-downs, and is some constant, k_1, times $(V_{inv} - V_{th})^2$. If V_1 is then changed by some small ΔV_1 to V_{init}, I_{pu2} remains constant but I_{pd2} changes immediately, producing a nonzero I_{g1}:

$$I_{g1} = I_{pu2} - I_{pd2} = k_1 \left[(V_{inv} - V_{th})^2 - (V_{inv} + \Delta V_1 - V_{th})^2 \right] .$$

For small ΔV_1, $I_{g1} = -2k_1(V_{inv} - V_{th})\Delta V_1$. More precisely, since I_{g1} is a function of (V_1, V_2), then near V_{inv},

$$\partial I_{g1}/\partial V_2 = -2k_1(V_{inv} - V_{th}).$$

Note that the pull-ups are not quite in saturation but are in the resistive region, and

$$\partial I_{g1}/\partial V_2 = -1/R_{pu},$$

where R_{pu} = effective resistance of the pull-up near V_{inv}. Noting that $I_{g1} = C_g dV_{g2}/dt$, we find that

$$dI_{g1}/dt = -2k_1(V_{inv} - V_{th})(dV_1/dt) - (1/R_{pu})(dV_2/dt) = C_g(d^2V_2/dt^2).$$

Evaluating the constants in this equation yields $-k_1(V_{inv} - V_{th}) = C_g/\tau_0$, where τ_0 is the saturation transit time of the pull-downs for t near zero. Assume a pull-up/pull-down Z ratio of 4:1, and consider the operating conditions near $t = 0$. Evaluating the effective resistance of the pull-ups in terms of the parameters of the pull-downs yields $1/R_{pu} \approx C_g/\tau_0$. Therefore,

$$-\frac{2}{\tau_0}\left(\frac{dV_1}{dt}\right) - \frac{1}{\tau_0}\left(\frac{dV_2}{dt}\right) = \frac{d^2V_2}{dt^2} .$$

Similarly,

$$-\frac{2}{\tau_0}\left(\frac{dV_2}{dt}\right) - \frac{1}{\tau_0}\left(\frac{dV_1}{dt}\right) = \frac{d^2V_1}{dt^2}.$$

Near time $t = 0$, dV_1/dt approximately equals $-dV_2/dt$, and therefore,

$$d^2V_1/dt^2 = -(1/\tau_0)\,dV_2/dt = (1/\tau_0)^2V_1 + \text{constant}. \qquad (1\text{--}17)$$

The solution to Eq. (1–17) is an exponential diverging from the equilibrium voltage V_{inv}, with a time constant $\tau_0/2$ equal to one half the pull-down delay time. Note that the solution given in Eq. (1–18) satisfies the conditions that $V(0) = V_{init}$ and that $V(t)$ is constant, if $V_{init} = V_{inv}$:

$$V_1(t) = V_{inv} + (V_{init} - V_{inv})\,e^{t/\tau_0}. \qquad (1\text{--}18)$$

The above analysis applies to cross-coupled circuits in the absence of noise. Noise unavoidably present in the circuit spreads the input voltage into a band from which such an unstable equilibrium can statistically be initiated. The width of the band is equal to the noise amplitude. Any timing condition that causes the input voltage to settle in the band has some probability of causing a balanced condition, from which the circuit may require an arbitrarily long time to recover. The time evolution of such a system is shown in Fig. 1.35, for several initial voltages near V_{inv}. The time for the cross-coupled system to reach one of its equilibria is thus logarithmic in the displacement from V_{inv} and is given approximately by Eq. (1–19):

$$t \approx \tau_0 \ln[V_{inv}/(V_{init} - V_{inv})]. \qquad (1\text{--}19)$$

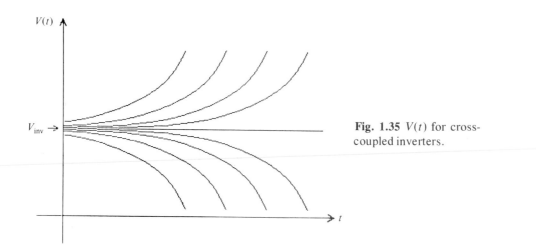

Fig. 1.35 $V(t)$ for cross-coupled inverters.

1.15   A FLUID MODEL FOR VISUALIZING MOS TRANSISTOR BEHAVIOR*

When designing circuits and systems, it is often useful to have some method for visualizing the physical behavior of the devices used as basic building blocks. This section develops such a method for the MOS transistor. Some readers of this text may be unfamiliar with the physics of semiconductor devices and would have difficulty visualizing what is going on inside an active semiconductor device, if device behavior were described in purely analytical terms. However, it is possible to construct a simple but very effective model of the behavior of certain charge-controlled devices, such as MOS transistors, charge-coupled devices (CCD's), and bucket-brigade devices (BBD's),[8] without referring to the details of device physics.

This model will be developed using two basic ideas: We think of electrical charge as though it were a fluid, and we mentally map the relevant electrical potentials into the geometry of a "container" in which the charge is free to move around. One can then apply one's intuitive understanding of, say, water in buckets of various shapes toward a visualization of what is going on inside the devices. Often a design guided by a good intuitive understanding of how a fluid would behave in the designed structure may show superior performance over designs based on complicated, but possibly inadequate, two-dimensional analytical modeling.

1.15.1   The MOS Capacitor

The basic element of MOS transistors or charge-transfer devices is the MOS capacitor. The notions of a fluid model will first be introduced using this elementary building block.

In physical space an MOS capacitor is a "sandwich" structure of a metal or polysilicon electrode on a thin insulator on the surface of a silicon crystal (Fig. 1.36a.) A suitable voltage applied to the electrode, i.e., positive for a p-type silicon substrate as used in nMOS, will repel the majority carriers in the substrate under the electrode, generating a depletion region that is at first free of any mobile charge carriers. Minority carriers, in this case electrons, can be either injected electrically into this area or generated by incident light and subsequently stored underneath the MOS electrode. Applying the notions of a fluid model, the same situation can be described as follows:

> The positive voltage applied to the MOS electrode generates a pocket in the surface potential of the silicon substrate. This can be visualized as a container, where the shape of the container is defined by the potential along the silicon surface, as plotted by the dashed line in Fig. 1.36(b). Note that in Fig. 1.36(b) *increasing positive potential* is plotted in the *downward direction*. The presence of minority charge carriers in an inversion layer changes the surface potential: an increase in this charge decreases the positive surface potential

*This section is contributed by Carlo H. Séquin, University of California, Berkeley.

under the MOS electrode. The potential profile in the presence of inversion charge is indicated by the solid line in Fig. 1.36(b). The area between the dashed and solid lines is hatched to indicate the presence of this charge. This representation shows charge sitting at the bottom of the container, just as a fluid would reside in a bucket. Of course the surface of the fluid (solid line) must be level in an equilibrium condition; if it were not, electrons would move under the influence of the potential difference until a constant surface potential is established.

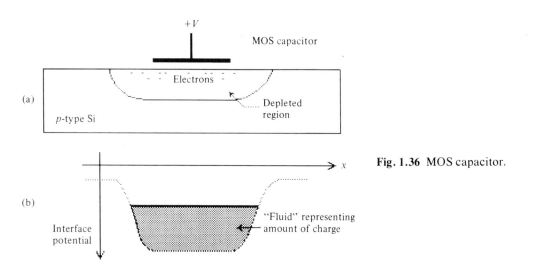

Fig. 1.36 MOS capacitor.

This model allows one to easily visualize the amount of charge present (hatched area), the fact that the charge tends to sit in the deepest part of the potential well, and the fact that the capacity of the bucket is finite and dependent on the applied electrode voltage. The higher this voltage, the deeper the bottom of the bucket and the more charge that can be stored. It should be kept in mind that this fluid model differs from physical reality insofar as the minority carriers in the inversion layer actually reside directly at the silicon surface.

1.15.2 The MOS Transistor

The same kind of model can be used to describe MOS transistor behavior. Figure 1.37(a) shows the physical cross section through an MOS transistor. Source and drain diffusions have been added to the simple MOS capacitor. For the moment we consider these two diffusions to be connected to two identical voltage sources, $V_{sb} = V_{db}$, which thus define the potential of the source and drain regions.

In the potential plot these diffusions are represented by exceedingly deep buckets, filled with charge carriers up to the levels of the potentials of the source

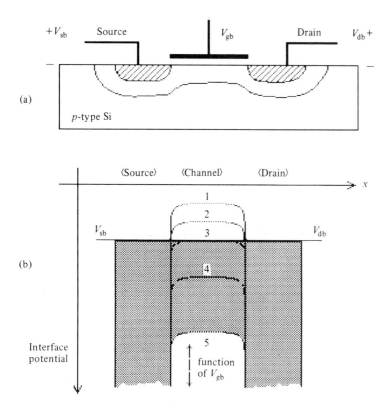

Fig. 1.37 MOS transistor.

and drain regions. Whether the MOS transistor is conducting, or is isolating the two diffused regions from one another, now depends on the potential underneath the MOS gate electrode. If the applied gate potential is chosen so that the potential underneath is less than V_{sb}, then there exists a potential barrier between source and drain regions (cases 1 and 2 in Fig. 1.37b). However, if the potential of an "empty bucket" under the gate electrode would be higher than V_{sb}, then the transistor is turned on (cases 4 and 5). Of course, in cases 4 and 5, carriers from the source and drain regions will spill underneath the gate electrode so that a uniform surface potential exists throughout the whole transistor. The conductivity of the channel area depends on the thickness of the inversion layer, which can readily be visualized in Fig. 1.37(b). Channel conductivity goes to zero at the turn-on threshold of the transistor (case 3), when the "empty bucket" potential under the gate electrode is equal to both the source and drain potential. Thus, the region under the gate can be viewed as a movable barrier of variable height that controls the flow of charge between the source and drain areas.

The same model enables us to visualize what happens when source and drain regions are biased to different potentials, as is usually the case in normal operation

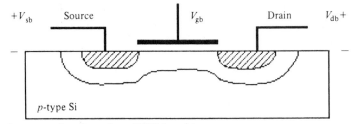

(a) Physical cross section through an MOS transistor.

(b) $V_{sb} = V_{db}$.

(c) $V_{db} = V_{sb} + \Delta V$.

Fig. 1.38 Various regimes of operation of an MOS transistor.

(d) $V_{db} > V_{sb}$.

(e) $V_{db} > V_{gb} - V_{th}$.

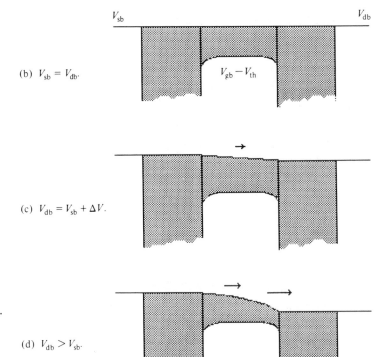

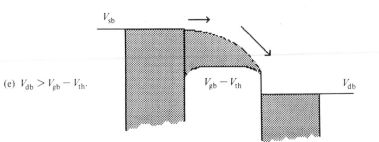

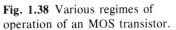

of MOS transistors. Figure 1.38(a) shows again a physical cross section through an MOS transistor, as a reference for the following figures. Part (b) reviews the case of equal source and drain potentials with the channel turned on fairly strongly, thus readily allowing charge to move between source and drain. Part (c) shows the situation when a small voltage difference, ΔV, has been applied between source and drain. Since the potential difference is maintained by external voltage sources, electrons will be forced to move from source to drain under the influence of the potential gradient, just as a liquid would flow from the higher to the lower level.

As the potential difference between source and drain is made larger, the variation in the "depth" of the fluid along the channel becomes significant (Fig. 1.38d). Continuity in the fluid requires that the charge move faster in the areas where the layer is thinner. This implies that the potential increases more rapidly closer to the drain region. With increasing drain potential the amount of charge flowing from source to drain per unit time increases, since the product of charge-layer depth and local gradient increases. However, there is a limit; once the drain potential exceeds the empty channel potential, the rate-of-charge flow will be limited by the drain-side edge of the barrier under the gate electrode. The MOS transistor has now reached saturation (Fig. 1.38e). The drain current density now is determined by the potential difference between the source and the empty channel and by the length of the channel (or the width of the barrier over which the charge has to flow); it is to first order independent of the drain voltage V_{db}.

Even in simple transistor circuits, the above fluid model helps one quickly develop a feeling for device and circuit operation. However, the real power of this intuitive model emerges when it is applied to complex structures where closed-form solutions describing charge motion can no longer be found. The empty potential under the various electrodes can first be plotted as in the above examples and the flow of charge then visualized using the analogy to the behavior of a fluid.

1.16   EFFECTS OF SCALING DOWN THE DIMENSIONS OF MOS CIRCUITS AND SYSTEMS

This section examines the effects on major system parameters resulting from scaling down all dimensions of an integrated system, including those vertical to the surface, by dividing them by a constant factor α. The voltage is likewise scaled down by dividing by the same constant factor α. Using this convention, all electric fields in the circuit will remain constant. Thus many nonlinear factors affecting performance will not change as they would if a more complex scaling were used.

Figure 1.39(a) shows a MOSFET of dimensions L, W, D, with $(V_{gs} - V_{th}) = V$. Figure 1.39(b) shows a MOSFET similar to that in part (a), but of dimensions $L' = L/\alpha$, $W' = W/\alpha$, $D' = D/\alpha$, and $V' = V/\alpha$. Refer now to Eqs. (1–1), (1–2), and (1–3). From these equations we will find that as the scale down factor α is increased, the transit time, the gate capacitance, and drain-to-source current of every individual transistor in the system scale down proportionally, as follows:

$$\tau \propto L^2/V, \quad \tau'/\tau = [(L/\alpha)^2/(V/\alpha)]/[L^2/V], \qquad \text{therefore, } \tau' = \tau/\alpha;$$

$$C \propto LW/D, \quad C'/C = [(L/\alpha)(W/\alpha)/(D/\alpha)]/[LW/D], \qquad \text{and } C' = C/\alpha;$$

$$I \propto WV^2/LD, \quad I'/I = [(WV^2/\alpha^3)/(LD/\alpha^2)]/[WV^2/LD], \qquad \text{and } I' = I/\alpha.$$

Switching power, P_{sw}, is the energy stored on the capacitance of a given device divided by the *clock period* (time between successive charging and discharging of the capacitance). A system's clock period is proportional to the τ of its smallest devices. As devices are made smaller and faster, the clock period is proportionally shortened. Also, the d.c. power, P_{dc}, dissipated by any static circuit equals I times V. Therefore, P_{sw} and P_{dc} scale as follows:

$$P_{sw} \propto CV^2/\tau \propto WV^3/DL \qquad \text{and} \qquad P'_{sw} = P_{sw}/\alpha^2;$$

$$P_{dc} = IV \qquad \text{and} \qquad P'_{dc} = P_{dc}/\alpha^2.$$

Both the switching power and static power per device scale down as $1/\alpha^2$. The average d.c. power for most systems can be approximated by adding the total P_{sw} to one half of the d.c. power that would result if all level-restoring logic pull-downs were turned on. The contribution of pass-transistor logic to the average d.c. power drawn by the system is due to the switching power consumed by the driving circuits that charge and discharge the pass-transistor control gates.

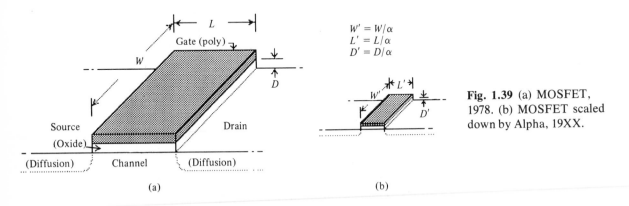

$$W' = W/\alpha$$
$$L' = L/\alpha$$
$$D' = D/\alpha$$

Fig. 1.39 (a) MOSFET, 1978. (b) MOSFET scaled down by Alpha, 19XX.

(a)   (b)

The *switching energy* per device, E_{sw}, is an important metric of device performance. It is equal to the power consumed by the device at maximum clock frequency multiplied by the device delay, and it scales down as follows:

$$E_{sw} \propto CV^2 \qquad \text{and} \qquad E'_{sw} = E_{sw}/\alpha^3.$$

Table 1.1 summarizes values of the important system parameters for current technology and for a future technology near the limits imposed by physical law:

Table 1.1

	1978	19XX
Minimum feature size	6 μm	0.3 μm
τ	0.3 to 1 ns	$\approx$0.02 ns
E_{sw}	$\approx 10^{-12}$ joule	$\approx 2 \times 10^{-16}$ joule
System clock period	$\approx$30 to 50 ns	$\approx$2 to 4 ns

A more detailed plot of the channel conductance of an MOS transistor near the threshold voltage is shown in Fig. 1.40. Below the nominal threshold, the conductance (1/R) is not in reality zero but depends on gate voltage and temperature as follows:

$$1/R \propto e^{(V_{gs} - V_{th})/(kT/q)},$$

where T is the absolute temperature, q is the charge on the electron, and k is Boltzmann's constant. At room temperature, $kT/q \approx 0.025$ volts. At present threshold voltages, as in the rightmost curve in Fig. 1.40, an off device is below threshold by perhaps 20 kT/q, that is, by about 0.5 volts, and its conductance is decreased by a factor of the order of ten million. Said another way, if the device is used as a pass transistor, a quantity of charge that takes a time τ to pass through the on device, will take a time on the order of $10^7 \tau$ to "leak" through the off device.

The use of pass-transistor switches to isolate and "dynamically store" charge on circuit nodes is common in many memory applications using 1978 transistor dimensions. However, if the threshold voltage is scaled down by a factor of perhaps 5, as shown in the leftmost curve in Fig. 1.40, then an off transistor is only 4 kT/q below threshold. Therefore, its conductance when off is only a factor of 100

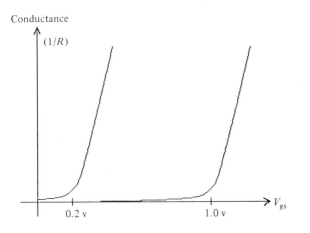
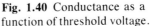

Fig. 1.40 Conductance as a function of threshold voltage.

or so less than when it is on. For such relatively large values of subthreshold conductance, charge stored dynamically on a circuit node by the transistor when on will safely remain on that node for only a few system clock periods. The charge will not remain on the node for a very large number of periods as it does in present memory devices using this technique. One way of possibly coping with this problem, as device dimensions and threshold voltages are scaled down, is to reduce the temperature of device operation.[9]

Suppose we scale down an entire integrated system by a scale-down factor of $\alpha = 10$. The resulting system will have one hundred times the number of circuits per unit area. The total power per unit area remains constant. All voltages in the system are reduced by the factor of 10. The current supplied per unit surface area is increased by a factor of 10. The time delay per stage is decreased by a factor of 10. Therefore, the power-delay product decreases by a factor of 1000.

This is a rather attractive scaling in all ways except for the current density. The delivery of the required average d.c. current presents an important obstacle to scaling. This current must be carried to the various circuits in the system on metal conductors, in order that the voltage drop from the off-chip source to the on-chip subsystems will not be excessive. Metal paths have an upper current density limit imposed by a phenomenon called metal migration (discussed further in Chapter 2). Many metal paths in today's integrated circuits are already operated near their current density limit. As the above type of scaling is applied to a system, the conductors get narrower but still deliver the same current on the average to the circuits supplied by them.

Therefore, it will be necessary to find ways of decreasing system current requirements to approximately a constant current per unit area relative to present current densities. In n-channel silicon gate technology, this objective can be partially achieved by using pass-transistor logic in as many places as possible and avoiding restoring logic except where it is absolutely necessary. Numerous examples of this sort of design are given later in this text. This design approach also has the advantages of tending to minimize delay per unit function and to maximize logic functions per unit area. However, when scaled down to submicron size, the pass transistors will suffer from the subthreshold current problem. It is possible that when the fabrication technologies have been developed to enable scaling down to submicron devices, a technology such as complementary MOS, which does not draw any d.c. current, may be preferable to the nMOS technology used to illustrate this text. However, even if this occurs, the methodology developed in the text can still be applied in the design of integrated systems in that technology.

The limit to the kind of scaling described above occurs when the devices created are no longer able to perform the switching function. To perform the switching function, the ratio of transistor on-to-off conductance must be $\gg 1$, and therefore the voltage operating the circuit must be many times kT/q. For this reason, even those circuits optimized for operation at the lowest possible supply voltages still require a VDD of ≈ 0.5 volts. Devices in 1978 operate with a VDD of approximately five volts and minimum channel lengths of approximately six microns. Therefore, the kind of scaling we have envisioned here will take us to devices with approximately one-half micron channel lengths and current densities approximately ten times what they are today. Power per unit area will remain constant over that range. Smaller devices might be built but must be used without lowering the voltage any further. Consequently the power per unit area will increase. Finally, there appears to be a fundamental limit[10] of approximately one-quarter micron channel length, where certain physical effects such as the tunneling through the gate oxide and fluctuations in the positions of impurities in the depletion layers begin to make the devices of smaller dimension unworkable.

REFERENCES

1. T. K. Young and R. W. Dutton, "MINI-MSINC: A Minicomputer Simulator for MOS Circuits with Modular Built-in Model," Stanford Electronics Laboratories, Technical Report No. 5013-1, March 1976.
2. L. Nagel and D. Pederson, "Simulation Program with Integrated Circuit Emphasis (SPICE)," 16th Midwest Symposium on Circuit Theory, Waterloo, Ontario, April 12, 1973.
3. W. M. Penney and L. Lau, eds., *MOS Integrated Circuits*, Princeton, N.J.: Van Nostrand, 1972, pp. 60–85.
4. R. C. Jaeger, "Comments on 'An Optimized Output Stage for MOS Integrated Circuits'," *IEEE J. Solid-State Circuits*, June 1975, pp. 185–186.
5. T. Kilburn; D. B. G. Edwards; and D. Aspinall, "A Parallel Arithmetic Unit Using a Saturated Transistor Fast-Carry Circuit," *Proc. IEE*, Pt. B, vol. 107, November 1960, pp. 573–584.
6. Staff of the Computation Lab, "Description of a Relay Calculator," Annals of the Harvard Computation Lab, vol. 24, Harvard University Press, 1949.
7. T. J. Chaney and C. E. Molnar, "Anomalous Behavior of Synchronizer and Arbiter Circuits," *IEEE Transactions on Computers*, April 1973, pp. 421–422.
8. C. H. Séquin and M. F. Tompsett, *Charge Transfer Devices*, New York: Academic Press, 1975.
9. F. H. Gaensslen; V. L. Rideout; E. J. Walker; and J. J. Walker, "Very Small MOSFETs for Low-Temperature Operation," *IEEE Transactions on Electron Devices*, March 1977.
10. B. Hoeneisen, and C. A. Mead, "Fundamental Limitations in Micro-electronics–I. MOS Technology," *Solid-State Electronics*, vol. 15, 1972, pp. 819–829.

2
INTEGRATED
SYSTEM
FABRICATION

The series of steps by which a geometric pattern or set of geometric patterns is transformed into an operating integrated system is called a *wafer fabrication process*, or simply a *process*. An integrated system in MOS technology consists of a number of superimposed layers of conducting, insulating, and transistor-forming materials. By arranging predetermined geometric shapes in each of these layers, a system of the required function may be constructed. The task of designers of integrated systems is to devise the geometric shapes and their locations in each of the various layers of the system. The task of the process itself is to create the layers and transfer into each of them the geometric shapes determined by the system design.

Modern wafer fabrication is probably the most exacting production process ever developed. Since the 1950s, enormous human resources have been expended by the industry to perfect the myriad of details involved. The impurities in materials and chemical reagents are measured in parts per billion. Dimensions are controlled to a few parts per million. Each step has been carefully devised to produce some circuit feature with the minimum possible deviation from the ideal behavior. The results have been little short of spectacular: chips with many tens of thousands of transistors are being produced for under ten dollars each. In addition, wafer fabrication has reached a level of maturity where the system designer need not be concerned with the fine details of its execution. The following sections present a broad overview sufficient to convey the ideas involved and in particular those relevant for system design. Our formulation of the basic concepts anticipates the evolution of the technology toward ever finer dimensions.

In this chapter we describe the patterning sequence and how it is applied in a simple, specific integrated system process: nMOS. A number of other topics are covered that are related to the processing technology or are closely tied to the properties of the underlying materials.

2.1 PATTERNING

The overall fabrication process consists of the *patterning* of a particular *sequence* of successive *layers*. The patterning steps by which geometrical shapes are transferred into a layer of the final system are very similar for each of the layers. The overall process is more easily visualized if we first describe the details of patterning one layer. We can then describe the particular sequence of layers used in the process to build up an integrated system, without repeating the details of patterning for each of the layers.

A common step in many processes is the creation of a silicon dioxide insulating layer on the surface of a silicon wafer and the selective removal of sections of the insulating layer exposing the underlying silicon. We will use this step for our patterning example. The step begins with a bare polished silicon wafer, shown in cross section in Fig. 2.1. The wafer is exposed to oxygen in a high-temperature furnace to grow a uniform layer of silicon dioxide on its surface (Fig. 2.2). After the wafer is cooled, it is coated with a thin film of organic "resist" material (Fig. 2.3). The resist is thoroughly dried and baked to ensure its integrity. The wafer is now ready to begin the patterning.

At the time of wafer fabrication the pattern to be transferred to the wafer surface exists as a *mask*. A mask is merely a transparent support material coated with a thin layer of opaque material. Certain portions of the opaque material are removed, leaving opaque material on the mask in the precise pattern required on the silicon surface. Such a mask, with the desired pattern engraved upon it, is brought face down into close proximity with the wafer surface, as shown in Fig. 2.4. The dark areas of opaque material on the surface of the mask are located where it is desired to leave silicon dioxide on the surface of the silicon. Openings in the mask correspond to areas where it is desired to remove silicon dioxide from the silicon surface. When the mask has been brought firmly into proximity with the wafer itself, its back surface is flooded with an intense source of ionizing radiation, such as ultraviolet light or low-energy x-rays. The radiation is stopped in areas where the mask has opaque material on its surface. Where there is no opaque material on the mask surface, the ionizing radiation passes through into the resist, the silicon dioxide, and silicon. While the ionizing radiation has little effect on the silicon dioxide and silicon, it breaks down the molecular structure of the resist into considerably smaller molecules.

We have chosen to illustrate this text using positive resist—in which case the resist material remaining after exposure and development corresponds to the opaque mask areas. Negative resists are also in common use. Positive resists are typically workable to finer feature sizes and are likely to become dominant as the technology progresses.

After exposure to the ionizing radiation, the wafer has the characteristics shown in Fig. 2.5. In areas exposed to the radiation, the resist molecules have been broken down to much lighter molecular weight than that of unexposed resist

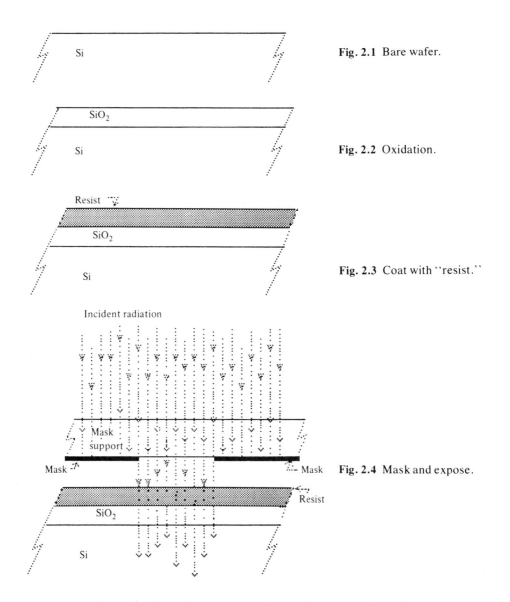

Fig. 2.1 Bare wafer.

Fig. 2.2 Oxidation.

Fig. 2.3 Coat with "resist."

Fig. 2.4 Mask and expose.

molecules. The solubility of organic molecules in various organic solvents is a very steep function of the molecular weight of the molecules. It is possible to dissolve exposed resist material in solvents that will not dissolve the unexposed resist material. The resist can be "developed" (Fig. 2.6) by merely immersing the silicon wafer in a suitable solvent.

Thus far, the pattern originally existing as a set of opaque geometries on the mask surface has been transferred as a corresponding pattern into the resist material on the surface of the silicon dioxide. The same pattern can now be transferred

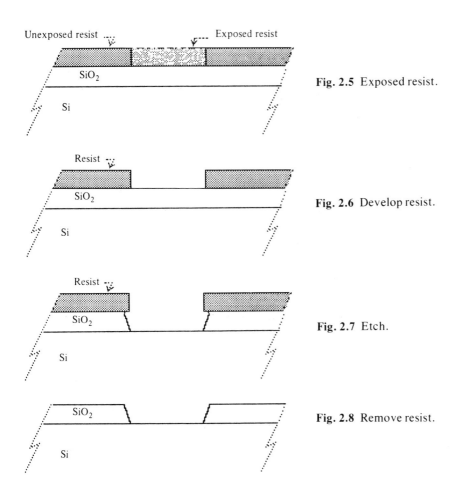

Unexposed resist .... Exposed resist

SiO₂

Si

Fig. 2.5 Exposed resist.

Resist

SiO₂

Si

Fig. 2.6 Develop resist.

Resist

SiO₂

Si

Fig. 2.7 Etch.

SiO₂

Si

Fig. 2.8 Remove resist.

to the silicon dioxide itself by exposing the wafer to a material that will etch silicon dioxide but will not attack either the organic resist material or the silicon wafer surface. The etching step is usually done with hydrofluoric acid, which easily dissolves silicon dioxide. However, organic materials are very resistant to hydrofluoric acid, and it is incapable of etching the surface of silicon. The result of the etching step is shown in Fig. 2.7.

The final step in patterning is removal of the remaining organic resist material. There are three techniques to remove resist materials: (1) using strong organic solvents that dissolve even unexposed resist material; (2) using strong acids, such as chromic acid, that actively attack organics; (3) exposing the wafer to atomic oxygen, which will oxidize away any organic materials present on its surface. Once the resist material is removed, the finished pattern on the wafer surface is as shown in Fig. 2.8. Notice that we have transferred the geometric pattern that originally existed on the surface of the mask directly into the silicon dioxide on the wafer surface. While a foreign material was present on the wafer surface during the patterning process, it has now disappeared and the only materials present are those that will be part of the finished wafer.

A similar sequence of steps is used to selectively pattern each of the layers of the integrated system. The steps differ primarily in the types of etchants used. Thus as we study the processing of the various layers, the reader need not visualize all the details of the patterning sequence for each layer but only recognize that a mask pattern for a layer can be transferred into a pattern in the material of that layer.

2.2 SCALING OF PATTERNING TECHNOLOGY

As discussed in Chapter 1, semiconductor devices could be at least an order of magnitude smaller in linear dimension than those typically manufactured in 1978 and still function correctly. The fundamental dimensional limitation is approximately a one-quarter micron channel length, corresponding to a length unit λ (to be discussed under design rules) of approximately 0.1 micron. This limitation appears to apply to both bipolar and MOS technologies. It has been possible for several years to create submicron lines using electron beam and x-ray techniques, and there is considerable research and development under way to bring these patterning technologies into general manufacturing use. It appears that there are no fundamental barriers preventing creation of patterns for ultimately small devices. (A detailed discussion of the techniques involved is given in Chapter 4.)

2.3 THE SILICON GATE *n*-CHANNEL MOS PROCESS

We now describe the particular sequence of patterned layers used to build up *n*MOS integrated circuits and systems. In color Plate 1, parts (a) through (f) illustrate a simple but complete sequence of patterning and processing steps that are sufficient to fabricate a complete integrated system. The example follows the fabrication of one simple circuit within a system, but all other circuits are simultaneously implemented by the same process. The example used is the basic inverter circuit. The top illustration in parts (a) through (f) shows the top view of the layers of the circuit layout. The lower illustration in each of those figures shows the cross section through the cut indicated by the downward arrows. (The vertical scale in these cross sections has been greatly exaggerated for illustrative purposes.)

The opening in the opaque material of the first mask is shown by the green outline in the top portion of Plate 1(a). This opening exposes all areas that will eventually be the diffusion level. It includes the sources and drains of all transistors in the circuit, together with the transistor gate areas, and any diffusion level circuit interconnection paths. This mask is used for the first step in the process, the patterning of silicon dioxide on silicon as described in the previous section. The resulting cross section is shown in the lower portion of Plate 1(a).

The second step in the process is to differentiate transistors that are normally on (depletion mode) from those that are normally off (enhancement mode). This is done by overcoating the wafer with resist material, exposing the resist material through openings in a second mask, and developing it in the manner shown in Plate 1(b). The patterning step leaves an opening in the resist material over the area to be selectively turned into depletion mode transistors. The actual conversion of the underlying silicon is then done by implanting ions of arsenic or antimony into the silicon surface. The resist material, where present, acts to prevent the ions from reaching the silicon surface. Therefore, ions are only implanted in the silicon area free of resist. The implanted layer, which causes a slight *n*-type conductivity in the underlying silicon, is shown by the yellow box in Plate 1(b). Once the depletion areas are defined, the resist material is removed from the surface of the wafer.

The wafer is then heated while exposed to oxygen, to grow a very thin layer of silicon dioxide over its entire surface. It is then entirely coated with a thin layer of polycrystalline silicon, usually called *polysilicon* (or *poly* for short). Note that the polysilicon layer is insulated everywhere from the underlying materials by the layer of thin oxide and additionally by thicker oxide in some areas. The polysilicon will form the gates of all the transistors in the circuit and will also serve as a second layer for circuit interconnections. A third mask is used to pattern the polysilicon by steps similar to those previously described, with the result shown in red in Plate 1(c). The leftmost polysilicon area will function as the gate of the pull-down transistor of the inverter we are constructing, while the square to the right will function as the gate of the depletion mode pull-up transistor.

Once the polysilicon areas have been defined, *n*-type regions can be diffused into the *p*-type silicon substrate, forming the sources and drains of the transistors and the first level of interconnections. This step is done by first removing the thin gate oxide in all areas not covered by the (red) polysilicon. The wafer is then exposed to *n*-type impurities such as arsenic, antimony, or phosphorus at high temperature for a sufficient period of time to allow these impurities to convert the exposed underlying silicon to *n*-type material. The areas of resulting *n*-type material are shown in green. Notice in the cross section of Plate 1(d) that the red polysilicon area and the thin oxide under it act to prevent impurities from diffusing into the underlying silicon. Therefore the impurities reach the silicon substrate only in areas not covered by the polysilicon and not overlain by the thick original oxide. In this way the active transistor area is formed in all places where the patterned polysilicon overlies the thin oxide area defined in the previous step. The diffusion level sources and drains of the transistors are automatically filled in between the polysilicon areas and extend up to the edges of the thick oxide areas. The major advantage of the silicon-gate process is that it does not require a critical alignment between a mask that defines the green source and drain areas and a separate mask that defines the gate areas. Rather, the transistors are formed by the intersection of the two masks, and the conducting *n*-type diffused regions are formed in all areas where the green mask is not covered by the red mask.

All the transistors of the basic inverter circuit are now defined. Connections must now be made to the input gate, between the gate and source of the pull-up, and to VDD and GND. These interconnections will be made with a metal layer that can make contact with both the diffused areas and the polycrystalline areas. However, in order to ensure that the metal does not make contact with underlying areas except where intended, another layer of insulating oxide is coated over the entire circuit. At the places where the overlying metal is to make contact with either the polysilicon or the diffused areas, the overlying oxide is selectively removed by the patterning process as previously described. The result of coating the wafer with the overlying oxide and then removing the oxide in places where contacts are desired is shown in Plate 1(e). In the top view, the black areas are those defined by openings in the contact mask, the fourth in the process's sequence of mask patterns. In the cross section, notice that in the contact areas all oxide has been removed down to either the polycrystalline silicon or the diffused area.

Once the overlying oxide has been patterned in this way, the entire wafer is coated with metal, usually aluminum, and the metal is patterned with a fifth mask to form the conducting areas required by the circuit. The top view in Plate 1(f) shows three metal lines running vertically, the leftmost connecting to the input gate of the inverter, the center one being ground, and the rightmost forming the VDD connection to the inverter. The peculiar structure formed by the metal square slightly to the right of center connects the polysilicon gate of the depletion mode pull-up transistor to its source and to the drain of the pull-down transistor. Rather than making two separate contacts from the metal line to the pull-up's polysilicon gate region and to the adjacent diffusion region, area can be conserved by coalescing the contacts into the compact arrangement shown. This geometrical arrangement is known as a *butting contact* and will be used extensively throughout the text.

In general, it is good practice to avoid placing contacts over active transistor area whenever possible. However, butting contacts in the location shown here (Plate 1(f)) reduce the area and simplify the geometry of the basic inverter and of many other circuits. (The authors have used this form of contact successfully in many systems implemented by a number of different commercial wafer fabrication lines.) A more conservative approach would be to place the butting contact adjacent to, rather than over, the active pull-up area. (See also Section 2.6.)

The inherent properties of the silicon-gate process allow the blue metal layer to cross over either the red polysilicon layer or the green diffused areas, without making contact unless one is specifically provided. The red polysilicon areas, however, cannot cross the green diffused areas without forming a transistor. The transistors formed by the intersection of these two masks can be either enhancement mode, if no yellow implantation is provided, or depletion mode, if such an implantation is provided. Hence, the enhancement mode transistors are defined by the intersection of the green and red masks; the depletion mode transistors are defined by the intersection of the green, red, and yellow masks.

If we wish to fabricate only a small number of prototype system chips and to have access to the metal level for the probing of test points, the wafer-fabrication sequence can be terminated at this step. However, when fabricating large numbers of chips of a debugged design, the wafer surface is usually coated with another layer of oxide. This step, called *overglassing*, provides physical protection for the devices in the system. A sixth mask is then used to pattern contact cuts in the overglassing at the locations of relatively large metal wire-bonding pads.

Each wafer contains many individual chips. The chips are separated by first scribing the wafer surface with a diamond scribe and then fracturing the wafer along the scribe lines. Each individual chip is then cemented in place in a package, and fine metal wire leads are bonded to the metal contact pads on the chip and to pads in the package that connect with its external pins. A cover is then cemented over the recess in the package that contains the silicon chip, and the completed system is ready for testing and use.

2.4 YIELD STATISTICS

Of the large number of individual integrated system chips fabricated on a single silicon wafer, only a fraction will be completely functional. Flaws in the masks, dust particles on the wafer surface, defects in the underlying silicon, etc., all cause certain devices to be less than perfect. With present design techniques, any single flaw of sufficient size will kill an entire chip.

The simplest model for the *yield*, or the fraction of the chips fabricated that do not contain fatal flaws, assumes (naively) that the flaws are randomly distributed over the wafer and that one or more flaws anywhere on a chip will cause it to be nonoperative. If there are N fatal flaws per unit area, and the area of an individual chip is A, the probability that a chip has n flaws is in the simplest case given by the Poisson distribution, $P_n(NA)$. The probability of a good chip is

$$P_0(NA) = e^{-NA}. \tag{2-1}$$

While this equation does not accurately represent the detailed behavior of real fabrication processes, it is a good approximate model for estimating the yield of alternative designs. The exponential is such a steep function that a very simple rule is possible: chips with areas many times $1/N$ will simply never be found without flaws. Areas must be kept less than a few times $1/N$ if one flaw will kill a system. Design forms may be developed in the future that will permit systems to work even in the presence of flaws. If such forms are developed, the entire notion of yield will be completely changed and much larger chips will be possible.

Once a wafer has been fabricated, each chip must be tested to determine if it is functional. Functional testing of simple combinatorial logic networks is straightforward and may be done completely. Complete functional testing of complex systems with internal sequencing is not possible in general, and most integrated system chips manufactured, even at 1978 levels of complexity, are not economically testable for even a small fraction of their possible internal states.

As time passes and the number of devices per chip increases, it will become important to consider including special functions in the design of integrated systems to improve their testability. The basic problem is to linearize an otherwise combinatorial problem. One approach to this is as follows:

1. Define the entire system as a set of register-to-register transfer blocks, that is, successive stages of storage registers with combinational logic between them.
2. Provide for reading and writing from the external world to/from each of the storage registers.

The storage locations are first tested independently for their ability to store data or control information. If all storage locations pass this test, each combinational logic block can be tested separately, by use of its input and output storage locations. Such a test becomes essentially linear in the number of components and may be accomplished in an acceptable time period, even for extremely complex systems. However, without access to the individual storage locations, testing rapidly becomes hopeless. For this reason even present day microprocessors are very incompletely tested. When one is used for a while, an apparently new and sudden malfunction may simply be the first occurrence of a particular state of control and data in the system and thus may represent the first time the device has been "tested" under those conditions.

From experience gained in testing memory parts, it is known that the behavior of one circuit can be influenced by the state of a nearby circuit. For example, a memory cell may be able to remember both a logic-1 and a logic-0 if its neighbor is at a logic-0, but may be able to retain only a logic-0 if its neighbor is at a logic-1. Failures of this type are dependent upon the data patterns present in the system and are known as *pattern-sensitive* failures. In a reasonable (or even an unreasonable) time, it is not possible to exercise even a minute fraction of all the combinations of bit patterns of many integrated systems. What we do instead is apply our knowledge of the physics of such failures and construct a *model* for possible failure modes. In the memory example, we may conclude that any flaw not visible optically will be unable to reach beyond the immediate locality of the cell involved. Hence, pattern sensitivity in the behavior of a particular cell may be introduced by other cells in the same row or column of an array of memory cells, or by diagonal nearest neighbors. A test for pattern sensitivity under this model is quite fast, being only slightly worse than a linear function of the number of devices on the chip.

In order to test for pattern-sensitive failures, we must construct a physical model for the possible failure mechanisms. The model will inevitably include the physical proximity of other signals. For this reason, any practical test for pattern-sensitive failures must be based on a knowledge of the physical location of the various elements of the subsystem being tested. The task of preparing such tests is thus greatly eased by regularity in the design and physical layout of a system.

2.5   SCALING OF THE PROCESSING TECHNOLOGY

In order to have a complete process for forming submicron transistors, it is necessary not only to make patterns in the resist material but to transfer these patterns to the underlying layers in the silicon and silicon dioxide. Traditionally, wet-etching processes have been used. However, wet-etching processes do not scale well into the submicron range.

Alternatives are now being developed which should prove workable. Etching with plasmas (i.e., glow discharges of gaseous materials resulting in free ions of great chemical activity) is already used in a number of advanced processing facilities. Very well controlled etching can be achieved in this way. It seems likely that no wet processing will be used in the construction of submicron devices. Ion implantation, an ideal method for achieving controlled doses of impurity ions in the silicon surface, is already a common production technique in essentially all MOS processing facilities.

Metal layers for submicron processes must be thicker in relationship to their width than that produced using today's commercial processing technology. A possible solution to this problem may be the use of a process known as ion milling for metal patterning. In this process, ions of modest energy sputter away any metal not covered with resist material, yielding much steeper sides on the patterned metal than could be produced using wet-etching processes.

It appears that the basic technological pieces exist to enable development of a complete patterning and wafer fabrication process at submicron dimensions. In reality, the ultimate submicron process will not emerge full-blown, but dimensions will gradually be reduced, as one after another of the myriad of technological difficulties are surmounted. The sketch we have given, and will cover in more detail in Chapter 4, is rather an artist's conception of the possibility of such an ultimate process. We do believe, however, that the evolution of this process is of fundamental importance to the entire electronics industry.

2.6 DESIGN RULES

Perhaps the most powerful attribute of modern wafer fabrication processes is that they are *pattern independent*. That is, there is a clean separation between the processing done during wafer fabrication and the design effort that creates the patterns to be implemented. This separation requires a precise definition to the designer of the capabilities of the processing line. The specification usually takes the form of a set of permissible geometries that may be used by the designer with the knowledge that they are within the resolution of the process itself and that they do not violate the device physics required for proper operation of transistors and interconnections formed by the process. When reduced to their simplest form, such geometrical constraints are called *design rules*. The constraints are of the form of minimum allowable values for certain *widths, separations, extensions,* and *overlaps* of geometrical objects patterned in the various levels of a system.

As processes have improved over the years, the absolute values of the permissible sizes and spacings of various layers have become progressively smaller. There is no evidence that this trend is abating. In fact, there is every reason to believe that at least another order of magnitude of shrinkage in linear dimensions is possible. For this reason we present a set of design rules in dimensionless form, as constraints on the allowable ratios of certain distances to a basic length unit. The basic unit of length measurement used is equal to the fundamental resolution of the process itself. This is the distance by which a geometrical feature on any one layer may stray from another geometrical feature on the same layer or on another layer, all processing factors considered and an appropriate safety factor added. It is set by phenomena such as over-etching, misalignment between mask levels, distortion of the silicon wafer (''runout'') due to high temperature processing, and overexposure or underexposure of resist. All dimensions are given in terms of this elementary distance unit, which we call the *length unit*, λ. In 1978 the length-unit λ is approximately 3 microns for typical commercial processes. One micron $(\mu m) = 10^{-6}$ meters.

The rules given below have been abstracted from a number of processes over a range of values of λ, corresponding to different points in time at different fabrication areas. They represent somewhat of a ''least common denominator'' likely to be representative of nMOS design rules for a reasonable period of time, as the value of λ decreases in the future.

A typical minimum for the line width of the diffused regions, W_d, is 2λ, as shown in Plate 2(a). The spacing required between two electrically separate diffused regions is a parameter that depends not merely upon the geometric resolution of the process but also upon the physics of the devices formed. If two diffused regions pass too closely, the depletion layers associated with the junctions formed by these regions may overlap and result in a current flowing between the two regions when none was intended. In typical processes a safe rule of thumb is to allow 3λ of separation, S_{dd}, between any two diffused regions that are unconnected, as shown in Plate 2(b). The width of a depletion layer associated with any diffused region depends upon the voltage on the region. If one of the regions is at ground potential, its depletion layer will of necessity be quite thin. In addition, some processes provide a heavier doping level at the surface of the wafer between the diffused areas in order to alleviate the problem of overlap of depletion layers. In cases where either very low voltage exists on both diffused regions or a heavily doped region has been implanted in the surface between the diffused areas, it is often possible to space diffused areas 2λ apart. However, this should not be done without carefully checking the actual process by which the design is to be fabricated.

The minimum for the width of polysilicon lines, W_p, is similarly 2λ. No depletion layers are associated with polysilicon lines, and therefore the separation of two such lines, S_{pp}, may be as little as 2λ. These rules are illustrated in Plate 2, parts (c) and (d).

We have so far considered the diffused and polysilicon layers separately. Another type of design rule concerns how the two layers interact with each other. Plate 2(e) shows a situation where a diffused line is running parallel to an independent polysilicon line, to which it is not anywhere connected. The only requirement here is that the two unconnected lines not overlap. If they did they would form an unwanted capacitor. Avoidance of this overlap requires a separation S_{pd} of only λ between the two regions, as shown in Plate 2(e). A slightly more complex situation is shown in Plate 2(f), where a polysilicon gate area intentionally crosses a diffused area, thereby forming a transistor. In order to make absolutely sure that the diffused region does not reach around the end of the gate and short-circuit the drain-to-source path of the transistor with a thin diffused area, it is necessary for the polysilicon gate to extend a distance E_{pd} of at least 2λ beyond the nominal boundary of the diffused area, as shown in Plate 2(f).

A composite of several of these design rules is shown in Plate 2(g). Note that the minimum width for a diffused region applies to diffused regions formed between a normal boundary of the diffused region and an edge of a transistor, as well as to a diffused line formed by two normal boundaries. This situation is illustrated in the lower left corner of Plate 2(g).

As we have seen in Plate 1(d), ion implantation in the region that becomes the gate of a transistor will convert the resulting transistor into the depletion mode type. It is important that the implanted region extend outward beyond all four boundaries of the gate region, as shown in Plate 2(h). To avoid any possibility that some small fraction of the transistor might remain in enhancement mode, the yellow ion implantation region should extend a distance E_{ig} of at least $1\frac{1}{2}\lambda$ beyond each edge of the gate region. The separation S_{ig} between an ion implantation region and an adjacent enhancement mode transistor gate region should also be at least $1\frac{1}{2}\lambda$. Both situations and their design rules are illustrated in Plate 2(h).

A contact may be formed between the metal layer and either the diffused level or the polysilicon level by means of the contact mask. A set of rules apply to the amount by which each layer must provide an area surrounding any contact to it, so that the contact opening will not find its way around the layer to something unintended below it. Since no physical factors apply here other than the relative registration of two levels, a very simple set of design rules results. Each level involved in a given contact must extend beyond the outer boundary of the contact cut by λ at all points, as illustrated by extension distances E_{dc}, E_{pc}, and E_{mc}, in Plate 3, parts (a), (b), and (f). The contacts themselves, like the minimum width lines in the other levels, must be at least 2λ long and 2λ wide (W_c). This situation is illustrated for the diffusion and polysilicon levels in Plates 3(a) and (b). When making contact between a large metal region and a large diffused region, many small contacts spaced 2λ apart should be used, as shown in Plate 3(c). Contact cuts to diffusion should be at least 2λ from the nearest gate region, as shown in Plate 3(c).

Note that a cut down to the polysilicon level does not penetrate the polysilicon. Thus one can in principle make a contact cut to poly over a gate region, and such contacts are permitted in these design rules. However, since such a cut must be 2λ wide and surrounded on all sides by 1λ of poly, it is not possible to make such a contact above a minimum-sized transistor's gate region. Also, as device dimensions scale down and the poly and thin oxide become ever thinner, such cuts might penetrate too far, and thus they may not be allowed in the design rules in the future.

When a direct connection is required between a polysilicon region and a diffused region, we normally use a construct known as the butting contact. The detailed geometric layout of the butting contact is shown in Plate 3(d). In its minimum-sized configuration, it is composed of a square region of diffusion 4λ on a side, overlapped by a 3λ by 4λ rectangle of polysilicon. A rectangular contact cut, 2λ by 4λ in size, is made in the center of this structure. The structure is then overlaid with metal, thus connecting the polysilicon to the diffusion. The rules involved in Plate 3(d) are identical to those given so far, with the addition of a minimum of one λ overlap, O_{pd}, of the diffused and polysilicon layers in the center area of the contact.

In considering the design rules for the metal layer, notice that this layer in general runs over much more rugged terrain than any other level, as can be seen by referring to the cross section of Plate 1(f). For this reason it is generally accepted practice to allow somewhat wider minimum lines and spaces for the metal layer than for the other layers. As a good working rule 3λ widths (W_m) and 3λ separations (S_{mm}) between independent metal lines should be provided, as shown in Plate 3(e).

The metal layer must surround the contact layer in much the same way that the diffused and polysilicon layers did. Since the resist material used for patterning the metal generally accumulates in the low areas of the wafer, it tends to be thicker in the neighborhood of a contact than elsewhere. For this reason, metal tends to be slightly larger after patterning in the vicinity of a contact than elsewhere. It is generally sufficient to allow only one λ of space around the contact region for the metal, as for the other two layers. The rule for metal surrounding contacts is shown in Plate 3(f). (Additional layout artifacts, such as alignment marks, that are associated with conveying a chip's layout through the processes of maskmaking and wafer fabrication, are given in Chapter 4. Included there are guidelines for sizing such macroscopic layout artifacts as scribe lines and wire-bonding pads. However, the design rules given here are sufficient for the layout of the functional circuitry within an n MOS integrated system.)

The above design rules are likely to remain valid as the length-unit λ scales down in size with the passage of time. Occasionally, for specific commercial fabrication processes, some one or more of these rules may be relaxed or replaced by more complex rules, enabling slight reductions in the area of a system. While these details may be important for certain competitive products such as memory systems, they have the disadvantage of making the system design a captive of the specific design rules of the process. Extensive redesign and checking is required to scale down such a design as the length-unit scales down. For this reason, we recommend use of the dimensionless rules given, especially for prototype systems. Designs implemented according to these rules are easily scaled and may have reasonable longevity.

2.7 ELECTRICAL PARAMETERS

By satisfying the constraints imposed by the design rules, designers may create circuit layout patterns with the knowledge that the appropriate transistors, lines, etc., produced by the wafer-fabrication process will be as originally specified in their layout patterns. To complete a design it is necessary to also know the electrical parameters of the transistors, diffused layers, polysilicon layers, etc., so that the performance of circuits can be evaluated. The resistances per square of the various layers and the capacitance per square micron with respect to underlying substrate are shown in Table 2.1. Note that the resistance of a square of material contacted along two opposite sides is independent of the size of the square and equals the resistivity of the material divided by its thickness. The tabulated values are typical of processes running in 1978. As the circuit dimensions are scaled *down* by dividing by a factor α, the parameters scale approximately as described in the table.

The relative resistance values of metal, diffusion, poly, and drain-to-source paths of transistors are quite different. Diffused layers and good polysilicon layers have more than one hundred times the resistance per square area of the metal layer. A fully turned-on transistor has approximately one thousand times the resistance of the diffused and polysilicon layers. The capacitances are not as wildly different as the resistances of the various layers. Compare the capacitances in Table 2.1 to the gate-to-channel capacitance, as a reference. The diffused areas typically have one fourth the capacitance per square micron. Polysilicon on thick oxide has approximately one tenth, and the metal layer slightly less than one tenth, of the gate-channel capacitance per square micron.

The relative values of the resistances and capacitances are not expected to vary dramatically as the processes evolve toward smaller dimensions, with the exception of the transistor resistance per square, which is independent of α.

Table 2.1 Typical MOS electrical parameters (1978).

Resistances		
Metal	≈ 0.03 ohms/$\square$	(Resistances/square scale *up* by α,
Diffusion	≈ 10 ohms/$\square$	as dimensions scale *down* by α,
Poly	$\approx 15\text{-}100$ ohms/$\square$	except that the transistor $R/\square$
Transistor	$\approx 10^4$ ohms/$\square$	is independent of α.)
Capacitances		
Gate-channel	$\approx 4 \times 10^{-4}$ pf/μm^2	(Capacitances/micron^2 scale
Diffusion	$\approx 1 \times 10^{-4}$ pf/μm^2	*up* by α, as dimensions
Poly	$\approx 0.4 \times 10^{-4}$ pf/μm^2	scale *down* by α.)
Metal	$\approx 0.3 \times 10^{-4}$ pf/μm^2	

One note of warning: There is a wide range of possible values of polysilicon resistance for different commercial processes. Polycrystalline silicon suffers from inordinately high resistances at the crystal grain boundaries if the doping level in the polysilicon itself is not held quite high. This disease does not affect the diffused layers. For this reason, any processing that tends to degrade the doping levels in the diffused and polysilicon layers affects the polysilicon resistance much more dramatically than it affects the resistance of the diffused area. In general it is difficult to design circuits that are optimum over the entire range of polysilicon resistivity. If a circuit is to be run on a variety of fabrication lines, it is desirable for the circuit to be designed in such a way that no appreciable current is drawn through a long, thin line of polysilicon. In an important example in Chapter 5, polysilicon lines are used as buses along which information flows. The timing of these buses can be dramatically affected by the resistance of the polysilicon. However, the protocol used on the buses has the polysilicon lines precharged during one period of a clock and then pulled low by the appropriate bus source during a following clock period. In this way the circuit is guaranteed to work independently of the resistance of the poly. However, it may be considerably slower in processes of high poly resistivity.

2.8 CURRENT LIMITATIONS IN CONDUCTORS

One limit that is not covered in Sections 2.6 and 2.7 is that associated with the maximum currents through metal conductors. There is a physical process called *metal migration* whereby a current flux through a metal conductor, exceeding a certain limit, causes the metal atoms to move slowly in the direction of the current. If there is a small constriction in the metal, the current density will be higher and therefore more metal atoms will be carried forward from that point, narrowing the point still more. Hence, metal migration is a destructive mechanism causing open circuits in the metal layer carrying heavy currents.

For metals like aluminum, this limit is a few times 10^5 amperes per square centimeter, that is, a few milliamperes per square micron of cross section. The limit does not interfere too drastically with the design of integrated systems in current MOS technologies. However, many metal conductors in present integrated systems are operated near their current limit, and currents do not scale well as the individual elements are made smaller. Applying the scaling rules developed earlier, we found that the power per unit area is independent of the scale-down ratio. However, the supply voltage decreases and therefore the current per unit area increases as the devices are scaled down. For this reason it will not be possible to use processes for very large-scale integrated systems where the metal thickness scales in the same way as do other dimensions in the circuit. Much work will likely be done to develop processes enabling fabrication of metal lines of greater aspect ratio than is now possible. (Metal lines in 1978 are $\simeq 1~\mu m$ thick.)

Short pulses of current are known to contribute much less to metal migration than steady direct current. Nanosecond pulses of currents two orders of magnitude higher than the d.c. limit given above may be carried in metal conductors without apparent damage. Therefore, switching current may not be as damaging to metal conductors as a steady current.

These effects strongly favor processes like CMOS that do not require static d.c. current and favor design methodologies that maximize system function per unit d.c. current.

2.9   A CLOSER LOOK AT SOME DETAILS

Thus far our discussion of fabrication has been a general one, adequate for readers whose primary interest is in the systems aspects of VLSI. The following sections involve a more detailed examination of the capacitance of several important structures and a discussion of the relative merits and scaling behavior of several common processes. We suggest that the reader just skim through these sections during the first reading of this text.

In Table 2.1 we gave typical capacitances for the various layers to the substrate. These capacitances are those that would be measured if the voltage on the particular layer were zero (relative to the substrate). The dependence upon voltage of the capacitances of the different layers may sometimes be important and we will now discuss how this dependence arises. For those wishing more background information on the concepts of device physics used in this text see Grove (1967), Cobbold (1970), Muller and Kamins (1977), and Richman (1973).

When a negative voltage is applied to an n-type diffused region relative to the p-type bulk silicon, the negative electrons are pushed out of the n-type layer into the bulk and a current flows (Fig. 2.9(a)). In integrated systems we are careful to never allow the voltage on the n-type diffused regions to be more negative than the p-type bulk. Diffused regions are biased positively with respect to the p-type bulk, resulting in a reversed biased p/n junction. With the exception of a small leakage current, the reverse biased p/n junction acts merely to isolate one diffused region from another. The p-type bulk of our integrated system has a small number (typically 10^{15} to 10^{16} per cubic centimeter) of impurity atoms. When a voltage is applied to an n-type diffused region, its influence is felt well out into the p-type bulk. Positive charge carriers in the p-type bulk are repelled from the positively charged n-type layer, thereby exposing negatively charged impurity ions. The region surrounding the n-type diffused layer that has been depleted of positive charge carriers is referred to as a *depletion layer* and is shown schematically in Fig. 2.9(b). As the voltage on the n-type layer is increased, charge carriers are pushed farther back from the junction between the n-type layer into the p-type bulk, widening the depletion layer and exposing more charged impurity ions. The charge thus induced in the depletion layer as the voltage on the n-type diffused

Fig. 2.9 (a) n-type diffusion in p-type bulk silicon. (b) Depletion layer.

region is increased is responsible for the capacitance of the n-type diffused region relative to the substrate.

We will now consider a unit area of the junction. The total charge in the depletion layer per unit area is proportional to the number per unit volume of impurity ions in the bulk (N), and the width, s_0, of the depletion layer:

$$\frac{\text{Total charge}}{\text{Area}} \propto N s_0.$$

The electric field in the region is proportional to the charge per unit area:

$$\text{Electric field} \propto \frac{\text{charge}}{\text{area}} \propto N s_0.$$

The voltage between the n-type diffused layer and the p-type bulk on the far side of the depletion layer is proportional to the electric field times thickness of the depletion layer and therefore to the density of negatively charged ions in the depletion layer times the square of the width of the depletion layer:

$$\text{Voltage} \propto \text{electric field} \times s_0 \propto N s_0^2.$$

The capacitance per unit area is just the charge per unit area divided by the voltage across the depletion layer. From the above equations, we find that the capacitance is proportional to the square root of the density of impurity atoms in the p-type bulk divided by the voltage:

$$\frac{\text{Capacitance}}{\text{Area}} = \frac{Q}{V} \propto \frac{1}{s_0} \propto \left(\frac{N}{V}\right)^{1/2}.$$

This relationship is plotted in Fig. 2.10. Notice that the capacitance tends toward infinity as the voltage across the junction tends to zero. It would seem that this large capacitance would be disastrous for the performance of our integrated systems. However, this proves not to be the case. When the p/n junction was formed, the n-type region had an excess of negative charge carriers while the p-type bulk had an excess of positive charge carriers. When the two were brought together to form the junction, there was no voltage to prevent charge carriers of either type from flowing over into the opposite region. The initial flow caused the n-type layer to become more positive than the p-type layer. The flow ceased when just enough voltage built up to stop it. In silicon the voltage required to prevent the flow of charge carriers in such a situation is approximately 0.7 volts. Thus the true voltage across the junction is the initial "built-in" voltage plus the voltage we apply in our circuit. The variation of the capacitance per unit area with applied voltage is shown in Fig. 2.11. An approximate equation that can be used to calculate the junction capacitance, C_j, per unit area of diffused layers as a function of the

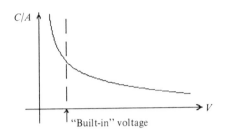

Fig. 2.10 C/A as a function of V.

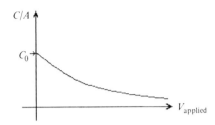

Fig. 2.11 C/A as a function of V_{applied}.

applied voltage is given by

$$C_j \approx 3 \times 10^{-12} \left(\frac{N}{V + 0.7} \right)^{1/2} \text{pf}/\mu\text{m}^2.$$

In this equation, N (the density of impurity ions in the p-type bulk) should be given in number per cm^3. The voltage is in volts and the capacitance per unit area is evaluated in picofarads per square micron. This equation is adequate for most design purposes.

Aside from the diffused regions, there are two other situations where the capacitance is of interest. The first is poly or metal over thick oxide and the second is the gate of an MOS transistor. We will discuss poly or metal over oxide first. Figure 2.12 illustrates once more the capacitance per unit area of a junction over the p-type bulk. If the poly or metal layer was laid on an oxide much thinner than the depletion layer, its capacitance would be nearly the same as that of the corresponding p/n junction. However, if an oxide is interposed whose thickness is of the order of the depletion layer thickness, the capacitance of the poly or metal line will be decreased. The formula which applies in this case is given by

$$\frac{1}{C_{\text{total}}} = \frac{1}{C_j} + \frac{1}{C_{\text{ox}}}.$$

A typical dependence is shown in Fig. 2.12. For an oxide thickness d, $C_{\text{ox}} = 3.5 \times 10^{-1}/d$, where d is given in angstrom units (10^{-4} microns) and the result is in picofarads per square micron as before.

The most spectacular voltage dependence of a capacitance in the technology we will be using is that of the gate of an MOS transistor. When the gate voltage, V_{gs}, is less than the threshold voltage, V_{th}, the capacitance of the gate to the bulk is just that given above for metal or poly over oxide, since the voltage on the gate merely depletes positive charge carriers back from the channel area. However, when the voltage on the gate reaches the threshold voltage of the transistor, nega-

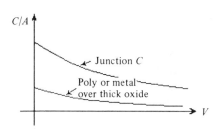

Fig. 2.12 Capacitance of poly or metal over thick oxide.

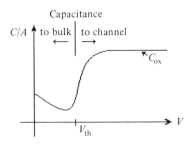

Fig. 2.13 MOS gate capacitance as a function of V.

tive charge is brought in under the gate oxide from the source of the transistor, and the capacitance changes abruptly from the small value associated with depleting charges in the bulk to the much larger oxide capacitance between the gate and the channel region. Further increase in voltage on the gate merely increases the amount of mobile charge under the gate oxide with no change in the width of the depletion layer underneath the channel. Hence, the character of the gate capacitance changes abruptly as the gate voltage passes through the threshold voltage.

The dependence of the total gate capacitance on gate voltage is shown in Fig. 2.13. The capacitance from channel to bulk is completely separate from the gate-to-channel capacitance. It is associated with the depletion layer underneath the channel region, and is almost identical to that of a diffused region of the same area. When the gate voltage is below threshold, the gate-to-channel capacitance disappears altogether leaving only the small parasitic overlap capacitances between the gate and the source and drain regions.

2.10 CHOICE OF TECHNOLOGY

Before proceeding to the chapters on system design, let us briefly examine some alternative technologies. Using the knowledge developed in these first two chapters, we will discuss the reasons for selecting nMOS as the single technology used to illustrate integrated systems in this text. Some of the factors that must be considered in choosing a technology include circuit density, richness of available circuit functions, performance per unit power, the topological properties of circuit interconnection paths, suitability for total system implementation, and general availability of processing facilities.

As the technology advances, more system modules can be placed on the same-sized chip. An ultimate goal is the fabrication of large scale systems on single chips of silicon. For this goal to be attained, any signal that is required in the system other than inputs, outputs, VDD, and GND must be generated in the technology on the chip. In other words, no subsystem can require a different technology for the generation of its internal signals. Thus such fully integrated systems cannot be implemented solely in a technology such as magnetic bubbles, since it cannot create the signals required for all operations in the on-chip medium.

We believe that for any silicon technology to implement practical large scale systems, it must provide two kinds of transistor. The rationale for this observation is as follows. In order to provide some kind of nonlinear threshold phenomenon there must be a transistor that is normally off when its control input is at the lowest voltage used in the system. Bipolar technologies use NPN transistors for this purpose. The nMOS technology uses n-channel enhancement mode devices. In addition to this transistor, a separate type of transistor must be supplied to allow the output of a driver device to reach the highest voltage in the circuit (VDD). In the bipolar technologies, PNP lateral devices are used to supply this function; in the n-channel technology, a depletion mode device is used; and in complementary MOS technology, a p-channel enhancement mode device is used. All three choices allow output voltages of drivers to reach VDD and thus meet the above criterion.

To date three technologies have emerged that are reasonably high in density and scale to submicron dimensions without an explosion in the power per unit area required for their operation. These are the n-channel MOS silicon gate process, the complementary MOS silicon gate process, and the integrated injection logic (I^2L) process.[1] Although present forms of I^2L technology lack the additional level of interconnect available in the silicon gate technologies, there is no inherent reason that such a level could not be provided. It is important to note that increasing the flexibility of interconnect enriches the types of functions that can be easily created. I^2L has the advantage over nMOS that the power per unit area (and hence the effective τ of its elementary logic functions) can be controlled by an off-chip voltage. The decision concerning at what point on the speed versus power curve to operate may thus be postponed until the time of application (or even changed dynamically) when using I^2L.

The nMOS scaling has been described previously. Any technology in which a capacitive layer on the surface induces a charge in transit under it to form the current control "transistor" will scale in the same way. Examples include Schottky Barrier Gate FET's (MESFET's), Junction FET's, and CMOS. Refer to Mead (1966) and Cobbold (1970) for descriptions of these alternative devices.

There are certain MOS processes (VMOS, DMOS) of an intermediate form in which the channel length is determined by diffusion profiles. While competitive at present feature sizes, these are likely to be interim technologies that will present no particular advantage at submicron feature sizes.

Scaling of the bipolar technology[2] is quite different from that of MOS technologies. For completeness, we include here a discussion of the scaling of bipolar devices, which may be of interest to those familiar with those technologies.

Traditionally, bipolar circuits have been "fast" because their transit time was determined by the narrow base width of the bipolar devices. In the 1950s, technologists learned how to form bipolar transistor base regions as the difference between two impurity diffusion profiles. This technique allowed very precise control of the distance perpendicular to the silicon surface and therefore permitted the construction of very thin base regions with correspondingly short transit times. Since current in a bipolar device flows perpendicular to the surface, both the current and the capacitance of such devices are decreased by the same factor as the device's surface dimensions are scaled down, resulting in no change in time performance. The base widths of high performance bipolar devices are already nearly as thin as device physics allows. For this reason, the delay times of bipolar circuits is expected to remain approximately constant as their surface dimensions are scaled down.

The properties of bipolar devices may be analyzed as follows. The collector current is due to the diffusion of electrons from emitter to collector. For a minority carrier density $N(x)$ varying linearly with distance x, from N_0 at the emitter to zero at the collector (at $x = d$), the current I per unit area A is

$$\frac{I}{A} = q(2D)\left(\frac{dN}{dx}\right) = \frac{q(2D)N_0}{d} = \frac{q(2kT/q)\mu N_0}{d} , \qquad (2\text{-}2)$$

where the diffusion constant $D = \mu kT/q$. The factor of two multiplies the diffusion constant in Eq. (2-2) because high-performance bipolar devices operate at high injection level (that is, where the injected minority carrier density is much greater than the equilibrium majority carrier density). The inherent stored charge in the base region is

$$\frac{Q}{A} = \frac{N_0 d}{2} . \qquad (2\text{-}3)$$

Therefore, the transit time is

$$\tau = \frac{Q}{I} = \frac{d^2}{4\mu kT/q} . \qquad (2\text{-}4)$$

The form of Eq. (2-4) is exactly the same as that for MOS devices (Eq. (1-1), with the voltage in the bipolar case being equal to $4 kT/q$ (at room temperature $kT/q = 0.025$ volts). A direct comparison of the transit times is shown in Table 2.2. At the smallest dimensions to which devices can be scaled, the base width of bipolar devices and the channel length of FET devices are limited by the same basic set of physical constraints and are therefore similar in dimension. The voltage on the FET devices must be many times kT/q to achieve the required nonlinearity. Hence at ultimately limiting small dimensions the two types of device have roughly equivalent transit times. At these limiting dimensions, choices between competing technologies will be made primarily on the grounds of the topological properties of their interconnects, the functional richness of their basic circuits, simplicity of

Table 2.2 Transit time: $\tau = (distance)^2/(mobility \times voltage)$

	MOSFET	MESFET, JFET	Bipolar
Distance	channel length	channel length	base width
Voltage	$\approx VDD/2$ (many kT/q)	$\approx VDD/2$ (many kT/q)	$4kT/q$
Mobility, in units of cm²/v-sec (Si)	≈ 800 (surface mobility)	≈ 1300 (bulk mobility)	≈ 1300 (bulk mobility)

process, and ability to control d.c. current per unit area. As supply voltages are scaled down to the 1-volt range, MOS devices become similar in most respects to other FET-type devices, and it is possible that mixed forms (MOS-JFET, MOS-MESFET, Bipolar-MESFET, etc.) may emerge as the ultimate integrated system technologies.

We have chosen to illustrate this text with examples drawn from the n-channel silicon gate depletion mode load technology. The reasons for this choice in 1978 are quite clear. In addition to meeting the required technical criteria we have described, this technology provides some important practical advantages to the student and to the teacher. It is the only high-density technology that has achieved universal acceptance across company and product boundaries. Readers wishing to implement integrated system designs may have wafers fabricated by essentially any wafer-fabrication firm, without fear that slight changes in the process or the vagaries of relationships with a particular firm will cut off their source of supply. It is also presently the highest density process available. This certainty of access to fabrication lines, the widespread knowledge of n MOS technology among members of the technical community, its density, and its performance similarity with bipolar technology in its ultimate scaling, are all important factors supporting its choice for this text on VLSI systems. However, the principles and techniques developed in this text can be applied to essentially any technology.

REFERENCES

1. F. M. Klaassen, "Device Physics of Integrated Injection Logic," *IEEE Transactions on Electron Devices*, March 1975, pp. 145–152, and cited papers by Hart & Slob, and by Berger & Weidmann.
2. B. Hoeneisen and C. A. Mead, "Fundamental Limits in Micro-electronics – II. Bipolar Technology," *Solid-State Electronics*, vol.15, 1972, pp. 891–897.

3

DATA AND CONTROL FLOW IN SYSTEMATIC STRUCTURES

3.1 INTRODUCTION

The process of designing a large-scale integrated system is sufficiently complex that only by adopting some type of regular, structured design methodology can one have hope that the resulting system will function correctly and not require a large number of redesign iterations. However, the methodology used should allow the designer to take full advantage of the architectural possibilities offered by the underlying technology.

In this chapter we present a number of examples of data and control flow in regularized structures. We discuss the way in which these structures can be assembled first into larger groups to form subsystems and then these subsystems assembled to form the overall system. The design methodology suggested in this chapter is but one of many ways in which integrated system design may be structured. The particular circuit form presented does tend to produce systems of very simple and regular interconnection topology and thus tends to minimize the areas required to implement system functions. Arrays of pass-transistor logic in register-to-register transfer paths are used wherever possible to implement system functions. This approach tends to minimize power dissipated per unit area and, with level restoration at appropriate intervals, tends to minimize the time delay per function. The methodology developed is applied in later chapters to the architecture and design of a data processing path and its controller, which together form a microprogrammed digital computer.

Computer architects, who usually design systems in a rather structured way using commercially available MSI and LSI circuit modules, are often surprised to discover how unstructured is the design within those modules. In principle one can use the basic NAND and NOR logic gates described in Chapter 1 to implement combinational logic, to build latches from these gates to implement data-storage registers, and then proceed to design integrated systems using traditional logic design methodology as applied to discrete devices. Integrated systems are often designed this way at the present time. However, it is unlikely that such unstructured approaches to system design can survive as the technology scales down towards maximum density VLSI.

There are historical reasons for the extensive use of random logic within integrated systems. The first microprocessors produced by the semiconductor industry were fairly direct mappings of early generation central processor architectures into LSI. A block diagram of the Intel 4004, the earliest microprocessor to see widespread commercial application, is illustrated in Fig. 3.1. The actual chip lay-

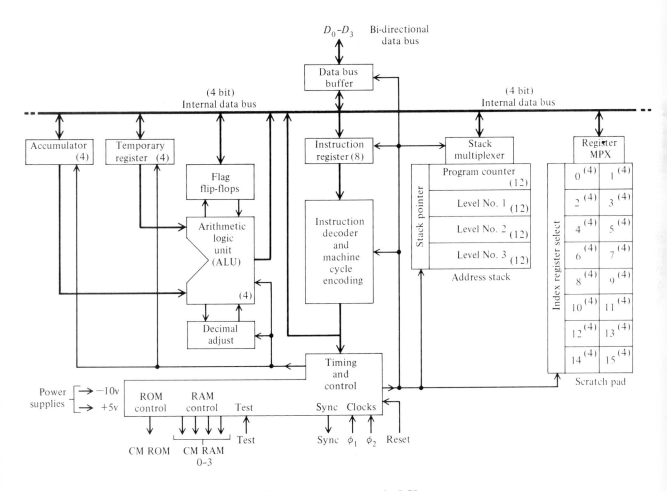

Fig. 3.1 Block diagram of the Intel 4004 Microprocessor, an early LSI system. (Reprinted with permission of Intel Corporation.)

out of the 4004 shown in Fig. 3.2 indicates the complexity of the LSI implementation of this simple central processing unit. Such LSI systems, using data paths and control functions appropriate in earlier component technologies, of necessity contained a great deal of random logic. However, the extensive use of random logic results in chip designs of very great geometrical and topological complexity, relative to their logical processing power.

To deal with such complexity, system design groups have often stratified the design problem into architecture, logic design, circuit design, and finally circuit layout, with specialists performing each of these levels of the design. Such stratification often precludes important simplifications in the realization of system functions.

Switching theory provides formal methods for minimizing the number of gates required to implement logic functions. Unfortunately, such methods are of little value in VLSI systems, since the area occupied on the silicon surface by circuitry is far more a function of the topological properties of the circuit interconnections than it is of the number of logic gates implemented. The minimum gate implementation of a function often requires much more surface area for its layout than does an alternative design using more transistors but having simpler interconnection topology.

There are known ways of structuring integrated circuit designs implemented using traditional logic design methods. A notable example is the *polycell* technique. In this technique, a group of standard cells corresponding to typical SSI or MSI functions are gathered into a library of functions. The logic diagram for the system to be implemented is used to specify which cells in the library are required. The cells are then placed into a chip layout and interconnections laid out between them by an automatic interconnection routing system. The polycell technique provides the logic designer who has limited knowledge of integrated systems with a means of implementing modest integrated circuit designs directly from logic equations. However, a heavy penalty is paid in area, power, and delay time. Such techniques, while valuable expedients, do not take advantage of the true architectural potential of the technology and do not provide insight into directions for further progress.

Switching theory not only yields the minimum number of gates to implement a logic function, but it also directly synthesizes the logic circuit design. Unfortunately, at the present time there is no general theory that provides us with a lower bound on area, power, and delay time for the implementation of logic functions in integrated systems. (Theoretical lower bounds for certain special structures and algorithms of interest are given in Chapter 8.)

In the absence of a formal theory, we can at best develop and illustrate alternative design methodologies that tend to minimize these physical parameters. Proposed design methodologies should, in addition, provide means of structuring system designs so as to constrain complexity as circuit density increases. We hope that the examples and techniques presented in this text will serve to clarify these issues and stimulate others to join in the search for more definitive results.[1]

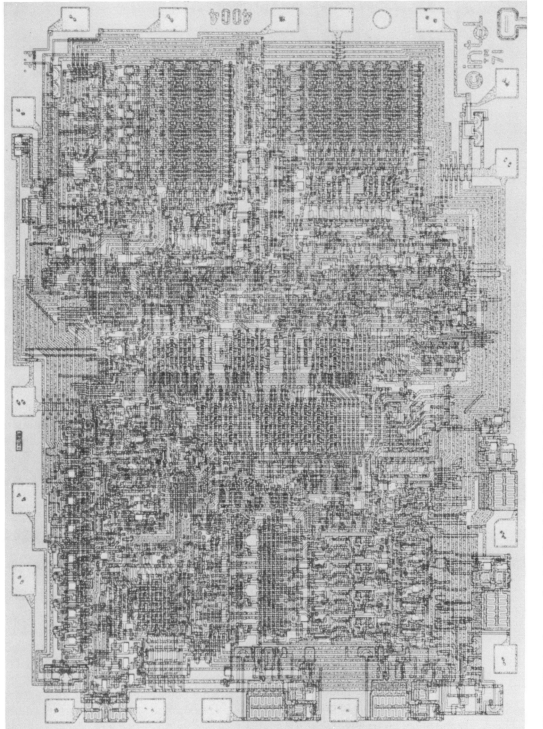

Fig. 3.2 Photomicrograph of a 4004 chip with pin designations. (Reprinted with permission of Intel Corporation.)

3.2 NOTATION

There are a number of different levels of symbolic representation for MOS circuits and subsystems used in this text. Plates 4(a), (b), (c), and (d) illustrate a NAND gate at several such levels. At times it may be necessary to show all the details of a circuit's *layout geometry* in order to make some particular point. For example, a clever variation in some detail of a circuit's layout geometry may lead to a significant compaction of the circuit's area without violating the design rules.

Often, however, a diagram of just the topology of the circuit conveys almost as much information as a detailed layout. Such *stick diagrams* may be annotated with important circuit parameters if needed, such as the length-to-width ratios shown in Plate 4(b). Many of the important architectural parameters of circuits and subsystems are a reflection of their interconnection topologies.

Alternative topologies often lead to very different layout areas after compaction. The discovery of a clever starting topology for a design usually provides far better results than does the application of brute force to the compression of final layout geometries. For this reason, many of the important structural concepts in this chapter and throughout the text will be represented for clarity by use of *colored* stick diagrams. The color coding in the stick diagrams is the same as in layout geometries: green symbolizes *diffusion* and *transistor channel region*; yellow symbolizes *ion implantation* for depletion mode transistors; red symbolizes *polysilicon*; blue symbolizes *metal*; black symbolizes a *contact cut*.

Later, through a number of examples in Chapter 4, we will present the details of procedures by which the stick diagrams are transformed into circuit layouts and then digitized for maskmaking. Note that if this topological form of representation were formalized, one might consider "compiling" such descriptions by implementing algorithms that "flesh out and compress" the stick diagrams into the final layout geometries,[2] according to the constraints imposed by the design rules.

When the details of neither geometry nor topology are needed in the representation, we may revert to the familiar *circuit diagrams* and *logic symbols*. At times we may find it convenient to *mix* several levels in one diagram, as shown in Plate 4(e). A commonly used mixture is (1) stick diagrams in portions where topological properties are to be illustrated, (2) circuit symbols for pull-ups, and (3) logic symbols, or defined higher level symbols, for the remaining portions of the circuit or system.

We will define logic variables in such a way that a *high voltage* on a signal path representing a variable corresponds to that variable being *true* (logic-1). Conversely, a *low voltage* on a signal path representing a logic variable corresponds to the variable being *false* (logic-0). Here *high voltage* and *low voltage* mean well above and well below the logic threshold of any logic gates into which the signal is an input. This convention simplifies certain discussions of logic variables and the voltages on the signal paths representing them. Thus when we refer to the logic variable β being *high*, we indicate simultaneously that β is *true* (logic-1) and is represented on the signal path named β by a *high voltage*, one well above the logic threshold. In Boolean equations and logic truth tables, we use the common notations of 1 and 0 to represent *true* and *false*, respectively, and by implication *high* and *low voltages* on corresponding signal paths.

3.3 TWO-PHASE CLOCKS

We will often make use of a particular form of "clocking" scheme to control the movement of data through MOS circuit and subsystem structures. By clocking scheme we mean a strategy for defining the times during which data is allowed to move into and through successive processing stages in a system, and for defining the intervening times during which the stages are isolated from one another.

Many alternative clocking schemes are possible, and a variety are in current use in different integrated systems.[3] The clocking scheme used in an integrated system is closely coupled with the basic circuit and subsystem structuring and has major architectural implications. For clarity and simplicity we have selected one clocking scheme, namely *two-phase, nonoverlapping clock signals*. This scheme is used consistently throughout the text and is well matched to the type of basic structures possible in MOS technology.

The two clock signals φ_1 and φ_2 are plotted as a function of time in Fig. 3.3. Both signals switch between zero volts (logic-0) and a voltage near VDD (logic-1), and both have the same period, T. Note that both signals are nonsymmetric and have nonoverlapping *high* times. The *high* times are somewhat shorter than the *low* times. Thus φ_2 is *low* all during each of those time intervals from when φ_1 rises, nears VDD, and then falls back to zero.

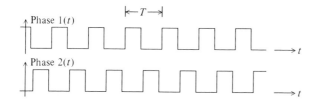

Fig. 3.3 Two-phase nonoverlapping clock signals.

Similarly, $\varphi_1 = 0$ all during each time interval when φ_2 is high. Therefore, at all times the logic AND of the two signals equals zero:

$$[\varphi_1(t)] \cdot [\varphi_2(t)] = 0, \qquad \text{for all } t.$$

For convenience, we will often use the following equivalence in our descriptions:

during φ_i is equivalent to *during the time period when* φ_i *is high.*

In the next section we will illustrate the use of these two clocking signals to move data through some simple MOS circuit structures. (A more detailed discussion of clocking requirements is given in Chapter 7.)

3.4   THE SHIFT REGISTER

Perhaps the simplest structure for enabling the movement of a sequence of data bits is the *serial shift register*, shown in the form of a circuit diagram in Fig. 3.4(a). The shift register is composed of level-restoring inverters coupled by pass transistors, with the movement of data controlled by applying clock signals φ_1 and φ_2 to the gates of alternate pass transistors in the sequence.

Data is shifted from left to right as follows. Suppose a logic signal X is present on the leftmost input to the shift register when clock signal φ_1 rises. Then, during the time when φ_1 is *high*, signal X will propagate through the pass transistor and be stored as charge on the input capacitance of the first inverter stage. For example, if signal X is *low*, then the inverter input gate capacitance will be discharged toward zero volts during the time when φ_1 is *high*. On the other hand, if X is *high*, the inverter input capacitance will charge up toward VDD $- V_{th}$ during φ_1.

When the clock signal φ_1 falls, the pass transistor becomes an open circuit, isolating the charge on the input of the inverter. The second clock phase is now initiated by the rise of φ_2. During the time interval when φ_2 is *high*, the logic signal X, now inverted, will flow through the second pass transistor onto the gate of the second inverter. This pattern can be repeated an arbitrary number of times to produce a shift register of any length.

Note that since the clock signals do not overlap, the successive pairs of stages of the shift register are effectively isolated from one another during the transfer of data between inverter pairs. For example, when φ_1 is *low* and φ_2 is *high*, all adjacent inverters connected by the φ_2-controlled pass transistors are in the process of transferring data from the left members to the right members of the pairs. All these pairs of inverters are isolated from each other by the intervening φ_1-controlled pass transistors that are all open circuits when φ_1 is *low*.

It is also important to note that the shortest period, T, we can use for clocks controlling such data transfers is determined by the time required to adequately charge or discharge the inverter input gate capacitance through the pass transistor and the preceding stage pull-up or pull-down. To this time must then be added an increment of time sufficient to ensure that the clocks do not overlap. For more complex systems, the minimum clock period may be estimated as a function of basic circuit parameters (as discussed in Chapter 1).

Figure 3.4(b) and Plate 5(a) illustrate the serial shift register using mixed notations. In Fig. 3.4(b), each inverter circuit diagram has been replaced by its logic symbol. In Plate 5(a), the pass transistor circuit symbols have been replaced by their stick diagrams. When visualizing the inverter as represented by its logic symbol in a circuit structure containing mainly stick diagrams, two points should be kept in mind:

1. The input to the inverter leads directly to the gate, and thus the gate capacitance, of the inverter's pull-down transistor. This input may be used to store a data bit by isolating the charge representing the bit with a pass transistor.

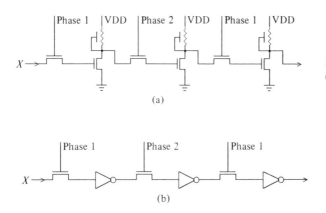

(a)

(b)

Fig. 3.4 (a) Shift register circuit diagram. (b) Shift register in mixed notation.

Note that the input path will end up on the poly level within the inverter. A contact cut may thus be required to connect the poly gate to the path, if it is metal or diffusion, on which the signal enters the inverter.

2. Since the connection between the source and gate of the inverter pull-up transistor requires a connection of all three conducting levels, the inverter output signal may easily be routed out on any one of the three levels.

Identical serial shift registers can be stacked next to each other and used to move a sequence of data *words*, as shown in Plate 5(b). The simple structure illustrated anticipates the elegant topological simplicity of many important MOS integrated subsystems. By connecting the successive inverter stages with diffusion paths, the pass transistors controlled by the clock signals are formed by simply running vertical clock lines in poly. The structure also anticipates another important point: topological simplification often results when control signals flow on lines that are at right angles to the direction of data flow. In this way, as many bits as necessary can be processed in parallel with the same control signals.

The example represented in Plate 5(b) is so rudimentary it is perhaps difficult to visualize the two clock signals as actually containing control information. Let us consider a slightly more complex example, the *shift-up register array* shown in Plate 5(c). In this structure, each data bit moving from left to right during φ_2 has two alternative pass transistor paths through which it can proceed to the next stage: a straight-through path and a path that shifts it up to the next higher row. If the control signal SH is *low*, then $[\varphi_2 \cdot \text{SH}']$ is *high*, and the straight-through pass transistor paths are used during φ_2. At the same time, $[\varphi_2 \cdot \text{SH}]$ is *low*, thus preventing data flow through the shift-up pass transistor paths. On the other hand, if SH is *high*, the straight-through pass transistors are off and the shift-up pass transistor paths are used during φ_2, resulting in the entire data word being shifted vertically as well as horizontally. Here the vertical control lines are run in metal, and the pass transistors are selectively formed by crossing the appropriate diffusion paths with short poly lines.

3.5 RELATING DIFFERENT LEVELS OF ABSTRACTION

In the discussions in this chapter, we will not have to make extensive calculations of the detailed electrical behavior of the devices and circuits involved in order to analyze the general behavior of digital logic constructed with these devices and circuits. Most of the examples presented in this chapter, and throughout the text, build upon the use of pass transistors to couple inverting logic stages as a means of structuring designs. The general results of Chapter 1 provide the solutions to most device and circuit problems encountered, such as ratio and delay calculations. In most cases, design concepts can be worked out using stick diagrams, and only at the stage of transforming the circuit topology into the detailed circuit layout geometry will any calculations need to be worked out, either by hand or with circuit simulation programs.

It is important to simplify our mental model of integrated circuitry, so as to more quickly and easily analyze or explain the function of a given circuit, and more easily visualize and invent new circuit structures without drifting too far away from physically realizable and workable solutions. Of course, it is a danger-ous practice to oversimplify our abstractions of electronic circuit behavior, and there are some nMOS circuits of deceptively simple appearance that have exceed-ingly complex behavior. However, throughout large portions of digital integrated systems that have circuit and subsystem designs structured as suggested in this text, an extremely simple mental model of device and circuit behavior will prove adequate to predict circuit and subsystem behavior.

Figure 3.5 illustrates a simple way of visualizing the operation of successive inverting logic stages coupled by pass transistors. Assume for the moment that any pass transistors in the paths between stages are on. To visualize the time behavior of an inverter, and the effect of the pull-up L/W to pull-down L/W ratio, imagine the flow of current from VDD to GND as the flow of a fluid and the inverter's two transistors as valves. The basis for thinking of the transistors in this

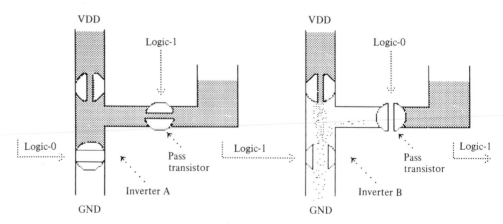

Fig. 3.5 A way of visualizing the operation of successive inverter stages.

manner is the fluid model of their internal behavior (as given in Chapter 1). Whether a transistor is on or off depends on the voltage, and thus on the charge, on its control gate and also on its threshold voltage. The upper ''valve'' is always open, since the pull-up transistor is always on. However, the valve corresponding to the pull-down transistor may be either open or closed, depending on the amount of charge on its gate.

In Fig. 3.5, the input to inverter A is a logic-0, so the pull-down of inverter A is off, and the lower valve is closed. Current is thus diverted to the large charge storage site corresponding to the gate of the pull-down of inverter B. At this level of diagram we have reverted to the common convention of positive charge flow from VDD to ground, rather than electron flow from ground to VDD. If sufficient positive charge has flowed onto this gate, corresponding to a high level of fluid in the tank representing the gate capacitance, then the pull-down of inverter B is turned on, and thus the lower valve of inverter B is open. If the lower valve in inverter B is much larger than the upper one, corresponding to a practical pull-up to pull-down ratio of size, then the pull-down of inverter B can sink all the source current provided by the pull-up. Also, if given sufficient time and if the connecting pass transistor is on, the pull-down can drain off any charge stored on the succeeding inverter's input gate. Thus we can visualize the sequence of inversions of a logic signal propagating through successive inverter stages as an alternation between high and low levels of fluid in the storage tanks. We can also visualize some of the time behavior of the signal propagation: the larger the gate capacitance, the longer it takes to build up enough charge to open the next stage, and the longer it takes to drain charge from the next stage to turn it off.

Figure 3.6 represents the same physical circuit modeled in Fig. 3.5, but on successively higher levels of abstraction. When analyzing circuit or logic diagrams

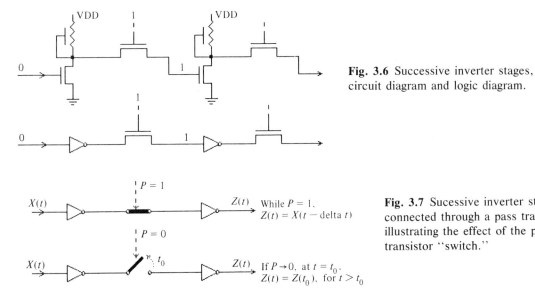

Fig. 3.6 Successive inverter stages, circuit diagram and logic diagram.

Fig. 3.7 Sucessive inverter stages connected through a pass transistor, illustrating the effect of the pass transistor ''switch.''

While $P = 1$, $Z(t) = X(t - \text{delta } t)$

If $P \rightarrow 0$, at $t = t_0$, $Z(t) = Z(t_0)$, for $t > t_0$

showing successive inverting logic stages, as in Fig. 3.6, one should keep the model of Fig. 3.5 in mind. Whether one is a novice or an expert in integrated system design, it is very helpful to compress the details of any given lower level of abstraction, so as to reduce the complexity of the problems presented at the next higher level and enable the mind to span problems of larger scope.

We are now able to visualize a very simple model for the pass transistor: it is in fact like a valve, or "switch," in the path between an inverter and the next charge storage site, i.e., the input gate of the next inverter. Figure 3.7 shows two inverters coupled by a pass transistor, with the pass transistor informally symbolized as a "switch." In the upper diagram, the pass transistor input is a logic-1, and so the "switch" is in the on position, resulting in the output Z being equal to the input X, after a suitable delay time Δt. Thus during the time the pass transistor gate input P is high, the output $Z(t)$ equals $X(t - \Delta t)$. Here Δt is some multiple of the transit time, τ, of the inverter pull-down transistor (as discussed in Chapter 1).

In the lower diagram, the pass transistor "switch" is moved to the off position since P is a logic-0. Therefore, according to our model, the valve in the path between the inverters is shut, and the charge, or lack of charge, is isolated in the storage site. Thus, once the pass transistor "valve" is shut, Z remains at a constant value, independent of changes in X. In other words, if $P \rightarrow 0$ at $t = t_o$, then $Z(t) = Z(t_o)$ for $t > t_o$.

These simple visualizations of the inverter and the pass transistor will carry us fairly far into LSI subsystem design. Several logic circuits in this chapter are drawn first in stick-diagram form and then informally sketched with pass transistors replaced with "switches"; this is both to clarify the behavior of the circuits involved and to further demonstrate the applicability of the model.

3.6   IMPLEMENTING DYNAMIC REGISTERS

Registers for the storage of data play a key role in digital system design. It is interesting to note that a group of adjacent inverters, with their gates isolatable by pass transistors, can be considered a form of temporary storage register. The arrangement, illustrated in Fig. 3.8, shows two levels of symbolism for a *dynamic register*. Such a register is very simple in structure. It consists of only three transistors per bit position: the pass transistor and the two transistors of the inverter. However, this dynamic form of register will preserve data only as long as charge

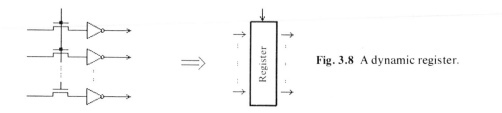

Fig. 3.8 A dynamic register.

can be retained on the inverter input gates. Typically, dynamic registers are used in situations where the updating control signals from the input gate are applied frequently. Dynamic registers are ideal in a clocked system in which they are reloaded every clock cycle, as in the shift register.

Suppose we wish to construct a simple register that can be loaded during the appropriate clock phase under the control of a *load* signal and will retain its information through an indefinite number of successive clock periods until it is reloaded using the *load* signal. A one-bit cell for such a register may be constructed using cross-coupled inverters in the configuration shown in Fig. 3.9. This register cell is still dynamic in form, since it uses charge storage on the gate of the first inverter to preserve its state. However, it need not be loaded on every successive φ_1, as was the simple register shown in Fig. 3.8. The pass transistor leading to it from the preceding stage is switched on only when *both* φ_1 and LD are *high*. On any following φ_1 when LD is *low*, the cell updates itself by the feedback path through the second pass transistor. Figure 3.10 illustrates a selectively loadable register composed of such cells. One important feature of this type of register is that it provides as output both the true and complemented forms of the stored data. This feature is often useful when the data are to be processed by a following network of combinational logic.

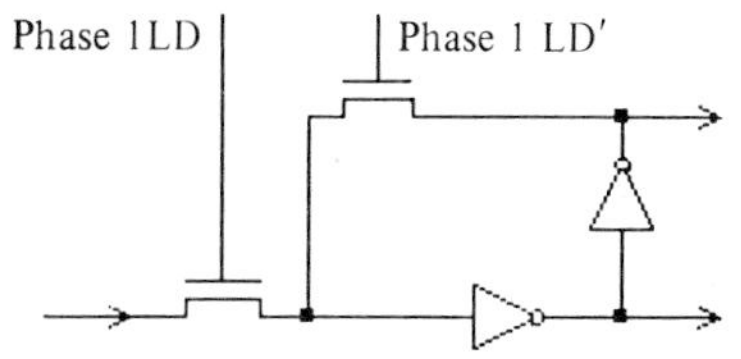

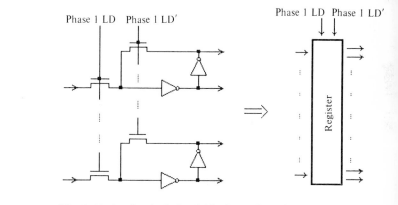

Fig. 3.9 A selectively loadable dynamic register cell.

Fig. 3.10 A selectively loadable dynamic register.

While there are more elaborate forms of dynamic and static registers, the above two forms are sufficient for many of the required data storage applications within integrated systems.

3.7 DESIGNING A SUBSYSTEM

The ideas used to construct simple dynamic registers in the preceding section may be applied to the construction of more sophisticated and interesting subsystems.

In this section we will describe the design of a *stack*. The methodology we use for this specific example we will find appropriate for a wide variety of functional subsystems. We first invent a "cell" that implements the most primitive function of the subsystem. This cell dictates a set of "timing" criteria necessary for its proper operation. The cell geometry together with the timing requirements dictates the design of control "circuits" that will surround an array of the basic cells. Once these control circuits are attached to the cell array, and the necessary "interconnections" are made, the entire assemblage constitutes a functional "module" with a well-defined "interface" to the next higher level of design. The interface consists of a functional specification, a geometrical specification, and a set of timing requirements for the control inputs, data inputs, and data outputs.

The stack subsystem is commonly called a *last-in, first-out* (LIFO) stack. It is also known as a *push-down* stack, although we will diagram it horizontally rather than vertically. It is a shift-register array with three basic operations: during each full clock period (1) we can *push* in a new data word at one end of the array, pushing all previously entered words one word position further into the array; or (2) we can leave all words in their current position; or (3) we can *pop* out a word from the end of the array, pulling all previously entered words back out by one word position.

Plate 6(a) shows the structure of one horizontal row of the stack. Here we have implemented a shift register that can perform the following three operations: shift data left to right, hold data in place, shift data right to left. There are four control signals used; two are active during φ_1 and two are active during φ_2. (The signals φ_1 and φ_2 are our familiar two-phase, nonoverlapping clock signals.)

In order for data to be shifted from left to right, the shift right control line (SHR) is driven *high* during φ_1, followed by driving the transfer right control line (TRR) *high* during φ_2. The bit of data appearing at the left is thus transferred by this operation onto the gate of the first inverter during φ_1, and thence to the gate of the second inverter during φ_2. In order for data to be held in place, the signal transfer left (TRL) is driven *high* during φ_1 and transfer right (TRR) is driven *high* during φ_2, causing the data to recirculate upon itself without shifting. Note that the data can be obtained at any time from the output of the first inverter. However, since new data may come to the gate of the first inverter during φ_1, the only safe time to take data out to the left is during φ_2. The transfer of data from right to left is caused by driving the shift left control (SHL) line *high* during φ_2, followed by driving transfer left (TRL) *high* during φ_1.

Plate 6(b) illustrates a possible topological structure of one horizontal row of the stack. There are two horizontal pathways on the diffusion level for shifting bits right or left. The two inverters for one stage of the row are nested between these paths. VDD, GND, and the four control lines run vertically in metal. The four pass transistors required for controlling the movement of data are conveniently implemented by short poly lines that cross the horizontal diffusion tracks at appropriate positions. Note that the entire row is composed of 180° rotations and repetitions of a basic cell containing one inverter.

In a typical implementation of the complete LIFO stack, a number of such rows run parallel to each other in the horizontal direction. The number of rows is equal to the width in bits of the data words involved. The control lines run vertically across the entire stack, perpendicular to the direction of data flow. For data words of any substantial width, the capacitive loading on the control signals would be sufficient to warrant use of super buffer drivers.

The stack as a whole may be controlled with only two logic signals: one signaling *push* and the other signaling *pop*. If neither of these two signals is activated, the data bits recirculate in place, awaiting the next active instruction.

Let us consider how to derive, from *push* and *pop*, the control signals for driving the four control lines SHR, TRR, SHL, TRL. A possible scheme is shown in Fig. 3.11. We use random logic for this purpose since only a few gates are required to control the large, regular array of circuit cells in the stack. The operation that determines what the stack will do during the subsequent clock phase is brought in on the path labeled OP. It is important to note in the following that only one signal path (OP) is required to bring in both *push* and *pop* logic signals, since these are active on mutually exclusive clock phases.

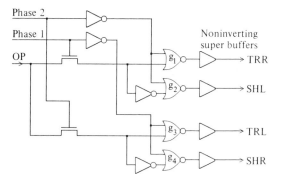

Fig. 3.11 Generating the stack control signals.

The control scheme is summarized in the timing diagrams in Fig. 3.12. Here we see that holding OP *high* during φ_2, followed by *low* during φ_1, implements *push*. Holding OP *low* during both φ_1 and φ_2 causes the data to recirculate in place. Holding OP *high* during φ_1, followed by *low* during φ_2, implements *pop*. Thus, the single signal path, OP, is sufficient to carry both stack control signals into the stack.

During φ_1, the OP signal is fed through the upper pass transistor into the inputs of the two NOR gates g_1 and g_2. The outputs for these NOR gates are *low* during this period, since φ_2' is *high*. If the incoming OP signal is *high* while φ_1 is *high*, then the lower input to NOR gate g_2 will be *low*. Thus when φ_2' falls *low*, the output of g_2 will go *high*, thereby driving SHL *high*. If the OP signal is instead kept *low* while φ_1 is *high*, then the output of the NOR gate g_1 will go *high* on the fall of φ_2', thereby driving TRR *high* during φ_2.

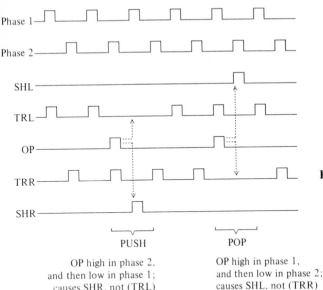

Phase 1

Phase 2

SHL

TRL

OP

TRR

SHR

PUSH

OP high in phase 2,
and then low in phase 1;
causes SHR, not (TRL)

POP

OP high in phase 1,
and then low in phase 2;
causes SHL, not (TRR)

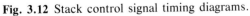

Fig. 3.12 Stack control signal timing diagrams.

During the period when φ_2 is *high* and either the shift left (SHL) or the transfer right (TRR) operation is being executed, the signal on the OP line is being stored on the corresponding input gates of the lower two NOR gates, g_3 and g_4. Thus, if OP is *high* while φ_2 is *high*, a logic-0 is stored on the input of the NOR gate g_4, and during the subsequent φ_1 *high* period, SHR will be driven *high*. Conversely, if OP is *low* while φ_2 is *high*, TRL will be driven *high* during the following φ_1 *high* period.

This kind of control scheme recognizes that there must be a lull period between any operation and its next occurrence. Control information is taken in during this period and set up for the subsequent operation. The scheme takes advantage of these lull periods, when possible, to perform other operations that can be done without conflict. It is an example of a fundamental design technique that can be extended to larger system structures.

When planning the overall architecture of a larger system, it is often useful to represent subsystems, such as the stack, using a higher level of symbolism. To be truly useful, such representations should, in addition to a functional definition, include the *topological* factors associated with the interconnection points of the subsystem and the *geometrical* factors of its shape and relative physical dimensions.

A system-level sketch of one particular implementation of the stack is shown in Fig. 3.13. Identical driver circuitry is placed along the top and bottom edges of the shift register array. The transfer-right and shift-left drivers that are set up during φ_1 (and active during φ_2) are placed along the top of the shift-register array. The transfer-left and shift-right drivers that are set up during φ_2 (and active during

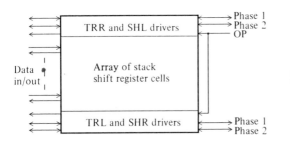

Fig. 3.13 Stack geometry and interconnect topology.

φ_1) are placed along the bottom of the array. The OP bit and the clock signals are required on both the top and the bottom of the shift-register array.

The integration of this subsystem into a larger integrated system design will require that the data-in and data-out paths be matched to those of subsystems to which the array is connected, and that the φ_1, φ_2, and OP signals be available at either the left or right side of the array. By using system-level representations that reflect as closely as possible the dimensions and locations of critical signals in all major subsystems, the interactions between topologies and dimensions of the subsystems can be assessed. The feasibility of an overall system architecture can thus be ensured prior to detailed design and layout.

3.8 REGISTER-TO-REGISTER TRANSFER

From an implementation point of view it is often desirable to combine logic-steering functions with the clocking of data into registers, since both require pass transistors as their elementary functional unit. An example is the shift-up register array; illustrated in Plate 5(c). From the next higher level system viewpoint, however, it is desirable to separate the two functions conceptually. In Fig. 3.14 we have shown some combination of inputs, X_0 through X_n going through some combination of pass transistors, *which may or may not have logic functions attached*, into the input gates of some inverting logic elements. This combination of pass transistors and logic elements is then abstracted into a register clocked on the phase during which the input pass transistors are turned on. Any logic function associated with the input pass transistors is considered part of the preceding combinational logic module. This viewpoint is an extension of the concept of dynamic register previously developed in Fig. 3.8.

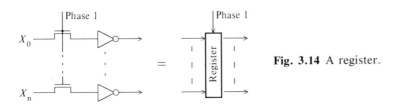

Fig. 3.14 A register.

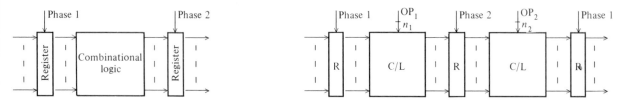

Fig. 3.15 A section of data path.   **Fig. 3.16** General form for a data path.

Using this notation, any processing function can be built up using blocks of the form shown in Fig. 3.15. Here we have a clocked input register, a block of strictly combinational logic *with no timing attached*, and an output register clocked on the opposite phase. In this case the inputs are stored in the input register during φ_1. They then propagate into and through the combinational logic (C/L), with the resulting outputs stored in the output register during φ_2. Any single data processing step can be viewed as a transfer from one such register to a second through a combinational logic block.

A sequence of such operations can be performed on a data stream by a series of such combinational blocks separated by registers, as shown in Fig. 3.16. Since different sets of data words in the stream may be operated on at the same time, but at different locations, this data path is a type of pipelined processing structure. Such pipelined processing structures offer the opportunity for improved processing bandwidth by performing many different operations concurrently. Notice that the throughput rate of such a pipeline system of register-to-register transfer operations is limited by the delay time through the slowest of the combinational logic blocks. If no registers had been interposed between the function blocks, and each operand set had been run separately through the entire sequence of combinational logic modules, the throughput rate would be much lower.

In line with the ideas developed earlier in this chapter, the detailed functions performed by the combinational logic modules may often be implemented in circuit structures of very simple and regular topology. Control signals will in general cross the data path at right angles to the direction of data flow. Figure 3.16 illustrates sets of such control inputs as n_1 lines carrying the control function OP_1 into the first C/L module, n_2 lines carrying OP_2 into the second, etc.

The idea of data being processed while passing through combinational logic interspersed between register stages in a sequence of register-to-register transfers is a basic and important concept in the hierarchy of digital system architecture. We have already described the implementation of registers. The next sections will describe some ways to implement combinational logic functions.

3.9   COMBINATIONAL LOGIC

Combinational logic modules contain no data storage elements. The outputs of a combinational logic module are functions only of the inputs to that module, provided that sufficient time has been allowed for those inputs to propagate through the module's circuitry.

In integrated systems, combinational logic design problems will typically fall within one of three general classes. The first is when a small amount of simple logic is required, for example, to derive control signals at the periphery of a system module (as in the stack control signal generation) or to implement a simple function within a single circuit cell (which may then be replicated in a regular array). In these cases, traditional logic design procedures using static NAND and NOR gates can be applied. Such designs involving a few gates are usually rather simple and can be produced by inspection rather than by use of formal minimization and synthesis procedures. Even in these simple cases, the minimum static logic gate implementation does not necessarily result in the most regular form, the minimum area, the minimum delay, or the minimum power. In fact, we often find alternative techniques to the use of static logic gates, which in specific instances lead to ''better'' designs by one of these measures than would minimum gate implementations. For example, Plate 7(a) shows a *selector* logic circuit in which one of the inputs S_0, S_1, S_2, S_3 is selected for output by the control variables A and B according to the function

$$Z = S_0 A'B' + S_1 A'B + S_2 AB' + S_3 AB.$$

The selector circuit is composed simply of poly paths crossing diffusion paths. Where depletion mode transistors are placed, the diffusion level path is always connected, thus placing control in the selectively located enhancement mode pass transistors, which function as simple switches. Figure 3.17 shows the circuit's paths from inputs to outputs using the ''switch'' abstraction for each of the pass transistors. For each possible combination of values of A and B, there is a path through the selector to Z from only one of the inputs S_i. For the specific inputs shown in the example in Fig. 3.17, the signal S_2 propagates through to Z since both A and B' are *high*. Note that no static power is consumed by the circuit, and the area occupied by the circuit is small since no contact cuts are required within it. (In Chapter 5 we describe a very general and powerful arithmetic logic unit (ALU)

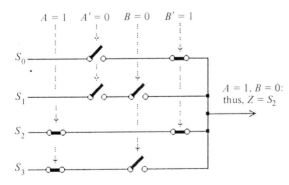

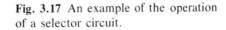

$A = 1, B = 0$:
thus, $Z = S_2$

Fig. 3.17 An example of the operation of a selector circuit.

that uses an array of such selector blocks to control a pass transistor carry network.)

The second general class of combinational logic design problems are those rather complex functions for which clever ways of structuring topologically regular implementations have been discovered. As an example, consider the implementation of a *tally* function with n inputs and $n + 1$ outputs. The kth output is to be *high* and all other outputs *low* if k of the inputs are *high*. The Boolean equations representing this function for the simple case of three inputs are

$$Z_0 = X_1' X_2' X_3';$$

$$Z_1 = X_1 X_2' X_3' + X_1' X_2 X_3' + X_1' X_2' X_3;$$

$$Z_2 = X_1 X_2 X_3' + X_1 X_2' X_3 + X_1' X_2 X_3;$$

$$Z_3 = X_1 X_2 X_3.$$

If this function were designed with random logic consisting of active pull-up static logic gates, it would result in a topological kludge. Plate 7(b) shows a topologically regular implementation of the tally function. A major portion of the function is implemented using a regular array of identical cells each containing only two pass transistors. The design is based on the idea of the shift-up register presented earlier. A *high* signal propagates through the array from the pull-up at the lower left. Whenever one of the variables X_i is *high*, the propagating *high* signal moves up to the next higher horizontal diffusion-level path. Thus the number of paths it moves up equals the number of inputs X_i that are *high*. Logic-0 signals propagate through the array from the ground points to all other outputs.

Figure 3.18 shows the paths from inputs to outputs for the tally circuit, using the "switch" abstraction for the pass transistors. The figure shows a specific example of a set of inputs controlling the pass transistors of the circuit. Since two of the inputs are *high*, the logic-1 signal is shifted up two rows and emerges at Z_2.

The tally function design can be easily expanded to handle more than three inputs by simply extending the array structure upward and to the right. However, remember that the delay through n pass transistors is proportional to n^2. Thus it may be necessary to insert level restoration prior to such extension. Similar comments apply to the extension of the selector circuit previously shown or to other pass-transistor logic arrays one might invent.

The electronic gates traditionally used in digital design are unilateral elements: they allow a logic signal to propagate in one direction only. It should be noted that the pass transistor is a bilateral circuit element. It permits the flow of current, and thus the passage of a logic signal, in either direction when its gate is *high*. While this property of the pass transistor is not necessarily of fundamental importance in integrated systems, it is an interesting and occasionally useful one.

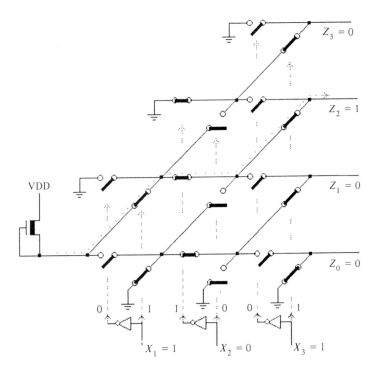

$Z_3 = 0$

$Z_2 = 1$

VDD

$Z_1 = 0$

$Z_0 = 0$

0 : : 1   1 : : 0   0 : : 1

$X_1 = 1$ $X_2 = 0$ $X_3 = 1$

Fig. 3.18 An example of the operation of a tally circuit, visualizing the location of the switches.

Early relay-switching logic used switching contacts that were bilateral elements. Interesting discussions of relay-switching logic are contained in Kohavi (1970) and Caldwell (1958). The tally array example just given is a basic *symmetric network* mapped directly into nMOS from relay-switching logic (Caldwell (1958), p. 241). The mathematics of switching universally used in digital systems today was proposed by Claude Shannon (1938). Shannon demonstrated that the calculus of propositions, based on the algebra of logic developed by Boole (1854), was directly applicable to relay-switching circuits.

A third combinational logic design situation occurs when a complex function must be implemented for which no direct mapping into a regular structure is known. Methods for handling this situation are the subject of the next section.

In the design methodology developed in this text, the combinational logic between stages in the register-to-register transfer paths is often done by operations on the *charge* moving between stages, using pass transistors to perform these operations. Many researchers at the present time are searching for alternative structures and techniques for performing elementary logic functions, including the use of charge-transfer devices.[4]

3.10 THE PROGRAMMABLE LOGIC ARRAY

On many occasions it is convenient to implement the combinational logic interspersed between register stages with regular structures of pass transistors.

However, we will often encounter important combinational logic functions that do not map well into such regular structures. In particular, combinational logic used in the feedback paths of finite-state machines is often highly complex and inherently irregular. Also, we may wish to delay binding the details of the logic functions used in sequencing a finite-state machine until most of the design is complete. If the combinational logic were implemented in an irregular structure, such changes could require a major redesign.

Fortunately, there is a way to map irregular combinational functions onto regular structures, using the *programmable logic array* (PLA) as described in this section. This technique of implementing combinational functions has a great advantage: functions may be significantly changed without requiring major changes of either the design or layout of the PLA structure.

One very general and regular way to implement a combinational logic function of n-inputs and m-outputs is to use a memory of 2^n words of m-bits each. The n-inputs form an address into the memory, and the m-outputs are the data contained in that address. Such a memory implements the full truth table for the output functions. Many systems are in fact built using memories as combinational logic elements. A common form of memory for this purpose is the *read-only memory* (ROM) where the data bits are permanently placed in the memory either by a mask pattern or by electrically altering the individual bit positions. There is one major difficulty with this approach: it is often the case that most of the possible input combinations cannot occur, due to the nature of the specific problem. Stated another way, many combinational logic functions require only a small fraction of all 2^n product minterms for a canonical sum of products implementation. In such cases, a ROM is very wasteful of area.

The programmable logic array (PLA) is a structure that has all the generality of a memory for implementing combinational logic functions. However, any specific PLA structure need contain a row of circuit elements only for each of those product terms that are actually required to implement a given logic function (see Kohavi (1970)). Since it does not contain entries for all possible minterms, it is usually far more compact than a ROM implementation of the same function. To achieve full compaction, the various output functions must be jointly minimized before the PLA layout pattern can be defined. However, such minimization is not essential. Less than full compaction increases the independence of the different entries, so that changes in function may require only local changes in the PLA.

Figure 3.19 illustrates the overall structure of a PLA. The diagram includes the input and output registers, in order to show how easily these are integrated into the PLA design. The inputs, stored during φ_1 in the input register, are run vertically through a matrix of circuit elements called the AND plane. The AND plane generates specific logic combinations of the inputs and their complements. The outputs of the AND plane leave at right angles to its inputs and run horizontally through another matrix called the OR plane. The outputs of the OR plane then run vertically and are stored in the output register during φ_2.

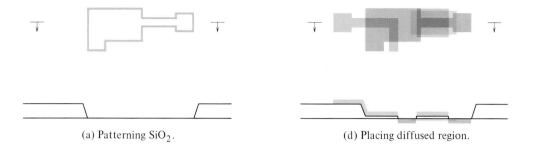

(a) Patterning SiO$_2$.

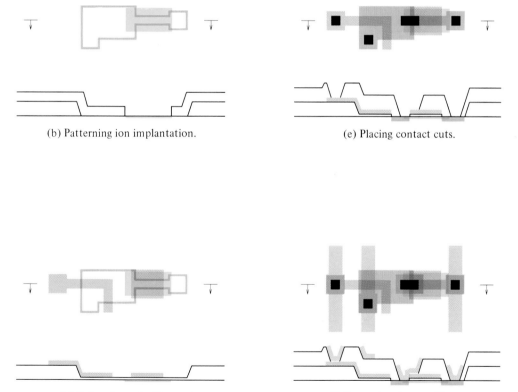

(b) Patterning ion implantation.

(c) Patterning polysilicon.

(d) Placing diffused region.

(e) Placing contact cuts.

(f) Patterning the metal layer.

PLATE 1 The *n*MOS process

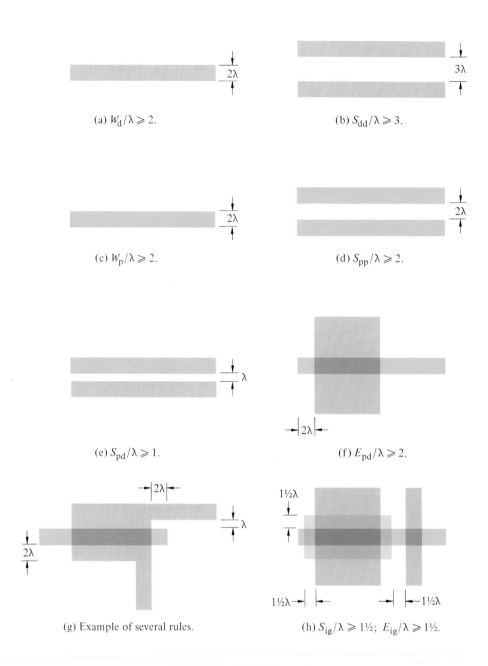

(a) $W_d/\lambda \geqslant 2$.

(b) $S_{dd}/\lambda \geqslant 3$.

(c) $W_p/\lambda \geqslant 2$.

(d) $S_{pp}/\lambda \geqslant 2$.

(e) $S_{pd}/\lambda \geqslant 1$.

(f) $E_{pd}/\lambda \geqslant 2$.

(g) Example of several rules.

(h) $S_{ig}/\lambda \geqslant 1\frac{1}{2}$; $E_{ig}/\lambda \geqslant 1\frac{1}{2}$.

PLATE 2 n MOS design rules

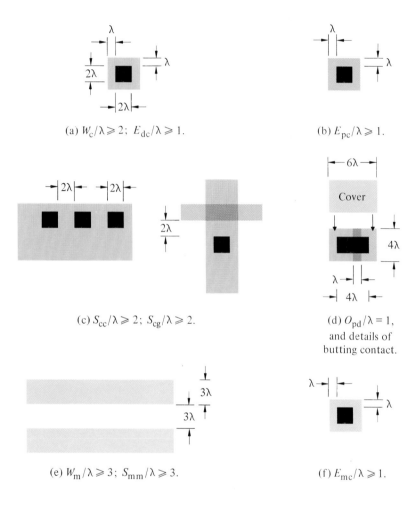

(a) $W_c/\lambda \geqslant 2$; $E_{dc}/\lambda \geqslant 1$.

(b) $E_{pc}/\lambda \geqslant 1$.

(c) $S_{cc}/\lambda \geqslant 2$; $S_{cg}/\lambda \geqslant 2$.

(d) $O_{pd}/\lambda = 1$, and details of butting contact.

(e) $W_m/\lambda \geqslant 3$; $S_{mm}/\lambda \geqslant 3$.

(f) $E_{mc}/\lambda \geqslant 1$.

PLATE 3 *n*MOS design rules (continued)

Scale in λ

H++++H
0 1 2 3 4 5 6

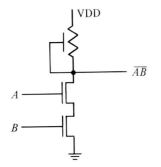

$\overline{AB}$

A

B

(a) NAND gate layout geometry.

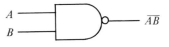

$Z = 4$

$\overline{AB}$

A —————— $Z = 1/2$

B —————— $Z = 1/2$

(b) NAND gate topology (stick diagram).

VDD

$\overline{AB}$

A

B

(c) NAND gate circuit diagram.

A ———⌐
B ———⌐ $\overline{AB}$

(d) NAND gate logic symbol.

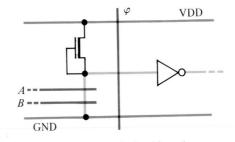

φ VDD

A
B

GND

(e) Example of mixed notation.

PLATE 4   Notation

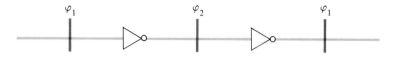

φ_1 φ_2 φ_1

(a) Shift register, more mixed notation.

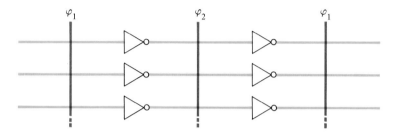

φ_1 φ_2 φ_1

(b) Array of shift registers.

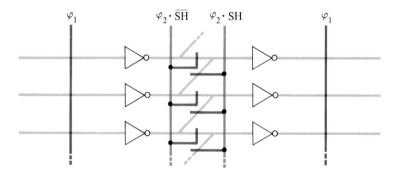

φ_1 $\varphi_2 \cdot \overline{SH}$ $\varphi_2 \cdot SH$ φ_1

(c) Shift–up register array.

PLATE 5 Shift register design

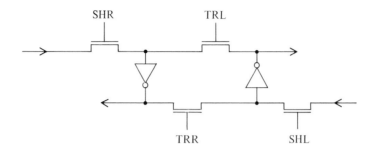

(a) One horizontal row of the stack. (SHR, TRL may be active only during φ_1; TRR, SHL may be active only during φ_2. See Fig. 3.15.)

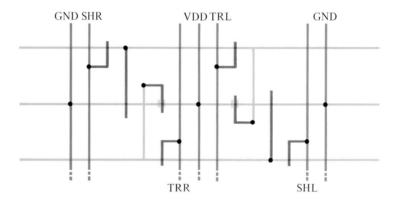

(b) Topology of one horizonal stack row.

PLATE 6 Stack cell design

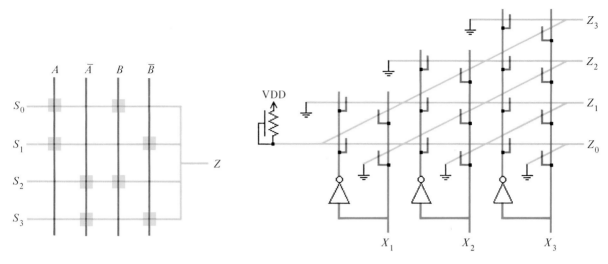

(a) Selector logic circuit.

(b) A tally circuit.

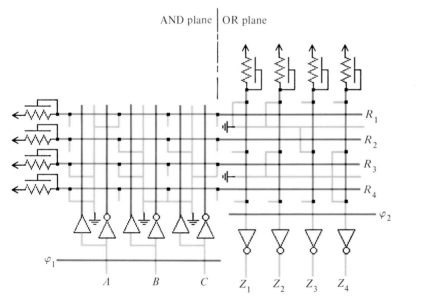

AND plane | OR plane

Product Terms

$R_1 = (A')' = A$
$R_2 = (B + C)' = B'C'$
$R_3 = (A + B + C')' = A'B'C$
$R_4 = (A + B' + C)' = A'BC'$

Outputs

$Z_1 = A$
$Z_2 = A + A'B'C$
$Z_3 = B'C'$
$Z_4 = A'B'C + A'BC'$

(c) Stick diagram of a PLA example.

PLATE 7 Several combinational logic functions

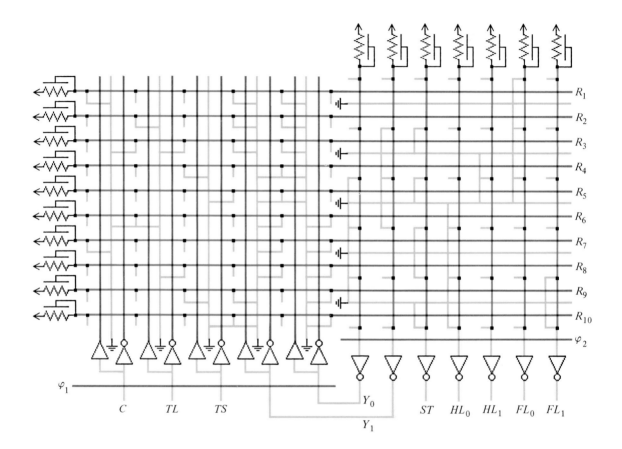

PLATE 8 PLA finite-state machine implementing the light controller

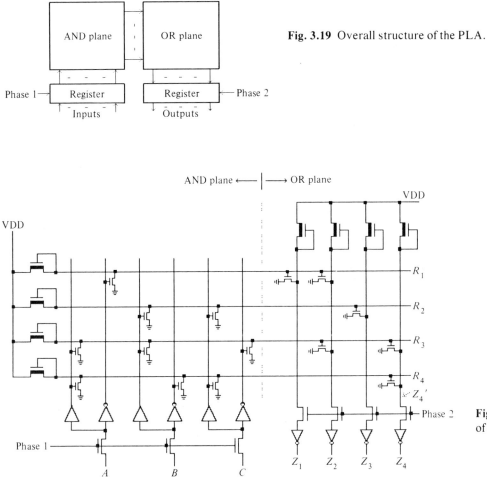

Fig. 3.19 Overall structure of the PLA.

Fig. 3.20 Circuit diagram of a PLA example.

The circuit diagram of a specific programmable logic array (Fig. 3.20) will help to clarify the structure and function of the AND/OR planes of the PLA. The input register bit for each input path is formed by a pass transistor clocked on φ_1 leading to both inverting and noninverting super buffers. The buffers drive two lines running vertically through the AND plane, one for the input term and one for its complement. The outputs of the AND plane are formed by horizontal lines with pull-up transistors at their leftmost end. The function of the PLA's AND plane is then determined by the locations and gate connections of pull-down transistors connecting the horizontal lines to ground.

Each output running horizontally from the AND plane carries the NOR combination of all input signals that lead to the gates of transistors attached to it. For example, the horizontal row labeled R_3 has three transistors attached to it in the AND plane, one controlled by A, one by B, and one by C'. If any of these inputs is *high*, then R_3 will be pulled down toward ground and will be *low*. Thus,

$$R_3 = (A + B + C')' = A'B'C.$$

Similarly,

$$R_4 = (A + B' + C)' = A'BC'.$$

The OR plane matrix of circuit elements is identical in form to the AND plane matrix, but rotated 90 degrees. Once again, each of its outputs is the NOR of the signals leading to the gates of all transistors attached to it. In Fig. 3.20, for example, both R_3 and R_4 lead to the gates of transistors leading from the output line Z'_4 to ground. If either R_3 or R_4 is *high*, Z'_4 will be *low*. Thus, $Z'_4 =$ NOR$(R_3, R_4) = (A'B'C + A'BC')'$. Up to this point the PLA implements the NOR-NOR *canonical form* of Boolean function of its inputs.

The output lines of the OR plane matrix are run into an output register formed by pass transistors (clocked on φ_2) leading into inverting drivers. Note that the output Z_4 at this point is $Z_4 = A'B'C + A'BC'$. This expression illustrates why the two PLA planes, each implementing the NOR function, are usually referred to as the AND plane and the OR plane. Following the output register, the outputs appear directly as the *sum of products canonical form* of Boolean functions of the PLA inputs, that is, as the OR of AND terms. Each horizontal line of the PLA carries one *product term*.

Plate 7(c) shows one possible layout topology for implementing the PLA in nMOS circuitry. The example is the same circuit illustrated in Fig. 3.20. The input lines crossing each plane are run in poly. The output lines from each plane are run in metal. Paths running to ground are placed between alternate poly lines, on the diffusion level. It is then a simple matter to form the pull-down transistors connecting the metal output lines to ground. They are selectively located diffusion lines under the appropriate input poly lines.

Although the PLA may implement a very irregular combinational function, the irregularity is confined to the irregular locations of pull-down transistors that "program" the function. The overall structure and topology of the PLA are very regular. Note that its overall shape and size is a function of the parameters: (1) the number of inputs, (2) the number of product terms, (3) the number of outputs, and (4) the length unit λ.

3.11   FINITE-STATE MACHINES

In many cases in the processing of data, it is necessary to know the outcome of the current processing step before proceeding with the next. Results of the current step may be used as inputs in the next step. The configuration shown in Fig. 3.21 can be used to implement a processing stage having this requirement. A typical

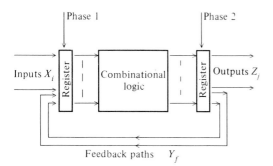

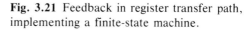

Fig. 3.21 Feedback in register transfer path, implementing a finite-state machine.

register-to-register transfer stage has been modified by simply feeding back some of its outputs to some of its inputs. This structure implements a form of sequential machine known as a *finite-state machine*.

The feedback signals form a binary number that may be regarded as identifying the *state* of the machine. The value of this number is stored, along with the external inputs, in the first register during φ_1. The combined inputs then propagate through the combinational logic. The resulting outputs are stored in the second register during φ_2. The falling edge of φ_2 must occur a sufficient time later to ensure that all signals have propagated through the combinational logic. Each complete machine cycle, consisting of φ_1 followed by φ_2, results in two new sets of outputs: (1) the external outputs that are typically used for controlling other units of the system, and (2) a new feedback number, which defines the *next state* of the machine. This process repeats during each clock period. The number of possible states is determined by the number of bits in the feedback path and is *finite*.

There are a number of ways of abstractly representing the states, the required state transitions, and the outputs of sequential machines under given input sequences. Possible representations include state diagrams, transition tables, Boolean or numerical difference equations, etc. A large body of theory has been developed concerning sequential machines. The serious reader will benefit from a further study of the results of switching theory on this subject (Dietmeyer (1971) and Kohavi (1970)).

Implementations of simple finite-state machines are used to produce the very lowest level of system control sequencing, since they can autonomously generate control sequences. The sequential machine having a finite number of states is a very important element in the hierarchy of fundamental concepts used in integrated system architecture.

The configuration shown in Fig. 3.21 implements a *synchronous* machine, since the feedback loop is activated only at times determined by the clock signals. In any clock period k, the output terms Z_j and the next state terms Y_f are valid during $\varphi_1(k)$. They are functions of the external inputs X_i and feedback terms Y_f that were valid during $\varphi_1(k-1)$.

If a sequential machine contains a feedback loop that is continuously active, then it may begin a response to a change in inputs or state at any time, rather than just at fixed clock times. Such a sequential machine is referred to as an *asynchronous* sequential machine. The analysis of asynchronous machines and their implementation is far more complex than that of synchronous ones. Great care must be exercised to avoid any difference in state sequencing and outputs under arbitrary differential delays of signals through the circuit paths of such machines (Dietmeyer (1971), Ch. 5). There will be only a few special cases where we use the asynchronous form of machine (Chapter 7), and these will be subject to detailed analysis.

Where finite-state machines are required within integrated systems, we will generally implement them in synchronous form. Synchronous machines are rather easy to implement correctly, and they fit naturally into the two-phase clocking scheme used for moving data around within our systems. However, the reader should carefully note that an implementation of a synchronous finite-state machine functions correctly only if the delays in the circuit paths are sufficiently short compared to the clock period. If we were to implement many copies of a particular machine, the probability of correct function for any given copy would be a function of both the clock period used and the distribution of differential delays in that copy's signal paths. Our estimate that a particular copy will function correctly is thus based in part on assumptions about the ratio of likely deviations in circuit delays to the clock period. (A discussion of delays in MOS circuits is given in Chapter 1.)

There is a very straightforward way to implement simple finite-state machines in integrated systems: we use the PLA form of combinational logic and feed back some of the outputs to the inputs, as illustrated in Fig. 3.22. The circuit's structure is topologically regular, has a reasonable topological interface as a subsystem, and is of a shape and size that are functions of the appropriate parameters. The function of this circuit is determined by the "programming" of its PLA logic. If, for example, early in a design cycle there is some uncertainty in the details of the desired sequencing of such a circuit, it is easy to provide layout space for extra unused inputs, product terms, or outputs as contingencies.

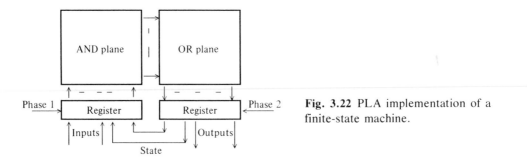

Fig. 3.22 PLA implementation of a finite-state machine.

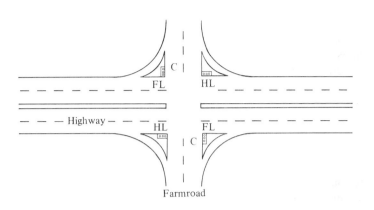

Fig. 3.23 A highway intersection.

3.11.1 An Example

The following simple example will help illustrate the basic concepts of finite-state machines and their implementation in n MOS circuitry. A busy highway is intersected by a little-used farmroad, as shown in Fig. 3.23. Detectors are installed that cause the signal C to go *high* in the presence of a car or cars on the farmroad at the positions labeled C. We wish to control traffic lights at the intersection, so that in the absence of any cars waiting to cross or turn left on the highway from the farmroad, the highway lights will remain green. If any cars are detected at either position C, we wish the highway lights to cycle through caution to red and the farmroad lights then to turn green. The farmroad lights are to remain green only while the detectors signal the presence of a car or cars, but never longer than some fraction of a minute. The farmroad lights are then to cycle through caution to red and the highway lights then to turn green. The highway lights are not to be interruptible again by the farmroad traffic until some fraction of a minute has passed.

A state diagram model of a finite-state machine to control the lights is sketched in Fig. 3.24. The diagram identifies four possible states of the machine

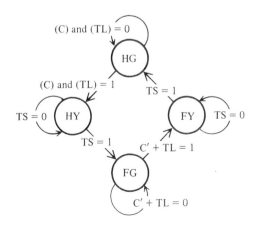

Fig. 3.24 Light controller state diagram.

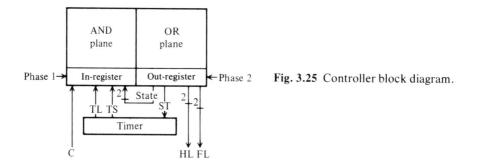

Fig. 3.25 Controller block diagram.

and indicates the input conditions that cause all possible state transitions. A block diagram of the PLA circuit implementing the machine is shown in Fig. 3.25. The circuit uses the signal C as an input and provides outputs HL and FL that encode the colors of the highway and farmroad lights it controls. (The input C can't normally be brought directly from a switch in the external world, but must be "conditioned" in some way. The issues surrounding the handling of such asynchronous inputs are considered in Chapter 7). Note that a timer is used to provide, as controller inputs, the short and long time-out signals (TS and TL) at appropriate times following a start timer (ST) signal output from the controller. This timer could be implemented as a synchronous digital counter in the same *n*MOS circuitry. Another abstract model describing the desired function of the controller is given in the state transition table (Table 3.1), which contains information similar to that in the state diagram.

Table 3.1 Transition table for the light controller.

In present state	If inputs* are	Next state will be	and outputs are		
			HL	FL	ST
Highway Green	(Cars)and(TimeoutL) = 0	Highway Green	Green	Red	No
	(Cars)and(TimeoutL) = 1	Highway Yellow	Green	Red	Yes
Highway Yellow	TimeoutS = 0	Highway Yellow	Yellow	Red	No
	TimeoutS = 1	Farmroad Green	Yellow	Red	Yes
Farmroad Green	(Cars)'or(TimeoutL) = 0	Farmroad Green	Red	Green	No
	(Cars)'or(TimeoutL) = 1	Farmroad Yellow	Red	Green	Yes
Farmroad Yellow	TimeoutS = 0	Farmroad Yellow	Red	Yellow	No
	TimeoutS = 1	Highway Green	Red	Yellow	Yes

*Inputs not listed = don't cares

The detailed sequencing of the machine under various input sequences is described by both the state diagram and transition table models of the controller. Consider starting in the state HG, where the highway lights are green. The

Table 3.2 Encoded state transition table for the light controller.

Stored during φ_1 in In-register				Stored during φ_2 in Out-register						
Inputs			Present state	Next state	Outputs					Product terms
C	TL	TS	Y_{p0}, Y_{p1}	Y_{n0}, Y_{n1}	ST	HL_0	HL_1	FL_0	FL_1	
0	X	X	0, 0 (HG)	0, 0 (HG)	0	0	0	1	0	R_1
X	0	X	0, 0 (HG)	0, 0 (HG)	0	0	0	1	0	R_2
1	1	X	0, 0 (HG)	0, 1 (HY)	1	0	0	1	0	R_3
X	X	0	0, 1 (HY)	0, 1 (HY)	0	0	1	1	0	R_4
X	X	1	0, 1 (HY)	1, 1 (FG)	1	0	1	1	0	R_5
1	0	X	1, 1 (FG)	1, 1 (FG)	0	1	0	0	0	R_6
0	X	X	1, 1 (FG)	1, 0 (FY)	1	1	0	0	0	R_7
X	1	X	1, 1 (FG)	1, 0 (FY)	1	1	0	0	0	R_8
X	X	0	1, 0 (FY)	1, 0 (FY)	0	1	0	0	1	R_9
X	X	1	1, 0 (FY)	0, 0 (HG)	1	1	0	0	1	R_{10}

machine remains in state HG as long as either no cars are detected or the long time-out has not occurred, in other words as long as (C)AND(TL) = 0. After the long time-out occurs, if any cars are detected, the machine restarts the timer and changes state to HY, where the highway lights are yellow. It remains in state HY only until the short time-out occurs, and then restarts the timer and changes to state FG, where the farmroad lights are green. It remains in state FG until either no cars are detected or the long time-out occurs, that is until (C)'OR(TL) = 1. Then it restarts the timer and changes to state FY, where the farmroad lights are yellow. It remains in state FY only until the short time-out occurs. It then restarts the timer and changes to state HG, the starting state.

The locations of transistors in the PLA light controller circuit can be determined by "hand assembling" the "program" specified in the "symbolic" transition table in Table 3.1, resulting in the encoded state transition of Table 3.2. First we assign codes to the states: state HG is encoded as $(Y_0, Y_1) = (0,0)$, HY as $(0,1)$, FG as $(1,1)$, and FY as $(1,0)$. Next we assign codes to the output light control signals: green is encoded as $(0,0)$, yellow as $(0,1)$, and red as $(1,0)$. We now form the encoded state transition table by constructing one row for each product term implied in Table 3.1. A row in Table 3.1 specifying a state transition as a function of a single input variable or single product term of input variables produces a single row in Table 3.2. A row in Table 3.1 specifying a state transition as a function of a sum or sum of products of input variables leads to a corresponding number of rows in Table 3.2.

Placement of the transistors within the PLA matrices follows directly from the encoded state transition table according to the three rules that follow. (Note that if all lines that control the transistors connecting a given product term line to ground are *low*, then that product term line will be *high*. Otherwise it will be *low*.):

1. For each logic-1 in the next state and output columns in the table, we run a diffusion path *from* the corresponding next state or output line in the PLA OR plane, *under* the corresponding product term line, *to* ground. This creates a transistor controlled by the product term line. Then, if that controlling product term line is ever *high*, the path to the output inverter will be *low*, and the output will be *high*. The output line will be *low* unless some product term line controlling it is *high*.

2. For each logic-1 in the input and present state columns in the table, we run a diffusion path *from* the corresponding product term line, *under* the corresponding *inverted* input or state line in the PLA AND plane, *to* ground. The transistor thus created is controlled by the *inverted* input or state line. Whenever that controlling line crossing the AND plane is *high*, the product term line will be *low*.

3. For each logic-0 in the input and present state columns in the table, we run a diffusion path *from* the corresponding product term line, *under* the corresponding *noninverted* input or state line in the PLA AND plane, *to* ground. The transistor thus created is controlled by the *noninverted* input or state line. Whenever that controlling line crossing the AND plane is *high*, the product term line will be *low*.

The PLA finite-state machine in Plate 8 is programmed from the transition table in Table 3.2, according to the rules above, and it implements the traffic light controller. Note that this LSI implementation does not exactly strain itself to meet the time response requirements of the control problem: it can run at a clock rate at least 10^7 times as fast as required. Also, note that the PLA controller is roughly $(150\lambda)^2$ in area. Using the 1978 value of $\lambda = 3\,\mu m$, this controller is $(450\,\mu m)^2 \approx$ 0.002 cm² in area. A PLA controller this size may contain over 150 transistors but occupies only 1/125 of the area of a typical 0.25 cm² silicon chip in 1978. By the late 1980s, as λ scales down toward its ultimate limits, such a controller will require only $\approx 1/25,000$ of the area of such a chip.

As we will see in later chapters, a data processing machine of any desired complexity can be created by interconnecting register-to-register data processing paths constructed along the lines of that shown in Fig. 3.16, such paths being controlled by finite-state machines implemented as shown in Fig. 3.22. The data paths form the "highways" for the movement of data under control of the finite-state machine "traffic controllers."

3.12   TOWARD A STRUCTURED DESIGN METHODOLOGY

The task of designing very complex systems involves managing, in some highly structured way, the space and time relationships between the various levels of system building blocks so that the entire system will function as intended when it is finished. The beginnings of a structured design methodology for VLSI systems can be produced by merging together in a hierarchy the concepts presented in this chapter. Designs are then done in a "top down" manner but with a full understanding by the architect of the successive lower levels of the hierarchy.

To begin, we plan our digital processing systems as combinations of register-to-register data transfer paths, controlled by finite-state machines. Then the geometric shapes, relative sizes, and interconnection topologies of all subsystem modules are collectively planned so all modules will merge together snugly, with a minimum of space and time wasted by random interconnect wiring. Storage registers are typically constructed by using charge stored on input gates of inverting logic. The combinational logic in the data paths is typically implemented using steering logic composed of regular structures of pass transistors. Most of the combinational logic in the finite-state machines is typically implemented using PLA's. All functioning is sequenced using a two-phase, nonoverlapping clock scheme.

When viewed in its entirety, a system designed in this manner is seen as a hierarchy of building blocks, from the very lowest level device and circuit constructs, up to and including the high-level system software and application programs in which the intended functions of the system are finally expressed. Individuals who understand the key concepts of each level in this hierarchy will recognize that the boundaries between levels are rather elastic ones. Each level of activity might best be optimized not on its own as a specialty but as it fits into an overall systems picture. For example, the activity "logic design" in integrated systems might best be conceptualized as the search for techniques and inventions that best couple the physical, topological, and geometric properties of integrated devices and circuits with the desired properties of digital VLSI systems. The search for alternative components for any given design hierarchy, and the search for alternative hierarchies, will be done best by those who span more than one specialty.

A particularly uniform view of such a system of nested modules emerges if we view every module at every level as a finite state machine or data path controlled by a finite-state machine. At the lowest level, elements such as the stack and register cells may be viewed as state machines with one feedback term (the output), two external inputs (the control signals), and a 1-bit state register. These rudimentary state machines are grouped in a structured manner to form portions of a state machine, or data path controlled by a state machine, at the next level of the hierarchy. Structured arrays of identical state machines often provide a mechanism for distributing processing among memory cells (Unger, 1958), thus enabling vast increases in processing bandwidth. Although in some cases the feedback paths are used in rather specialized ways, the state-machine metaphor still provides a precise description of module behavior. The entire system may thus be viewed as a giant hierarchy of nested machines, each level containing and controlling those below it. (A detailed quantitative treatment of certain hierarchically organized machines is given in Chapter 8.)

(In Chapters 5 and 6 we will apply the design methodology developed in this chapter to the design of a digital computer system. A 1-chip implementation of the data path portion of this computer system is illustrated in the frontispiece. Consistent use of the described design methodology resulted in a design of great regularity, short delay times, low power consumption, and high logical processing capability. As we will see in Chapter 4, regular designs, with small numbers of basic circuit cell types replicated in two dimensions to form subsystems, also have significant implementation advantages over less structured designs.)

REFERENCES

1. I. E. Sutherland and C. A. Mead, "Microelectronics and Computer Science," *Scientific American*, September 1977, pp. 210–228.
2. J. D. Williams, "Sticks—A New Approach to LSI Design," M.S.E.E. thesis, Dept. of Electrical Engineering and Computer Science, M.I.T., June, 1977.
3. W. M. Penney and L. Lau, eds., *MOS Integrated Circuits*, Princeton, N.J.: Van Nostrand, 1972, Chapter 5.
4. C. H. Séquin and M. F. Tompsett, *Charge Transfer Devices*, New York: Academic Press, 1975, Chapter VIII.

4
IMPLEMENTING INTEGRATED SYSTEM DESIGNS: FROM CIRCUIT TOPOLOGY TO PATTERNING GEOMETRY TO WAFER FABRICATION

This chapter presents the basic concepts involved in implementing integrated system designs, from the system designer's point of view. Tools are described that help the designer produce the geometrical layout patterns for each layer of an integrated system, given the logic, circuit, or topological level design of the system. Procedures are described for encoding the layout patterns and then using the encoded layouts in the patterning and fabrication processes to implement the integrated system. In addition, we discuss how design tools and implementation procedures are likely to evolve, under the influence of increased complexity of design and predictable changes in the technologies of implementation.

To enable groups of readers to actually design moderate-sized LSI systems, we include descriptions of easily constructed design tools and procedures for organizing and implementing LSI multiproject chips. In each case, the tools are described as part of a complete system of design and implementation procedures, some of which are performed manually while others are machine assisted. Those experienced in software system design will recognize that construction of the machine-assisted portions of the systems is fairly straightforward. Contrary to what many think, designing your own LSI projects, merging them onto collaborative multiproject chips, and having these implemented by commercial maskmaking and wafer-fabrication firms is now well within the computational and financial reach of most industrial R&D groups and university EE/CS departments.

We are firm believers in *learning by doing* and hope that the information provided in this chapter will both help and encourage many groups of readers to try their hand at building design tools and designing integrated systems. Such firsthand experience will lead to a deeper understanding of the remaining material in this text.

4.1 INTRODUCTION

An overview of the stages of integrated system design, layout, and implementation is given in Fig. 4.1. The designer first transforms the circuit and topological level designs into a geometrical layout of the system, using procedures described later in this chapter. In order to optimize the layout, perform various design checks, and discover errors, the designer usually "iterates" several times between design and layout. The result is a set of *design files* that describe the *layout*. The files are in a particular representation called an *intermediate form*, which efficiently and unambiguously describes the layout geometry.

   The design files are then converted to files for driving the chosen patterning mechanism. At present, design files are commonly converted to *pattern generator*

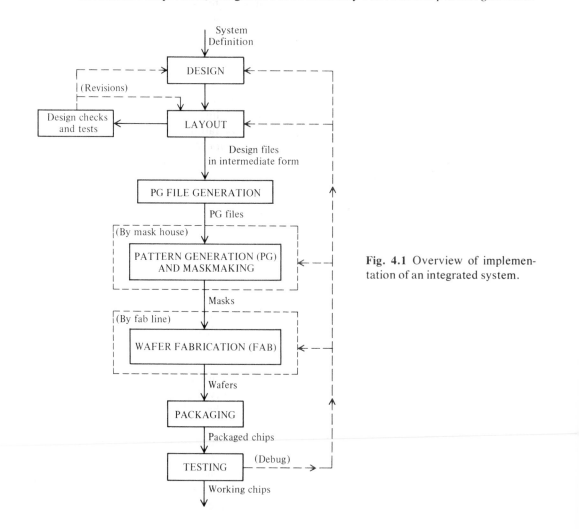

Fig. 4.1 Overview of implementation of an integrated system.

(PG) *files*, for use by a maskmaking firm for driving an optical pattern generator, the first step of maskmaking. By a sequence of photolithographic steps, the mask house produces a set of *masks*, which a commercial wafer fabrication firm then uses to pattern silicon *wafers*. Each finished wafer contains an array of system chips. The wafers are then diced into separate chips, which are packaged and tested to yield working systems.

From the system designer's point of view, maskmaking and fabrication can be visualized as a film-processing service: the designer produces the "artwork" (design files), from which the mask house makes "negatives" (masks), which are then run on a fab line to produce "prints" (wafers). The maskmaking–fabrication sequence is *function, design, and layout independent*: the mask and fab firms do not require detailed information about the integrated systems they fabricate. If the original layouts satisfy the design rules, and satisfy a few constraints imposed by patterning and fabrication, then these processes will yield correctly patterned wafers.

One need not closely bind a system's design to the detailed processing specifications of particular mask and fab firms. Various firms will differ somewhat in the minimum value of the length unit λ they can successfully process. The transit time of the transistors fabricated will vary from one fab line to another, as will the resistance per square and capacitance per unit area of fabricated features. However, well-structured and relatively process-independent nMOS designs will function correctly if scaled to a value of λ appropriate for the chosen fabrication facilities and operated using an appropriate system clock period.

We next examine some of the present implementation procedures a bit more closely, to set the stage for sections on design and layout. Those later sections will be clearer if one can visualize how the design files are to be used during patterning and fabrication.

4.2 PATTERNING AND FABRICATION

On completion of design and layout, the system design is contained in system layout files in intermediate form. Prior to fabrication, a final *check plot* of the layout is usually generated by converting the design files to files for driving a graphics plotter. Check plots are used for visually checking for violations of design rules and other design errors. Once the designers have done as much visual checking as they are going to do, the system layout files are converted to pattern generator (PG) files, to be sent to the maskmaking facility. Figure 4.2 summarizes the sequence of patterning and fabrication procedures that follows and also identifies the artifacts passed on at each step in the sequence.

Maskmaking begins with *pattern generation* to produce *reticles*. Present pattern generators are projector-like systems containing (1) a precisely movable stage, (2) an aperture of precisely variable rectangular size and angular orientation, and (3) a light source, all program controllable by a computer system. To produce a reticle, a photographic plate is mounted on the stage, and then the PG

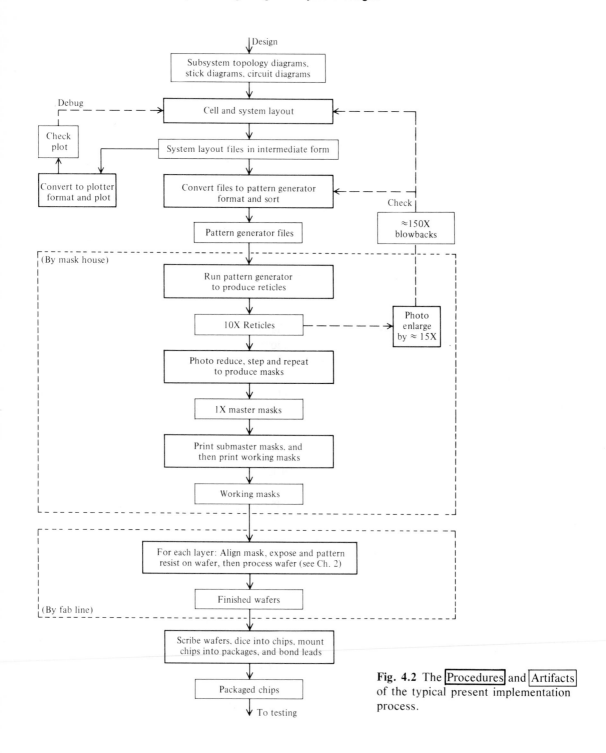

Fig. 4.2 The Procedures and Artifacts of the typical present implementation process.

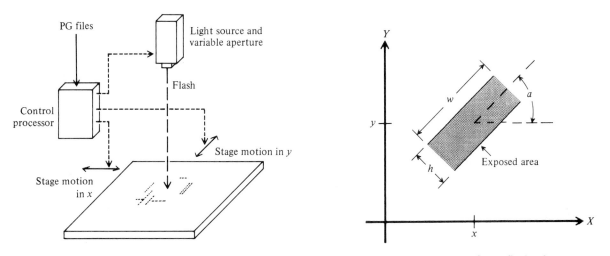

Fig. 4.3 The function of a pattern generator.

Fig. 4.4 Parameters of one flash of a pattern generator.

file for a particular system layer is used to direct the "flashing" of a sequence of rectangular exposures, of particular sizes and orientations, onto a sequence of coordinate locations on the plate, as illustrated in Fig. 4.3.

The PG file contains a sequence of entries, each of which describes a rectangle.[1] A typical representation uses five numbers for each rectangle: the x,y coordinates of its center, and its height, width, and angular orientation, as shown in Fig. 4.4. One can now visualize the nature of the conversion from intermediate form to PG files: the layout of each layer must be decomposed into its equivalent as a set of rectangles, each having (x,y,h,w,a) values "flashable" by the particular pattern generator, and these rectangles must be sorted into an efficient flashing sequence for that pattern generator.

When the flashing sequence is completed, the plate is developed, yielding the reticle (Fig. 4.5). Each reticle is a photographic master copy (much like a photo

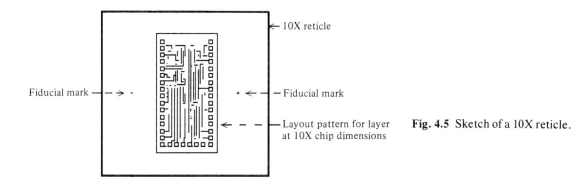

Fig. 4.5 Sketch of a 10X reticle.

negative) of the layout of one system layer, usually at a scale ten times (10X) the final system chip size. Photo enlargements of reticles, called *blowbacks*, may be obtained from the mask house, to provide a further level of checking of the design layout, PG file conversion, and pattern generation. At the current value of $\lambda = 3$ microns, blowbacks at approximately 100 to 150 times actual chip dimensions have sufficient detail to enable visual checking of the smallest features. Blowbacks of reticles may also be obtained in the form of color transparencies, to enable inspection of superposed overlays of various layers.

Once the 10X reticles have been generated, a 1X *master mask* is made from each reticle using a *photorepeater*, often called a "step and repeat" camera. The photorepeater exposes a photographic plate held on a movable stage, as in the pattern generator. In this case, however, each plate exposure is a 10:1 photo reduction of the reticle pattern. Between exposures the stage is moved by a precise *x,y* stepping distance. This process is repeated until a complete array of 1X chip patterns for one layer of the system has been exposed. The plate is then developed to produce a 1X master mask. Figure 4.6 sketches such a mask made from the reticle shown in Fig. 4.5. Note that as each reticle is inserted in the photorepeater, the position and angular orientation of the reticle pattern is carefully adjusted by microscopic examination of two *fiducial marks* on the reticle. These marks are placed as part of the pattern generation process and have the same precise position relative to the chip pattern origin on each of the system's reticles, thus assuring that all mask levels produced with the photorepeater will accurately register with each other.

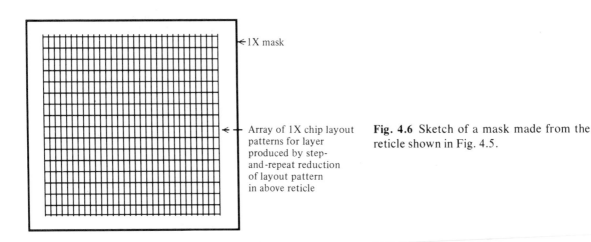

←1X mask

← Array of 1X chip layout patterns for layer produced by step-and-repeat reduction of layout pattern in above reticle

Fig. 4.6 Sketch of a mask made from the reticle shown in Fig. 4.5.

A succession of contact prints is made from each master mask to yield a number of *working masks*, sometimes called *working plates*, for each system layer. These are the actual masks used in wafer fabrication. During the contact-printing step of the typical wafer fabrication procedure, the working plates sometimes become worn or damaged, so several are usually made for each layer.

The wafer fabrication facility uses the working plates, in the sequence of patterning and process steps described in Chapter 2, to produce finished wafers. The fab line requires no detailed information about the design or mask patterns of the integrated system being fabricated. However, several auxiliary patterns are normally included in the mask patterns, some of which are replicated on each chip and are examined during wafer fabrication: (1) *alignment marks*, which are used to accurately overlay successive masks with previous patterning steps, (2) *line-width testers*, sometimes called *critical dimensions* (C/D's), which are lines in each mask layer of stated width that may be examined during maskmaking and fabrication to control dimensional tolerances, and (3) a few simple *test transistors* and their associated probe pads, which may be electrically tested prior to packaging to verify that the wafer fabrication process was successful.

The finished wafers are divided into chips and packaged by the sequence of steps sketched in Fig. 4.7. The surface of the wafer is marked along the *scribe lines* (the boundary lines between chips) with either a diamond-tipped scribe or a diamond-edged saw blade. The wafer is then fractured along these lines into single chips. Each individual chip is then cemented into the cavity of a package. Fine wires are bonded between the contact pads on the chip and the leads of the package, and a cover is cemented over the cavity; the system is then ready for functional testing.

From the preceding we see that once a system's design files have been produced, all the remaining implementation procedures are design and layout independent, and largely automatic. However, the many extraneous parameters, pat-

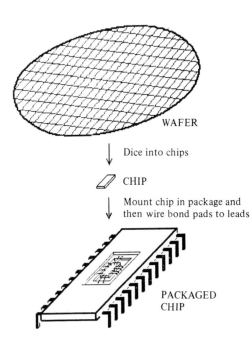

WAFER

↓ Dice into chips

CHIP

↓ Mount chip in package and then wire bond pads to leads

PACKAGED CHIP

Fig. 4.7 The packaging sequence.

terns, and constraints involved in maskmaking and fabrication must be carefully thought through and defined in order to guarantee successful implementation within a reasonable turnaround time. The PG files must be correctly sorted and formatted for the chosen pattern generator. The 10X pattern of the chip must fit within the largest reticle that the pattern generator can produce. The photore-peater used will determine the shape, size, and location of the fiducial marks on the reticle. The size, surface material, and photographic polarity, either positive (clear background field with opaque features) or negative (dark field with transparent features), of the working plates will be a function of the fabrication facility to be used. Each fab line also typically prescribes its own patterns for the alignment marks and test transistors to be included along with the system in the mask patterns.

While many designs may be scalable and have some longevity, the parame-ters, patterns, and constraints of maskmaking and fabrication are changing rapidly as the technologies evolve. This constant change complicates interactions with mask and fab firms. Later we describe procedures for implementating moderate-sized LSI systems as part of multiproject chips. Such chips are collaborative ef-forts of many designers, facilitating the merger of many projects into one maskmaking and wafer fabrication run. In this way the procedural overhead in-volved may be shared.

4.3 HAND LAYOUT AND DIGITIZATION USING A SYMBOLIC LAYOUT LANGUAGE

A simple and common method of producing system layouts is to draw them by hand. This is typically done on a one lambda grid using the familiar color codes to identify various system layers. Once the layout has been hand drawn it can then be *digitized*, or translated into machine-readable form, by encoding it into a sym-bolic layout language. This method, hand layout and digitization using a symbolic layout language, is quite practical for generating design files for highly structured system designs. Be warned, however, that implementing irregular structures using these primitive procedures is a difficult and tedious task.

If a system has only a few cell types that are replicated over and over, and otherwise has little "random wiring," one need draw only a single copy of each cell type, and then make reproductions or equivalent-sized outlines of these cell drawings. All these cell reproductions may then be patched together to plan and build up the overall layout. Similarly, only one symbolic digitization need be made for each cell type. The replication of cells in various orientations and locations in the system layout can then be easily described using the symbolic layout language. In a sense, the ease with which a system's layout can be described using a primi-tive layout language provides a measure of the regularity of its design. The OM2 Data Chip pictured in the frontispiece was laid out and digitized in this way, using only the simplest machine aids.

The function of a symbolic layout language, in its simplest form, is similar to that of a macroassembler. The user defines *symbols* (macros) that describe the layout of basic system cells. The locations and orientations of instances of these symbols are described in the language, as a function of appropriate parameters. These symbolic descriptions may then be mechanically processed in a manner similar to the expansion of a macro assembly language program, to yield the intermediate form description of the system layout, which is analogous to machine code for generating output files. (An example intermediate form is described in Sec. 4.5.) The intermediate form files may be processed to yield the PG files, each layer being a machine-encoded collection of rectangles encoded as $[x,y,h,w,a]$ values. The generation of PG files is analogous to the loading and execution of machine code to produce output files: it is a process of "unrolling" and fully instantiating all symbol descriptions into a sequence and format suitable for a particular output device. Definition of simple layout languages and the construction of their assemblers is fairly straightforward. The reader may define and implement layout languages by using the macro assembler or higher level language facilities of any commonly available computer system (Donovan, 1972; Freeman, 1975).

The following example will clarify the concepts and procedures of hand layout and symbolic layout description: We wish to create an array of shift registers consisting of parallel horizontal rows of inverters coupled by clocked pass transistors, as in color Plate 5(b). Plate 9(a) sketches the stick diagram of one row of the array. The entire array can be constructed from one basic cell containing an inverter, the pass transistor following it, VDD and GND buses crossing through on metal, and a clock line passing through on poly. Plate 9(b) shows a hand sketch of the layout of the basic shift register cell, SRCELL, on a 1λ grid, subject to the design rules given in Section 2.6. Since the inverters are coupled by pass transistors, the inverter pull-up/pull-down ratio is $\approx 8:1$ (see Sec. 1.12). Also, while the 4λ wide metal lines could be 1λ narrower in between the contact regions, the cell size would not decrease. As an exercise, the reader might check for design rule violations and also for ways of further shrinking the cell size.

The SRCELL layout shown in Plate 9(b) is composed using only rectangles placed at orientations that are integer multiples of 90°. The illustrations and descriptions in this section are considerably simplified by the use of such constrained layout constructions and yet they still illustrate the general principles involved. Were completely arbitrary shapes used, the SRCELL could be made somewhat smaller and still satisfy the design rules. Interestingly, experience has shown that the simple extension of including rectangles at orientations that are integer multiples of 45° enables most cell layouts to reach within a few percent of the minimum area achievable using arbitrary shapes. There is a clear trade-off here: the inclusion of increasingly complex geometrical objects in a layout will tend to reduce the minimum achievable layout area but will also increase the computational complexity of the associated machine aids.

We can informally characterize a simple layout language by examining Fig. 4.8, which contains a description of the layout of an array of SRCELL's using such

```
SCALE  LAMBDA=3.0MICRON;
;
SYMBOL    START, SRCELL;
    BOX      DIFF,X=3,Y=0,LX=4,LY=4,NY=2,IY=19;
    BOX      DIFF,X=2,Y=3,LX=6,LY=9;
    BOX      DIFF,X=8,Y=8,LX=3,LY=2;            INVERTER OUTPUT
    BOX      DIFF,X=9,Y=10,LX=2,LY=1;
    BOX      DIFF,X=9,Y=11,LX=7,LY=2;
    BOX      DIFF,X=16,Y=9,LX=4,LY=4;
    BOX      DIFF,X=4,Y=12,LX=2,LY=7;
    BOX      IMPL,X=2.5,Y=9.5,LX=5,LY=10;       PULLUP IMPLANT
    BOX      POLY,X=0,Y=5,LX=10,LY=2;           CELL INPUT
    BOX      POLY,X=12,Y=0,LX=2,LY=26;          CLOCKLINE
    BOX      POLY,X=16,Y=5,LX=5,LY=2;           CELL OUTPUT
    BOX      POLY,X=16,Y=7,LX=4,LY=3;
    BOX      POLY,X=2,Y=11,LX=6,LY=7;
    BOX      CUTS,X=4,Y=1,LX=2,LY=2,NY=2,IY=19;
    BOX      CUTS,X=17,Y=8,LX=2,LY=4;
    BOX      CUTS,X=4,Y=9,LX=2,LY=4;
    BOX      METL,X=0,Y=0,LX=21,LY=4,NY=2,IY=19;   VDD & GND
    BOX      METL,X=3,Y=8,LX=4,LY=6;
    BOX      METL,X=16,Y=7,LX=4,LY=6;
SYMBOL    END;
;
    DRAW    SRCELL,NX=4,NY=2,IX=21,IY=38,X=0,Y=0;
    DRAW    SRCELL,MIRRORX,NX=4,IX=21,X=0,Y=42;
;
END;
```

Fig. 4.8 Symbolic description of a shift register array.

a language. The language describes layouts as collections of BOXes on various layers. BOX statements describe each of these boxes by specifying their layer, the X,Y coordinates of their lower left corner, and then their lengths, LX in the x-direction, and LY in the y-direction. The use of a box corner to encode its location simplifies the encoding task. BOX statements may describe arrays of identical boxes, with the array's lower left corner origin at X,Y, by including optional parameters that specify the number NX and replication interval IX in the x-direction, and NY and IY in the y-direction. Dimensions are given in the length unit, λ. A SCALE statement defines the value of λ for this particular layout as $\lambda = 3.0$ microns.

In Fig. 4.8, the SRCELL is first described as a macro, or SYMBOL. The reader can verify that the collection of BOXes in the definition of the SYMBOL SRCELL, when ORed together, produces the layout shown in Plate 9(b). This

SRCELL is then replicated a number of times in various layout locations according to parameters in several DRAW statements.

Each DRAW statement describes the placement of an array of cells as follows: The cell described by the named SYMBOL definition is considered to be drawn at the origin. It is then *mirrored* (about the *x*- and/or *y*-axis), and/or *rotated* (by 0°, 90°, 180°, or 270°) about the origin, as specified by MIRROR or ANGLE transformations. The cell thus positioned may then be *replicated NX* times at distance intervals *IX* in the *x*-direction, and that row of cells may then be *replicated NY* times at intervals *IY* in the *y*-direction. The resulting array of cells is then *translated* a distance *X,Y* from the origin and placed into the layout.

The "program" in Fig. 4.8 describes an array of 3 rows by 4 columns of SRCELL's. After machine assembly of this program, the resulting design file can be used to generate check plots, which may be inspected to detect errors made in encoding the layout. A check plot of one SRCELL is given in Fig. 4.9(a), and we see that the cell has been correctly digitized. A set of stipple patterns is used in this check plot to encode the different system layers (see Fig. 4.9(b)). If available, color check plots are much better: color check plots can be made denser and still be readable, and association of colors with layers and functions is more easily

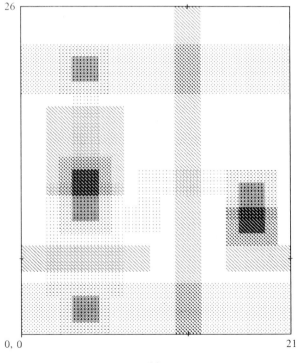

26

0, 0   21

(a)

Fig. 4.9 (a) Check plot of the SRCELL. (Dimensions in lambda. Implant layer not shown.) (b) Check plot of stipple codes.

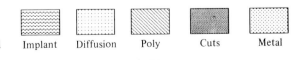

Implant   Diffusion   Poly   Cuts   Metal

(b)

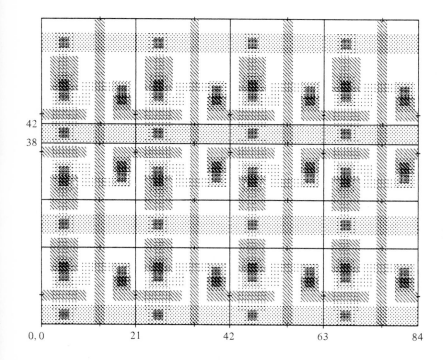

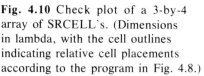

Fig. 4.10 Check plot of a 3-by-4 array of SRCELL's. (Dimensions in lambda, with the cell outlines indicating relative cell placements according to the program in Fig. 4.8.)

made and subject to fewer errors in practice. Note: the implant layer hasn't been plotted in Fig. 4.9(a) so that the other layers may be more easily seen.

A check plot of the complete 3 by 4 array of cells is given in Fig. 4.10 (again the implant layer is not plotted). Although Fig. 4.10 is of insufficient scale to check details within the cells, it enables us to check for correct relative placement of the SRCELL's. The individual cell outlines are included to indicate the nature of the placement of the central row of the array. By mirroring the central row prior to its placement, that row is able to share VDD and GND with the other two rows, thus reducing the overall array size. There is one column of cells per 21 lambda in the x-direction and one row of cells per 19 lambda in the y-direction. It is *very impor-tant* to note that the *outcome of each DRAW statement is determined by the order in which any mirror, rotate, replicate, and translate operations occur* (see Sec. 4.5 on the Caltech Intermediate Form, and also Newman and Sproull (1979)). Any permutation in the order of these operations may lead to a completely different result.

In Chapter 3 we found that the PLA is a useful subsystem structure, often used to implement finite-state machines and combinational logic. We now present a worked-out example of a PLA's layout, to further clarify symbolic layout description. An examination of the PLA stick diagrams shown in Plate 7(c) and Plate 8 reveals that a PLA can be constructed using six basic cell types and a slight amount of "random wiring." Once these six basic cells have been laid out by hand

and symbolically digitized, it is easy to construct symbolic descriptions of different-sized PLA's having various numbers of inputs, product terms, and outputs.

The digitized layouts of four of these basic cells are check-plotted in Fig. 4.11. The AND and OR planes of the PLA are constructed as arrays of the 14λ by 14λ PLAcellpair cell plotted in Fig. 4.11(a), which contains two poly and two metal signal lines, and one ground line on the diffusion layer. Diffusion paths may be added in any of four locations in such cells to form transistors and thus program the PLA. The connection between the AND and OR planes is made using the PLAconnect cell plotted in Fig. 4.11(b): these cells change the signal paths from the metal to the poly layer. The pull-up transistors to be placed at the edges of the AND and OR planes are implemented by the Pulluppair cell shown in Fig. 4.11(c). The ground return paths, to be connected to the diffusion lines crossing the planes, are implemented by the PLAground cell shown in part (d). The PLA-

(*cont. p.* 106)

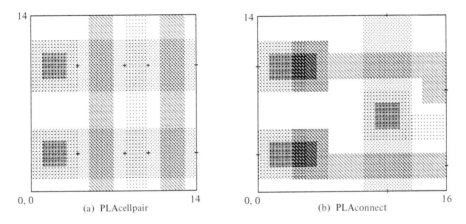

(a) PLAcellpair

(b) PLAconnect

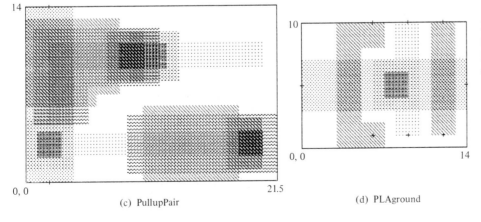

(c) PullupPair

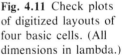

(d) PLAground

Fig. 4.11 Check plots of digitized layouts of four basic cells. (All dimensions in lambda.)

```
SCALE  LAMBDA=3.0MICRON;
;
;        PLA  CELL  DEFINITIONS:
;
SYMBOL    START,PLACELLPAIR;                            [SEE FIGURE 4.11A.]
   BOX    DIFF,X=0,Y=1,LX=4,LY=4,NY=2,IY=7;
   BOX    DIFF,X=8,Y=0,LX=2,LY=14;                      DIFF  TO  GND
   BOX    POLY,X=5,Y=0,LX=2,LY=14,NX=2,IX=6;
   BOX    CUTS,X=1,Y=2,LX=2,LY=2,NY=2,IY=7;
   BOX    METL,X=0,Y=1,LX=14,LY=4,NY=2,IY=7;            METL  TO  PULLUPS
SYMBOL    END;
;
SYMBOL    START,PLACONNECT;                             [SEE FIGURE 4.11B.]
   BOX    DIFF,X=0,Y=1,LX=4,LY=4,NY=2,IY=7;
   BOX    DIFF,X=9,Y=4,LX=4,LY=4;
   BOX    DIFF,X=13,Y=4,LX=3,LY=2;
   BOX    POLY,X=6,Y=1,LX=10,LY=2,NY=2,IY=8;
   BOX    POLY,X=3,Y=1,LX=3,LY=4,NY=2,IY=7;
   BOX    POLY,X=14,Y=7,LX=2,LY=2;
   BOX    CUTS,X=1,Y=2,LX=4,LY=2,NY=2,IY=7;
   BOX    CUTS,X=10,Y=5,LX=2,LY=2;
   BOX    METL,X=9,Y=0,LX=4,LY=14;                      GND
   BOX    METL,X=0,Y=1,LX=6,LY=4,NY=2,IY=7;
SYMBOL    END;
;
SYMBOL    START,PULLUPPAIR;                             [SEE FIGURE 4.11C.]
   BOX    IMPL,X=8.5,Y=0.5,LX=13,LY=5;
   BOX    IMPL,X=0.5,Y=4.5,LX=5,LY=8;
   BOX    IMPL,X=0.5,Y=7.5,LX=11,LY=5;
   BOX    DIFF,X=0,Y=1,LX=4,LY=4;
   BOX    DIFF,X=4,Y=2,LX=16,LY=2;
   BOX    DIFF,X=2,Y=5,LX=2,LY=4;
   BOX    DIFF,X=2,Y=9,LX=18,LY=2;
   BOX    DIFF,X=9,Y=8,LX=4,LY=4;
   BOX    POLY,X=10,Y=0,LX=8,LY=6;
   BOX    POLY,X=18,Y=1,LX=2,LY=4;
   BOX    POLY,X=8,Y=8,LX=2,LY=4;
   BOX    POLY,X=0,Y=7,LX=8,LY=6;
   BOX    POLY,X=0,Y=6,LX=6,LY=1;
   BOX    CUTS,X=1,Y=2,LX=2,LY=2,NX=2,IX=17;
   BOX    CUTS,X=8,Y=9,LX=4,LY=2;
   BOX    METL,X=0,Y=0,LX=4,LY=14;                      VDD
   BOX    METL,X=7,Y=8,LX=6,LY=4;
   BOX    METL,X=17,Y=1,LX=3,LY=4;
```

                                                            (cont.)

Fig. 4.12 Symbolic description of a 5-input, 10-pterm, 8-output PLA.

```
SYMBOL    END;
;
SYMBOL    START,PLAGROUND;                              [SEE FIGURE 4.11D.]
   BOX    DIFF,X=8,Y=1,LX=2,LY=9;
   BOX    DIFF,X=6,Y=3,LX=4,LY=4;
   BOX    POLY,X=3,Y=0,LX=2,LY=10;
   BOX    POLY,X=5,Y=0,LX=2,LY=2,NY=2,IY=8;
   BOX    POLY,X=11,Y=1,LX=2,LY=9;
   BOX    CUTS,X=7,Y=4,LX=2,LY=2;
   BOX    METL,X=0,Y=3,LX=14,LY=4;                      GND
SYMBOL    END;
;
SYMBOL    START,PLAINPUT;
;
[ insert symbol definition; size: 14 wide by ≈35 high ]
;
SYMBOL    END;
;
SYMBOL    START,PLAOUTPUT;
;
[ insert symbol definition; size: 14 wide by ≈41 high ]
;
SYMBOL    END;
;
;         LAYOUT 5-INPUT,10-PTERM,8-OUTPUT PLA:
;                   [SEE FIGURE 4.13]
;
   DRAW   PLACELLPAIR,NX=5,NY=5,IX=14,IY=14,X=0,Y=0;
   DRAW   PLACONNECT,NY=5,IY=14,X=70,Y=0;
   DRAW   PULLUPPAIR,NY=5,IY=14,X=-19,Y=0;
   DRAW   PLAGROUND,NX=5,NY=2,IX=14,IY=79,X=0,Y=-10;
   DRAW   PLACELLPAIR,ANGLE=270,NX=4,NY=5,IX=14,IY=14,X=86,Y=14;
   DRAW   PULLUPPAIR,ANGLE=270,NX=4,IX=14,X=86,Y=89;
   DRAW   PLAGROUND,ANGLE=270,NY=5,IY=14,X=141,Y=14;
   DRAW   PLAINPUT,NX=5,IX=14,X=0,Y=-44;
   DRAW   PLAOUTPUT,NX=5,IX=14,X=86,Y=-41;
   BOX    DIFF,X=70,Y=-15,LX=4,LY=4;
   BOX    CUTS,X=71,Y=-14,LX=2,LY=2;
   BOX    METL,X=70,Y=-15,LX=4,LY=4;
   BOX    METL,X=-19,Y=70,LX=4,LY=9;                    VDD
   BOX    METL,X=-19,Y=79,LX=105,LY=4;                  VDD
   BOX    METL,X=82,Y=83,LX=4,LY=6;                     VDD
   BOX    METL,X=142,Y=85,LX=9,LY=4;                    VDD
   BOX    METL,X=151,Y=-40,LX=4,LY=129;                 VDD
```

(cont.)

```
BOX     METL,X=142,Y=-40,LX=9,LY=4;              VDD
BOX     METL,X=-19,Y=-15,LX=19,LY=4;             VDD
BOX     METL,X=-19,Y=-11,LX=4,LY=11;             VDD
BOX     METL,X=70,Y=71,LX=9,LY=4;                GND
BOX     METL,X=70,Y=-7,LX=9,LY=4;                GND
BOX     METL,X=70,Y=-24,LX=9,LY=4;               GND
BOX     METL,X=79,Y=-45,LX=4,LY=45;              GND
BOX     METL,X=83,Y=-21,LX=3,LY=4;               GND
BOX     METL,X=142,Y=-21,LX=2,LY=4;              GND
BOX     METL,X=144,Y=-21,LX=4,LY=21;             GND
BOX     POLY,X=-4,Y=-43,LX=4,LY=2;               PH1
BOX     POLY,X=142,Y=-7,LX=15,LY=2;              PH2
;
[ insert the PLA "program", using BOXes on the diffusion
            layer to form transistors in the PLAcellpair cells ]
;
[ insert the PLA's input, output, clock, and power connections ]
;
END;
```

Fig. 4.12 Continued.

ground cell is structured so that rows of the cell may be inserted at intervals within AND planes and so that columns of the cell may be inserted at intervals within OR planes, to provide proper ground returns in large PLA's. The two other cell types required are the input drivers and output inverters: these cell layouts are left as exercises for the reader. The cells shown in Fig. 4.11 have been collectively planned so as to fit on a 14λ pitch surrounding the PLA's planes. Figure 4.12 contains a symbolic description of each of these cell types and a description of a moderate-sized PLA constructed from these cells.

A check plot of the PLA described in Fig. 4.12 is given in Fig. 4.13. This check plot has been simplified to include only the outlines of the basic cells, plus the additional wiring necessary to complete the PLA. The dimensions and orientations of the cells may be found by comparing these outlines with the cell details shown in Fig. 4.11. Note that in Fig. 4.11 some of the *connection points*, where paths leave or enter at cell edges or where internal connections may be later inserted, are tagged with tick marks. Cell placements and orientations in the check plot may be visualized by locating and identifying the appropriate connection point marks. A comparison of the check plot with the symbolic description above will clarify the function of the various DRAW statements. To assist in this comparison, the origin cell of the array of cells produced by each DRAW statement has been marked in Fig. 4.13 with its cell name. Note that this PLA layout could contain the PLA example presented in Plate 8.

Symbolic layout languages are easy to define and may be primitive or sophisticated, according to the requirements of the user. The function of the assembler

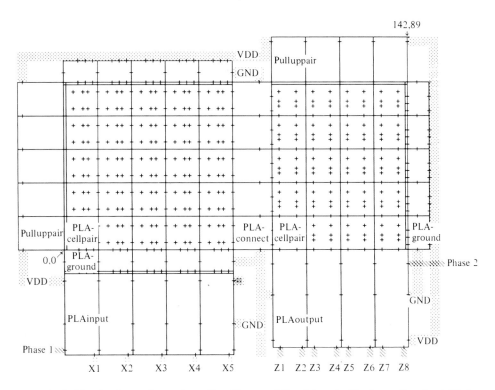

Fig. 4.13 Check plot, using only cell outlines, of the 5-input, 10-pterm, 8-output PLA. (Dimensions in lambda. Symbol labels on origin cells of the DRAW statements shown in Fig. 4.12.)

for such a language is simply to scan and decode the statements and translate them into design files in intermediate form. Conversion of design files into check plot or pattern generator output files is straightforward for the above simple language, since we have used only boxes with a severe constraint on angular orientations. MIRROR and ANGLE transformations are easily handled: x- and y-coordinates of symbols and boxes are simply replaced by $\pm x$ or $\pm y$, according to the specific parameters, during the instantiation of symbols and drawing of boxes prior to their replication and translation into the layout output file.

The effectiveness of the above language could be further increased by constructing an assembler capable of handling nested symbols. Through the use of nested symbols, system layouts may be described in a hierarchical manner, leading to very compact descriptions of structured designs. At the lowest level, one might define symbols for such small but commonly encountered structures as the various forms of contacts. Boxes and these simple symbols could then be used to construct cells such as those in the PLA example (Fig. 4.12). The PLA could be constructed with these cells and then defined as a symbol to be used in a larger design. An example of the sort of function one might add to create a much more sophisticated language, and language processor, would be the capability of generating the layout description of a PLA from the collection of basic cells, as a function of its input, product term, and output size parameters and logic function parameters.

Figure 4.14 summarizes the procedures and artifacts of hand layout, and layout description and digitization using a layout language. By studying Fig. 4.14 and thinking back over the material and examples of this section, one can visualize a

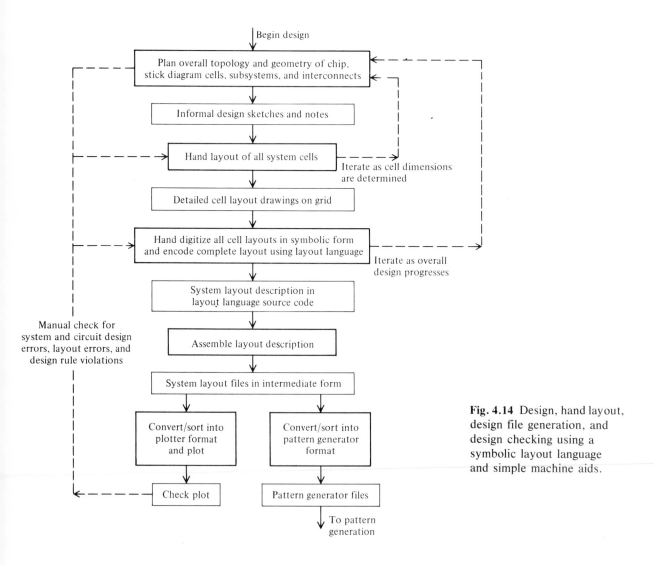

Fig. 4.14 Design, hand layout, design file generation, and design checking using a symbolic layout language and simple machine aids.

complete, though primitive, sequence of steps sufficient to prepare a design for implementation. These procedures are entirely adequate for preparing small LSI projects for implementation. The procedures may also be used for those larger integrated systems which have highly structured designs.

The primary obstacle that these primitive procedures place in the path of the system designer is the sheer time and effort it takes to get through the loop to a new check plot each time a small design change is made. The enthusiasm aroused by a sudden insight, such as the conception of a completely new topological possibility for an important system cell, can be dampened by the tedious tasks of hand layout and box digitization required before one can really see the full effect of the idea on the overall system layout.

Though often supported by large batch mode CAD systems for containing, modifying, check plotting, and simulating designs, the majority of LSI layout now done in industry begins with hand layout. Digitization is usually simplified by the use of digitizing tables, which are much like graphics plotters in reverse: a new section of a design, laid out by hand, is placed on the table and digitized by tapping switches while manually following the outlines of the cell's boxes with a pointer. Although this is less tedious than digitization using a layout language, it is still time-consuming and hardly interactive.

The next section describes an interactive graphics layout system that enables the system designer to quickly sketch new layout ideas and see their effect immediately.

4.4 AN INTERACTIVE LAYOUT SYSTEM*

Computing hardware of sufficient power to support highly interactive graphics has in the past been quite expensive, and this has inhibited the widespread application of interactive computing techniques. However, because of expected advances in VLSI technology, we are rapidly approaching the day when many will have access to personal computers with computing power rivaling today's medium to large-scale systems. It will be more difficult to provide effective software for these systems than it will be to build the computers themselves.[2] In this section we describe a highly interactive layout system that runs on a modest personal computer, rather than on an expensive, limited access, centralized system. This system was developed anticipating the work environment of the future, in which most "knowledge" workers will have personal computers as part of their normal office equipment.

ICARUS[3] (Integrated Circuit ARtwork Utility System) is a software system that enables the user to create and modify an integrated system layout directly on a CRT display screen. ICARUS was conceived with the idea that the designer

---

*This section is contributed by Douglas Fairbairn, Xerox PARC, Palo Alto, California, and James Rowson, California Institute of Technology.

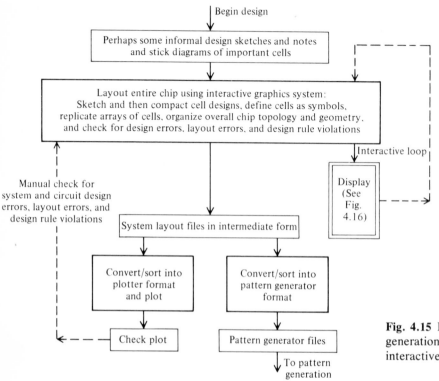

Fig. 4.15 Design, layout, design file generation, and design checking using an interactive graphics layout system.

would create and edit a layout at the display, without doing any more than a rough sketch or ''stick diagram'' before beginning work. Creating and moving items is fast and easy enough so that the designer can truly sketch on the screen. Once the layout is basically correct, the items can be moved or modified to arrive at the most compact layout.

The user is required to remember very little about the available commands or their use because the commands themselves are displayed on the screen and the system prompts the user for additional information as it is needed. The system can format and output check plots to matrix-type printers or raster-scan laser printers. ICARUS design files can be used to create standard pattern generation files from which masks can be made. An overview of design and layout procedures using the system is given in Fig. 4.15. It is instructive to compare this with Fig. 4.14, which presents equivalent steps for hand layout.

All the software to accomplish these various steps runs on a small experimental minicomputer known as the Alto. This machine was designed by researchers at Xerox PARC as a general purpose personal computer suitable for both text and graphics applications. No additional, special hardware is used by ICARUS. The

ICARUS system is programmed in BCPL, an ALGOL-like high-level language. There are about 30K words of compiled code in the system, of which half is in memory at any given time. At minimum, the Alto memory has 64K 16-bit words. A 2.5 Mbyte cartridge disk drive is an integral part of the system. The user interacts with the system through an unencoded keyboard (software definable keys) and with a pointing device called a mouse (Newman and Sproull, 1979). A cursor is controlled on the screen by moving the mouse around on a small area of the user's desk. A bit map display with a resolution of 600×800 dots is used for output, and printers for doing check plots are available through an in-house computer network.

The ICARUS display features two windows that provide a flexible working view of the layout, as shown in Fig. 4.16. The upper window is normally used for viewing a large piece of the layout at small magnification, and the lower window is used for looking at a smaller section in more detail. The magnifications of the windows may be set independently.

In addition to the two windows, there are various menus and status lines presented in the display. The menu on the left is the *command menu*. The menu under the upper window is the *parameter menu*. Under the parameter menu is the *stipple menu*, containing the mask level codes. Rectangles at a given level are stippled with the pattern for that level. The patterns were chosen so that, where necessary, one pattern could be seen through the other to verify that appropriate layers are overlapping properly. Current drawing coordinates and the status of system memory space are displayed to the right of the stipple menu.

The user interface is implemented principally through the display, the mouse, and five conveniently located keys on the keyboard. Frequently used commands are given using only one or two simple hand operations and can be done without glancing away from the display. These characteristics, coupled with rapid display redrawing, enhance the system's interactiveness.

The internal data representation in ICARUS is based on three types of items: rectangles, symbols, and text strings. The organization of these items into memory data structures and the typical run-time memory space allocation are illustrated in Fig. 4.17.

Rectangles are created with the aid of the mouse. They may have angular orientations that are integer multiples of $45°$. They can be moved, copied, or deleted using the mouse and one key. As items are created, they are added to an item list in main memory. Each rectangle is stored as six words in memory: the first word is the pointer to the next item, the second specifies what layer it is on, what type of item it is, etc. The third through sixth words specify the minimum and maximum x- and y-coordinates. The items are kept in order of increasing values of minimum x-coordinate, so that the display may be quickly redrawn.

When a symbol is defined by the user, the items that are contained within it are stored on the disk, while a pointer, the name, and the bounding box for the symbol are placed in main memory. Symbols can be nested to any level. Once a

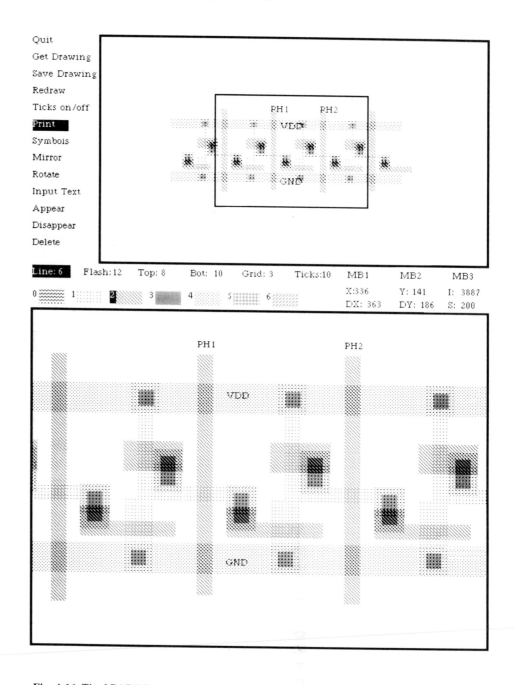

Fig. 4.16 The ICARUS display. Two views of a layout in progress.

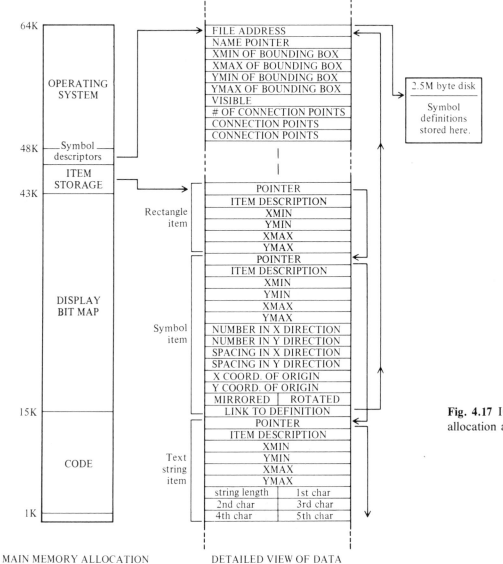

	FILE ADDRESS
	NAME POINTER
	XMIN OF BOUNDING BOX
	XMAX OF BOUNDING BOX
	YMIN OF BOUNDING BOX
	YMAX OF BOUNDING BOX
	VISIBLE
	# OF CONNECTION POINTS
	CONNECTION POINTS
	CONNECTION POINTS

64K OPERATING SYSTEM

48K Symbol descriptors

ITEM STORAGE

43K

DISPLAY BIT MAP

15K

CODE

1K

2.5M byte disk

Symbol definitions stored here.

Rectangle item
POINTER
ITEM DESCRIPTION
XMIN
YMIN
XMAX
YMAX

Symbol item
POINTER
ITEM DESCRIPTION
XMIN
YMIN
XMAX
YMAX
NUMBER IN X DIRECTION
NUMBER IN Y DIRECTION
SPACING IN X DIRECTION
SPACING IN Y DIRECTION
X COORD. OF ORIGIN
Y COORD. OF ORIGIN
MIRRORED | ROTATED
LINK TO DEFINITION

Text string item
POINTER
ITEM DESCRIPTION
XMIN
YMIN
XMAX
YMAX
string length | 1st char
2nd char | 3rd char
4th char | 5th char

MAIN MEMORY ALLOCATION   DETAILED VIEW OF DATA

Fig. 4.17 ICARUS memory allocation and data structure.

symbol definition has been created, one is free to define *symbol instances*, which are references to that definition. The symbol instance may be a command to draw one copy of the symbol at a certain location, or a whole array. The size of the symbol instance, which resides in main memory, is the same in both cases. The use of symbols wherever possible tends to preserve main memory space. Rather large systems can be designed using ICARUS, if the systems are well structured and make extensive use of symbols. This is true even when using a minimum-sized 64K memory, which leaves little space for layout data.

Text is used for identifying data and control lines and is merely a memory aid to the user. There is no attempt to make use of the text or other information in the drawing for connectivity or other types of checking.

Operations more complex than those such as Draw and Move are implemented through the use of menus as shown in Fig. 4.16. The desired command is chosen by pointing at it with the cursor and clicking a mouse button. The selected command is then inverted to white-on-black video to identify its selection, which the user then confirms with a key on the keyboard. At this point, the system prompts the user with instructions presented in the display area that normally holds the stipple menu. The instructions lead the user through the individual steps required, for example, to Mirror or Rotate a group of items.

Operations on symbols are defined in a secondary menu that can be reached by selecting the command "symbols" on the primary menu. The secondary menu offers commands such as Define symbol, Draw symbol, List the names of the symbols in the symbol library, or Expand symbol. This last command is used to modify a symbol that is already defined, the modified symbol definition immediately updating all symbol instances that point to it.

Various system parameters are displayed in the parameter line directly below the top window. Values such as the default line-width for the currently selected layer, the magnification of the top and bottom windows, and the spacing of the tick marks are all displayed. The parameter values can be changed at any time by selecting the desired one and typing the new parameter value on the keyboard. The X,Y layout coordinates of the point last clicked with the mouse are displayed at the right of the screen. The DX,DY distances between the last two clicks are also displayed. This feature provides a convenient "ruler" for measuring distances on the layout.

By the way, the design in progress shown in Fig. 4.16 has a number of bugs in it. Can you find them all?

The construction of an interactive layout system such as ICARUS is a relatively straightforward task for one who is experienced in interactive computer graphics (Newman and Sproull, 1979), given a display-oriented, minicomputer system and effective, systems-building software. A first version of ICARUS was constructed in three man-months, and a mature version was produced in an additional five man-months.

ICARUS has been used internally in Xerox to lay out many integrated system projects and to organize a number of multiproject chips. Among the users were a number of individuals previously unfamiliar with integrated circuit layout, who nevertheless successfully completed LSI projects with up to 10,000 transistors. We find that the interactive nature of such a system not only aids the experienced designer but also enhances the learning process for the novice. We believe that such interactive, personal design systems greatly enhance the creative ability of the designer by enabling easy generation and examination of many more design alternatives per unit time than would be the case with centralized, noninteractive design systems.

However, there is more to integrated system design than circuit layout. Design rules must be checked, logic transfer functions tested, and, in certain cases, circuit transfer functions computed to determine delays and predict system performance. We believe that the direction in which to search for further improvements in design tools is in the replacement of the primitive ICARUS-type of data structure with one that allows design functions other than just layout to also interactively operate upon the same data base.

4.5 THE CALTECH INTERMEDIATE FORM FOR LSI LAYOUT DESCRIPTION*

The Caltech Intermediate Form (CIF Version 2.0) is a means of describing graphic items (mask features) of interest to LSI circuit and system designers. Its purpose is to serve as a standard machine-readable representation from which other forms can be constructed for specific output devices such as plotters, video displays, and pattern-generation machines. The intermediate form is not intended as a symbolic layout language: CIF files will usually be created by computer programs from other representations, such as a symbolic layout language or an interactive design program. Nevertheless, the form is a fairly readable text file, in order to simplify combining files and tracing difficulties.

The basic idea of the form is to specify literally every geometric object in the design using ample precision. Use of this form provides participating design groups easy access to output devices other than their own, enables sharing designs with others, allows combining several designs to form a larger chip, and the like. It is not necessary for all participating groups to implement the entire set of features of CIF, as long as their programs and documents contain warnings about unimplemented functions; nevertheless, the syntax must be correctly interpreted by all programs that read CIF, to assure a reasonable result.

CIF thus serves as the common denominator in the descriptions of various integrated system projects. No matter what the original input methods are (hand layout and coding, or a design system), the designs will be translated to CIF as an intermediate, before being translated again to a variety of formats for output devices or other design aids.

This section is divided into four parts: a description of the syntax of the form, a description of the semantics, an explanation of the transformations used, and a discussion of the conversion of wires to boxes.

4.5.1 Syntax

A CIF file is composed of a sequence of characters in a limited character set. The file contains a list of commands, followed by an end marker; the commands are separated with semicolons. Commands and their forms are as follows.

---

* This section is contributed by Robert F. Sproull, Carnegie-Mellon University, and Richard F. Lyon, Xerox PARC, Palo Alto, California.

Command	Form
Polygon with a path	P path
Box with length, width, center, and direction (direction defaults to (1,0) if omitted)	B integer integer point point
Round flash with diameter and center	R integer point
Wire with width and path	W integer path
Layer specification	L shortname
Start symbol definition with index, a, b (a and b both default to 1 if omitted)	DS integer integer integer
Finish symbol definition	DF
Delete symbol definitions	DD integer
Call symbol	C integer transformation
User extension	digit userText
Comments with arbitrary text	(commentText)
End marker	E

A more formal definition of the syntax is given below. The standard notation proposed by Niklaus Wirth[4] is used: production rules use equals = to relate identifiers to expressions, vertical bar | for or, and double quotes `` '' around terminal characters; curly brackets { } indicate repetition any number of times including zero; square brackets [ ] indicate optional factors (i.e., zero or one repetition); parentheses ( ) are used for grouping; rules are terminated by a period. Note that the syntax allows blanks before and after commands, and blanks or other kinds of separators (almost any character) before integers, etc. The syntax reflects the fact that symbol definitions may not nest.

```
cifFile                  = { { blank } [ command ] semi } endCommand { blank }.
command                  = primCommand | defDeleteCommand |
                             defStart Command semi { { blank } [ primCommand ] semi } defFinishCommand.
primCommand              = polygonCommand | boxCommand | roundFlashCommand | wireCommand |
                             layerCommand | callCommand | userExtensionCommand | commentCommand.
polygonCommand           = ``P'' path.
boxCommand               = ``B'' integer sep integer sep point [ sep point ].
roundFlashCommand        = ``R'' integer sep point.
wireCommand              = ``W'' integer sep path.
layerCommand             = ``L'' { blank } shortname.
defStartCommand          = ``D'' { blank } ``S'' integer [ sep integer sep integer ].
defFinishCommand         = ``D'' { blank } ``F''.
defDeleteCommand         = ``D'' { blank } ``D'' integer.
callCommand              = ``C'' integer transformation.
userExtensionCommand     = digit userText.
commentCommand           = ``('' comment Text'')'' .
endCommand               = ``E''.
```

transformation   = { { blank } (``T`` point | ``M`` { blank } ``X`` | ``M`` { blank } ``Y`` | ``R`` point) }.

path   = point { sep point }.
point   = sInteger sep sInteger.

sInteger   = { sep } [`` − ``] integerD.
integer   = { sep } integerD.
integerD   = digit { digit }.

shortname   = c [c] [c] [c].
c   = digit | upperChar.
userText   = { userChar }.
commentText   = { commentChar } | commentText ``(``commentText``)`` commentText.

semi   = { blank } ``;`` { blank }.
sep   = upperChar | blank.
digit   = ``0`` | ``1`` | ``2`` | ``3`` | ``4`` | ``5`` | ``6`` | ``7`` | ``8`` | ``9``.
upperChar   = ``A`` | ``B`` | ``C`` | . . . | ``Z``.
blank   = any ASCII character except digit, upperChar, ``−``, ``(``, ``)``, or ``;``.
userChar   = any ASCII character except ``;``.
commentChar   = any ASCII character except ``(``or``)``.

4.5.2 Semantics

The fundamental idea of the intermediate form is to describe unambiguously the geometry of patterns for LSI circuits and systems. Consequently, it is important that all readers and writers of files in this form have exactly the same understanding of how the file is to be interpreted. Many of the decisions in designing the file format were made to avoid ambiguity or small but troublesome errors: floating point numbers are avoided; there are no iterative constructs, though there may be in future additions to CIF.

A simple file format might include only primitive geometric constructs, such as polygons, boxes, flashes, and wires. Unfortunately, the geometric description of a chip with hundreds of thousands of rectangles on it would require an immense file of this sort. Consequently, we have made provision for defining and calling symbols; this should reduce the size of the file substantially.

It is important that programs processing CIF files operate cautiously, maintaining a constant vigilance for mistakes or entries that will not be processed properly. The description below mentions implementation suggestions or cause for caution displayed inside brackets[ ].

Measurements

The intermediate form uses a right-handed coordinate system shown in Fig. 4.18, with x increasing to the right and y increasing upward. (Directions and distances

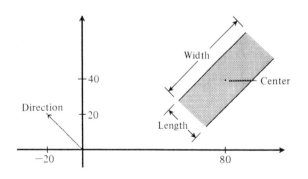

Fig. 4.18 Box representation in intermediate form.

are always interpreted in terms of the front surface of the finished chip, not in terms of the various sizes and mirrorings of the intermediate artifacts.) The units of distance measurement are hundredths of a micron (μm); there is no limit on the size of a number. [Programs reading numbers from CIF files should check carefully to be sure that the number does not overflow the number of bits in the internal representation used, and should specify their own limits, if any.]

Directions

Rather than measure rotation by angles, CIF uses a pair of integers to specify a "direction vector." This eliminates the need for trigonometric functions in many applications and avoids the problem of choosing units of angular measure. The first integer is the component of the direction vector along the x-axis; the second integer along the y-axis. Thus a direction vector pointing to the right (the $+x$-axis) could be represented as direction (1 0), or equivalently as direction (17 0); in fact, the first number can be any positive integer as long as the second is zero. A direction vector pointing northeast (i.e., rotated 45 degrees counterclockwise from the x-axis) would have direction (1 1), or equivalently (3 3), and so on. [A (0 0) direction vector may be defaulted to mean the $+x$-axis; a warning should be generated.]

Geometric primitives

The various primitives that specify geometric objects are not intended to be mutually exclusive or exhaustive. CIF may be extended occasionally to accommodate more exotic geometries. At the same time, it is not necessary to use a primitive just because it is provided. Notice in the examples below that lower case comments and other characters within a command are treated as blanks, and that blanks and upper case characters are acceptable separators.

Boxes.   Box Length 25 Width 60 Center 80,40 Direction −20,20; (or B25 60 80 40 −20 20;)

The fields that define a box are shown graphically in Fig. 4.18. Center and direction (optional, defaults to +x-axis) specify the position and orientation of the box, respectively. Length is the dimension of the box parallel to the direction, and Width is the dimension perpendicular to the direction.

Polygons.   Polygon A 0,0 B 10,20 C −30,40; (or P0 0 10 20 −30 40;)

A polygon is an enclosed region determined by the vertices given in the path, in order. For a polygon with n sides, n vertices are specified in the path (the edge connecting the last vertex with the first is implied; see Fig. 4.19). [Programs that try to interpret polygons may place various restrictions on their paths; no set of constraints has been generally accepted, and no program currently exists for converting completely general polygons to pattern generator output.]

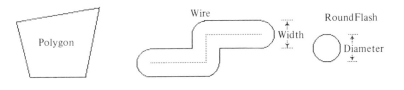

Fig. 4.19 Other items in the intermediate form.

Flashes.   RoundFlash Diam 200 Center −500,800; (or R200 −500 800;)

The diameter of a flash is sufficient to specify its shape, and the center specifies its position (see Fig. 4.19). [Some programs may substitute octagons, or other approximations, for round flashes.]

Wires.   Wire Width 50 A 0,0 B 10,20 C −30,40; (or W50 0 0 10 20 −30 40;)

It is sometimes convenient to describe a long, uniform-width run by the path along its centerline. We call this construct a wire (see Fig. 4.19). An ideal wire is the locus of points within one half-width of the given path. Each segment of the ideal wire therefore includes semicircular caps on both ends. Connecting segments of the wire is a transparent operation, as is connecting new wires to an existing one: the semicircular overlap ensures a smooth connection between segments in a wire and between touching wires. [For output devices that have a hard time constructing circles, we approximate the ideal wire with squared-off ends. Notice that squared-off ends work nicely for segments meeting at right angles, but cause problems if wires or wire segments are connected at arbitrary angles. A way to circumvent this problem is to convert, prior to output, any wires in a file into connected sets of boxes of appropriate length, width, angle, and center position.

The width of each box is the same as the width of the wire. The length of the boxes must be adjusted to minimize unfilled "wedges" and overlapping "ears." An algorithm for constructing boxes from a wire description is given in Section 4.5.4. If the wire is specified within a symbol definition, the approximation need be computed only once and can then be used each time the symbol is instantiated.]

Layer specification.   Layer ND nmos diffusion; (or LND;)

Each primitive geometry element (polygon, box, flash, or wire) must be labeled with the exact name of a fabrication mask on which it belongs. Rather than cite the name of the layer for each primitive separately, the layer is specified as a "mode" that applies to all subsequent primitives, until the layer is set again (layer mode is preserved across symbol calls, which are discussed later).

The argument to the layer specification is a short name of the layer. Names are used to improve the legibility of the file, and to avoid interfering with the various biases of designers and fabricators about numbers (one person's "first layer" is another's "last"). [The intention of the layer specification command is to label locally the layer for a particular geometry. It is therefore senseless to specify a box, wire, polygon, or flash if no layer has been specified. In order to detect this error, the command LZZZZ is implicitly inserted at the beginning of the file, and as the first command of a symbol definition (DS; see below). Any attempt to generate geometric output on layer ZZZZ will result in an error.]

It is important that layer names be unique, so that combining several files in intermediate form will not generate conflicts. The general idea is that the first character of the name denotes the technology, and the remainder is mnemonic for the layer. At present, the following layers are defined:

ND	nMOS	Diffusion
NP	nMOS	Polysilicon
NC	nMOS	Contact cut
NM	nMOS	Metal
NI	nMOS	depletion mode Implant
NB	nMOS	Buried contact
NG	nMOS	overGlass openings

New layer names will be defined as needed.

[Programs that read CIF will want to check to be sure that layer names used do in fact correspond to fabrication masks being constructed. However, the file may cite layer names not used in a particular pass over the CIF file. It would be helpful for the program to provide a list of the layer names that it ignored.]

Symbols

Because many LSI layouts include items that are often repeated, it is helpful to define often-used items as "symbols." This facility, together with the ability to "call" for an instance of the symbol to be generated at a specific position, greatly reduces the bulk of the intermediate form.

The symbol facilities are deliberately limited, in order to avoid mushrooming difficulties of implementing programs that process CIF files. For example, symbols have no parameters; calling a symbol does not allow the symbol geometry to be scaled up or down; there are no direct facilities for iteration. The main reason for symbol facilities is to limit the file size; if the symbol mechanism is not adequate for some application, the desired geometry can still be achieved with less use of symbols and more use of explicit geometrical primitives. [Symbols need not be used at all; this eliminates the need for intermediate storage for symbol definitions, but results in larger design files. Machines that must process a fully instantiated representation of a layer (such as pattern generators) might accept only CIF files without symbol definitions, to reduce the cost of implementation. Therefore, it would be useful to have a program that would convert general CIF files to fully instantiated CIF files, and then maybe sort them by layer, location, or whatever.]

The ability to call for iterations (arrays) of symbols is not provided in CIF Version 2.0. This is primarily due to the difficulty of defining a standard method of specifying iterations, without introducing machine-dependent computation problems. It is still possible to achieve a great deal of file compaction by defining several layers of symbols (e.g., cell, row, double-row, array). However, the ability to iterate symbol calls is a likely prospect for a future addition to CIF.

Defining symbols.   Definition Start #57 A/B = 100/1; . . . ; Definition Finish; (or DS57 100 1; . . . ;DF;)

A symbol is defined by preceding the symbol geometry with the DS command, and following it with the DF command. The first argument of the DS command is an identifying symbol number (unrelated to the order of listing of symbol definitions in the file).

The mechanism for symbol definition includes a convenient way to scale distance measurements. The second and third arguments to the DS command are called a and b, respectively. As the intermediate form is read, each distance (position or size) measurement cited in the various commands (polygons, boxes, flashes, wires, and calls) in the symbol definition is scaled to $(a*distance)/b$. For example, if the designer uses a grid of 1 micron, the symbol definition might cite all distances in microns and specify $a = 100$, $b = 1$. Or the designer might choose lambda (characteristic fabrication dimension) as a convenient unit. This mechanism reduces the number of characters in the file by shrinking the integers that specify dimensions and may improve the legibility of the file (it provides neither scaling nor the ability to change the size of a symbol called within the definition).

Definitions may not nest. That is, after a DS command is specified, the terminating DF must come before the next DS. The definition may, however, contain calls to other symbols, which may in turn call other symbols.

There is only one restriction on the placement of symbol definitions in the file: a symbol must be defined before its instantiation becomes necessary. This constraint can be satisfied by placing all symbol definitions first in the file, followed by calls on the symbols. In fact, it is often convenient to have the file consist exclusively of symbol definitions and only one call on a symbol. This call will be the last command in the file before the end command. [If a file redefines a symbol that already exists, the previous definition is discarded: a warning message should be generated. When several people contribute to a design, some symbol management is therefore necessary; see *Deleting symbol definitions* below.]

Calling symbols. Call Symbol #57 Mirrored in X Rotated to −1,1 then Translated to 10,20; (or C57 MX R−1 1 T 10 20;)

The C command is used to call a specified symbol and to specify a transformation that should be applied to all the geometry contained in the symbol definition. The call command identifies the symbol to be called with its "symbol index," established when the symbol was defined.

The transformation to be applied to the symbol is specified by a list of primitive transformations given in the call command. The primitive transformations are

T point	Translate the current symbol origin to this point.
M X	Mirror in x, that is multiply x-coordinate by −1.
M Y	Mirror in y, that is, multiply y-coordinate by −1.
R point	Rotate symbol's x-axis to this direction

Intuitively, each coordinate given in the symbol is transformed according to the first primitive transformation in the call command, then according to the second, etc. Thus "C1 T500 0 MX" will first add 500 to each x-coordinate from symbol 1, then multiply the x-coordinate by −1. However, "C1 MX T500 0" will first multiply the x-coordinate by −1, and then add 500 to it: the order of application of the transformations is therefore important. In order to implement the transformations, it is not necessary to perform each primitive operation separately; the several operations can be combined into one matrix multiplication (see Sec. 4.5.3 on transformations).

Symbol calls may nest; that is, a symbol definition may contain a call to another symbol. When calls nest, it is necessary to "concatenate" the effects of the transformations specified in the various calls (see Sec. 4.5.3). [There is no sensible way in which a symbol may be invoked recursively (i.e., call itself, either directly or indirectly). Programs that read the intermediate form should check that no recursion occurs. This can be achieved by retaining a single flag with each symbol to indicate whether the symbol is currently being instantiated; the flags are initialized to "false." When a symbol is about to be instantiated, we check the flag; if it is "true," we have detected recursion, so we print an error message and do not perform the call. Otherwise, we mark the flag "true," instantiate the symbol as specified, and mark the flag "false" when the instantiation is complete.]

Layer settings are preserved across symbol calls and definitions. Thus, in the sequence

LNM;
R6 20 0;
C 57 T45 13;
DS 114;
. . . .;
DF;
LNM;
R3 0 0;

the second LNM is not necessary, regardless of the specification of symbols 57 and 114.

Deleting symbol definitions. Delete Definitions greater than or equal to 100; (or DD100;)

The DD command signals the program reading the file that all symbols with indices greater than or equal to the argument to DD can be "forgotten"—they will not be instantiated again. This feature is included so that several intermediate form files can be appended and processed as one. In such a case, it is essential to delete symbol definitions used in the first part of the file, both because the definitions may conflict with definitions made later and because a great deal of storage can usually be saved by discarding the old definitions.

The argument to DD that allows some definitions to be kept and some deleted is intended to be used in conjunction with a standard "library" of definitions that a group may develop. For example, suppose we use symbol indices in the range 0 to 99 for standard symbols (pull-up transistors, contacts, etc.) and want to design a chip that has two student projects on it. Each project defines symbols with indices 100 or greater. The CIF file will look like this:

(Definitions of library symbols);
DS 0 100 1;
(. . .definition of symbol 0 in library);
DF;
DS 1 100 1;
(. . .definition of symbol 1);
DF;
(. . .remainder of library);

(Begin project 1);
DS100 100 1;
(. . .first student's first symbol definition);
DF;
. . .

```
DS109 100 1;
( . . .first student's main symbol definition);
DF;
C109 T403 − 110; (call on first student's main symbol);

DD100; (Preserve only symbols 1 to 99);

(Begin project 2);
DS100 100 1;
( . . .second student's first symbol definition);
DF;
. . .
DS113 100 1;
( . . .second student's main symbol definition);
C1 T−3 45; (Call on library symbol, still available);
DF;
C113 T401 0; (call on second student's main symbol);

E
```

User expansion.   3'SYMBOL.LIBRARY'; 5:NONSTANDARD DESIGN RULES: LAMBDA = 4.0;

Several command formats (any command starting with a digit) are reserved for expansion by individual users; the authors of the intermediate form agree never to use these formats in future expansions of the standard format. For example, private expansions might provide for (1) requesting that another file be "inserted" at this point in the processing, thus simplifying the use of symbol libraries; (2) inserting instructions to a preprocessor that will be ignored by any program reading only standard intermediate form constructs; or (3) recording ancillary information or data structures (e.g., circuit diagrams, design-rule check results) that are to be maintained in parallel with the geometry specified in the style of the intermediate form.

Comments.   (HISTORY OF THIS DESIGN:);

The comment facility is provided simply to make the file easier to read. [It is possible to deactivate any number of commands by simply enclosing them within a pair of parentheses, even if they already include balanced parentheses.]

End Command.   End of file.

The final E signals the end of the CIF file. [Programs that read CIF should give either an error message if the file ends without an End command or a warning if more text other than blanks follows the E.]

4.5.3   Transformations*

When we are expanding a symbol, we need to apply a transformation to the specification of an item in the symbol definition to get the specification into the coordinate system of the chip. There are three sorts of measurements that must be transformed: distances (for widths, lengths), absolute coordinates (for "points" in all primitives), and directions (for boxes).

Distances are never changed by a symbol call, because we allow no scaling in the call. Thus a distance requires no transformation.

A point (x,y) given in the symbol is transformed to a point (x',y') in the chip coordinate system by a 3×3 transformation matrix T:

$$[x'\ y'\ 1] = [x\ y\ 1]T$$

The matrix T is itself the product of primitive transformations specified in the call: $T = T_1 T_2 T_3$, where T_1 is a primitive transformation matrix obtained from the first transformation primitive given in the call, T_2 from the second, and T_3 from the third (of course, there may be fewer or more than three primitive transformations specified in the call). These matrices are obtained using the following templates for each kind of primitive transformation:

$$T\ a\ b \qquad T_n = \begin{array}{ccc} 1 & 0 & 0 \\ 0 & 1 & 0 \\ a & b & 1 \end{array}$$

$$M\ X \qquad T_n = \begin{array}{ccc} -1 & 0 & 0 \\ 0 & 1 & 0 \\ 0 & 0 & 1 \end{array}$$

$$M\ Y \qquad T_n = \begin{array}{ccc} 1 & 0 & 0 \\ 0 & -1 & 0 \\ 0 & 0 & 1 \end{array}$$

$$R\ a\ b \qquad T_n = \begin{array}{ccc} a/c & b/c & 0 \\ -b/c & a/c & 0 \\ 0 & 0 & 1 \end{array} \quad \text{where } c = \sqrt{a^2 + b^2}$$

Transformation of direction vectors $(x\ y)$ is slightly different from the transformation of coordinates. We form the vector $[x\ y\ 0]$, and transform it by T into the new vector $[x'y'0]$. The transformed direction vector is simply $(x'y')$. [Note that some output devices may require rotations to be specified by angles, rather than by

---

*For further information on this subject, see Newman and Sproull (1979).

direction vectors. Conversion into this form may be delayed until necessary to generate the output file. Then we calculate the angle as arctan(y/x), applying care when $x=0$.]

Nested calls require that we combine the transformations already in effect with those specified in the new call. Suppose we are expanding a symbol a, as described above, transforming each coordinate in the symbol to a coordinate on the chip by applying matrix *Tac*. Now we encounter, in a's definition, a call to b. What is to happen to coordinates specified in b? Clearly, the transformations specified in the call will yield a matrix *Tba* that will transform coordinates specified in symbol b to the coordinate system used in symbol a. Now these must be transformed by *Tac* to convert from the system of symbol a to that of the chip. Thus, the full transformation becomes

$$[x'\ y'\ 1] = [x\ y\ 1]Tba\ Tac.$$

The two matrices may be multiplied together to form one transformation *Tbc* = (*Tba Tac*) that can be applied to convert directly from the coordinates in symbol b to the chip. This procedure can be carried to an arbitrary depth of nesting.

To implement transformations, we proceed as follows: we maintain a "current transformation matrix" T, which is initialized to the identity matrix. We use this matrix to transform all coordinates. When we encounter a symbol call, we

1. "Push" the current transformation and layer name on a stack.

2. Set layer name to *ZZZZ*.

3. Collect the individual primitive transformations specified in the call into the matrices T_1, T_2, T_3, etc.

4. Replace the current transformation T with $T_1\ T_2\ T_3\ \ldots\ T$ (i.e., premultiply the existing transformation by the new primitive transformations, in order).

5. Now process the symbol, using the new T matrix.

6. When we have completed the symbol expansion, "pop" the saved matrix and layer name from the stack. This restores the transformation to its state immediately before the call.

4.5.4 Decomposing Wires Into Boxes

The following algorithm for decomposing wires into boxes was developed by Carver Mead and first implemented at Caltech by Ron Ayres; it was further modified to be consistent with the use of direction vectors, to allow more general path lengths, and to avoid use of trigonometric functions (see Fig. 4.20). [Note that this decomposition covers more area than the locus of points within $w/2$ of the path for small angles of bend, but less area for sufficiently sharp bends; in particular, if a path bends by 180 degrees (reverses), it will have no extension past the point of reversal (it is missing a full semicircle). Other decompositions are possible and may better approximate the correct shape.]

Let the wire consist of a path of n points $p_1, \ldots, p_n$.
Let w represent the width of the wire.

```
IF n = 1 THEN
    {MAKEFLASH[Diameter ← w, Center ← p₁]; "single-point gets a flash";
    DONE;};
i ← 1;
OldExtension ← w/2; "initial end of wire"
Segment ← p₂ − p₁; "Segment is a vector (a point)"
"LoopConditions:"
FOR pᵢ, pᵢ₊₁ in path UNTIL pᵢ₊₁ is last DO
"calculate the box for the segment from pᵢ to pᵢ₊₁:"
IF pᵢ₊₁ is last THEN { Extension←w/2; "final end of wire"}
    ELSE
    {"compute Extension for intermediate point:"
    Next Segment←pᵢ₊₂ − pᵢ₊₁; "next vector in path"
    T ← MATRIX[    X[Segment],   −Y[Segment],
                   Y[Segment],    X[Segment] ];
    "T transforms Segment to +x axis."
    Bend ← MULTIPLY[ NextSegment, T ]; "relative direction vector"
    "if Bend is (0 0), delete pᵢ₊₁, reduce n, and start over"
    Extension←w/2∗( ABS[Y[Bend]]/( LENGTH[Bend] + ABS[X[Bend]]) );
};
MAKEBOX [{ Length←LENGTH[Segment] + Extension + OldExtension;},
         { Width←w;},
         { Center←(pᵢ + pᵢ₊₁)/2 + ( Segment / LENGTH[Segment] )∗
                             (Extension − OldExtension)/2; },
          { Direction ← Segment; "careful, may be zero vector" } ];
i ←i + 1;
OldExtension ← Extension;
Segment ← NextSegment; "next vector in path"
ENDLOOP;
DONE;
```

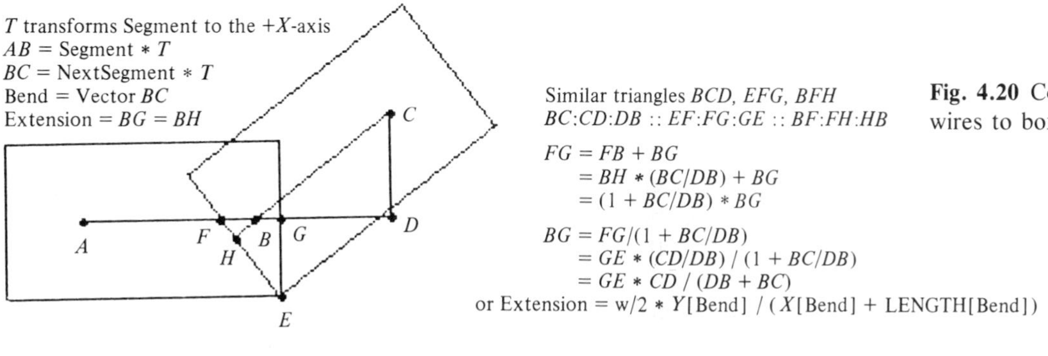

T transforms Segment to the $+X$-axis
AB = Segment $\ast T$
BC = NextSegment $\ast T$
Bend = Vector BC
Extension = $BG = BH$

Similar triangles BCD, EFG, BFH
$BC:CD:DB :: EF:FG:GE :: BF:FH:HB$

$FG = FB + BG$
$\quad = BH \ast (BC/DB) + BG$
$\quad = (1 + BC/DB) \ast BG$

$BG = FG/(1 + BC/DB)$
$\quad = GE \ast (CD/DB) / (1 + BC/DB)$
$\quad = GE \ast CD / (DB + BC)$
or Extension = w/2 $\ast Y$[Bend] / (X[Bend] + LENGTH[Bend])

Fig. 4.20 Converting wires to boxes.

4.6 THE MULTIPROJECT CHIP

Insight into integrated system design is most quickly gained by actually carrying through to completion several LSI design projects, each of increasing scope. A large, complex VLSI system could be quickly and successfully developed by designers able to easily implement and test prototypes of its subsystems. The separate subsystems can be implemented, tested, debugged, and then merged together to produce the overall system layout. However, such activities are practical only if a scheme exists for carrying out implementation with minimum turnaround time and low procedural overhead per project.

In this section we describe procedures for organizing and implementing many small projects by merging their layouts onto one *multiproject chip*, so that each designer of a small project or subsystem need not carry the entire procedural burden involved in maskmaking and fabrication. We also include a collection of practical tips and hints that may prove useful to those undertaking their first projects or organizing their first multiproject chips. While the details in this section are specific to present maskmaking and fabrication technology, they nevertheless give a feeling for the sort of things that must be done to implement projects in general. In a later section we discuss how multiple project implementation might be done in the future.

Figure 4.21 contains a photomicrograph of a Caltech class project chip containing 15 separate student projects. The individual projects were simply merged together onto one moderate sized chip layout, approximately 3 mm by 4 mm, and implemented simultaneously as one chip type. Most of these projects are prototypes of digital subsystems designed using the methodology of this text. By implementing a small "slice" of a prototype subsystem array, one can verify that its design, layout, and implementation are correct, and measure its power and delay characteristics as yielded by the particular fabrication process, thus gaining almost as much information as would be obtained by implementing the full array.

Following fabrication, the wafers containing such multiproject chips are scribed, diced, and then divided up among the participants. The typical minimum fabrication run yields about 10 to 20 wafers, each $\approx$7.5 to 10 cm in diameter. Thus even a minimum run provides a few thousand chips, and each participant ends up with many chips. Participants may then each package their chips, bonding the package leads to the contact pads of their individual project. Since most such projects are relatively small in area, yields are unusually high: if a project's design and layout have been done correctly, most of the corresponding chips will work.

Organizing a multiproject chip involves (1) creating the layout of a *starting frame*, into which the various projects are to be merged, (2) gathering, relocating, and merging the project layouts into the starting frame to create one design file and generating from this the PG files for the overall project chip, and (3) documenting various parameters and specs to be used during maskmaking and fabrication.

The starting frame contains all the auxiliary portions of the chip layout: scribe lines, alignment marks, line width testers (critical dimension marks), and test patterns. The starting frame may contain fiducial marks on each mask level if these

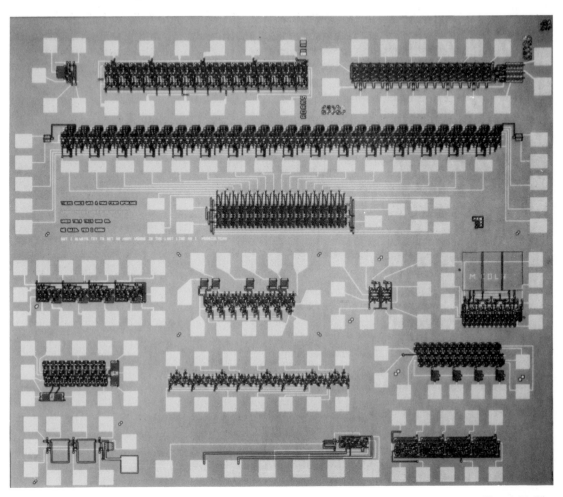

Fig. 4.21 Photomicrograph of a Caltech class project chip.

are not to be placed by the mask house, and in some cases it may contain a parity mark on each level to mark the appropriate reticle side and orientation during step-and-repeat reduction. [A tip: placing a mask level name somewhere within the chip's scribe line boundary on each level helps prevent the fatal error of level interchange during project merging, maskmaking, or fabrication.]

The contents of the starting frame must be carefully worked out to meet the requirements and constraints of the chosen mask house and fab line. The important factor of turnaround time for the entire mask and fab sequence may be reduced to some extent by repeatedly using a relatively standard starting frame, which then becomes familiar to all those involved. Some typical 1978 values for the time involved: 3 to 5 weeks for maskmaking, and then 3 to 4 weeks for fabrication; longer if large work queues exist at the mask or fab firms. It is interesting to note that almost all of the implementation time involves waiting in various queues within commercial facilities that are usually optimized for high thruput. The actual physical processing times required for maskmaking, fabrication, and packaging add up to only a few days. Were an implementation facility to optimize its structure and procedures for fast turnaround, implementation times of less than one week should be easily achievable.

When a multiproject chip is scheduled, a tentative chip partition for each project can be negotiated among the participants. Project design and layout can then proceed, with iterations on the space allocation being done right up till the final merging. The gathering and merging of project layout files into one design file is simplified if they are in a *common intermediate form*. Projects may then be relocated to their respective partitions of the chip, displayed, plotted, or otherwise checked, using minimum and consistent software operating upon manageably sized files. When the project chip appears to be correctly organized, pattern generator (PG) files are produced and written on a mag tape to be sent to the mask house.

An alternative to the merging of projects at the intermediate form level is the relocation and merging of their PG files. However, the PG files for major designs, containing fully instantiated artwork, become unwieldy in size even at today's complexity. The PG file merging scheme is workable for projects of small to moderate size and does provide a contingency plan for including projects having alien intermediate forms. If designs are relocated and merged at the PG level, additional software should be provided for displaying or plotting the chip at that level, so that merging errors may be spotted. [A tip: it is a good idea in any case to have some bounds checking to prevent stray items of one project from clobbering another.]

A thought: the interface between design groups and mask houses would be cleaner if design files in a common intermediate form (such as CIF) rather than PG files were used to transmit designs to the patterning process. Files would be much smaller. The use of data links would be eased. The process, involving patterning mechanism dependent optimization, to convert and sort design files into PG files, would be appropriately located: in association with the particular patterning mechanism.

4.6.1 Examples of Multiproject Chips

The above concepts and some further possibilities may be clarified by examining the details of some specific examples. Figure 4.22 illustrates a collaborative Xerox PARC/Caltech multiproject chip set (organized by D. Fairbairn, D. Johannsen, R. Lyon, J. Rowson, S. Trimberger). The figure was produced as a software blowback from the PG file of the metal level of this chip set. Projects in the set ranged in scope from the test of a few cells of an experimental, low-power shift register (by C. Séquin, U. C. Berkeley, and R. Lyon, Xerox PARC) up to a complete content addressable cache memory system (by D. Fairbairn).

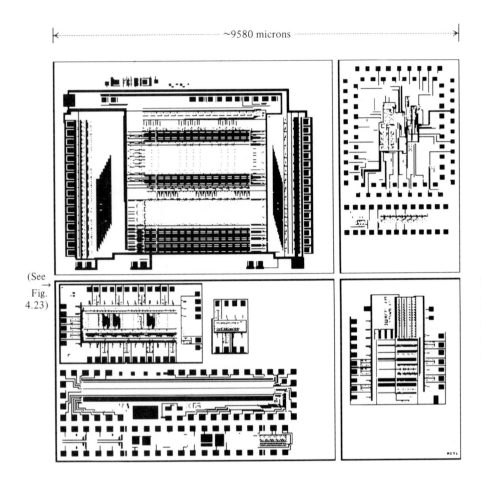

~9580 microns

(See
Fig.
4.23)

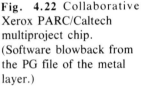

Fig. 4.22 Collaborative Xerox PARC/Caltech multiproject chip. (Software blowback from the PG file of the metal layer.)

Although several of the projects in the set are fairly large, all were individually designed to yield chip sizes packagable in standard 40-pin packages, which can hold chips up to ≈ 7 mm square. The pattern generator at the intended mask house was a GCA/D.W.Mann 3600, and the photorepeater was a Mann 3696. Together, this equipment can produce 10X reticles having field sizes as large as 10 cm square and can reduce, step, and repeat these at a maximum of 10 mm (x,y)-intervals onto masks. Therefore, the 3600/3696 can provide masks for square chips up to 10 mm $(10,000\ \mu)$ on a side. A 10 mm square chip can hold the patterns of several normal-sized chips. By including *interior scribe lines* in the starting frame, as indicated in Fig. 4.22, one reticle set can be patterned on the Mann 3600 to contain a number of different chips, each of which may contain more than one project. When masks are made, each reticle is photorepeated at intervals in x and y corresponding to its outer dimensions minus some scribe-line overlap. In the example in Fig. 4.22, the x and y stepping distances were both ≈ 9700 microns. Fabricated wafers are scribed and diced on all scribe lines, including the interior ones, to yield chips of typical sizes. One of the projects, on the lower left chip in Fig. 4.22, is an experimental charge coupled device array (by R. Davies). The CCD's rode along on this chip set to obtain working masks for use in a completely different process technology (triple-poly) from the standard nMOS the other projects used.

Figure 4.23 provides a higher magnification PG file software blowback of the region near the center of the left scribe line of the chip set. Alignment marks and line width testers (C/D's) were placed in this region, as noted in the figure. Software blowbacks of individual mask levels, more closely resembling the reticles and masks than would a composite design checkplot of all levels, are useful in conveying such location information to the mask and fab houses. Parity marks were not needed on the reticles for this project chip set. Fiducial marks were placed on the reticles by the mask house. Since the software converting the design files to PG files had just been constructed prior to organizing this chip set, reticle blowbacks were requested before proceeding further with maskmaking to verify that everything through pattern generation had worked correctly.

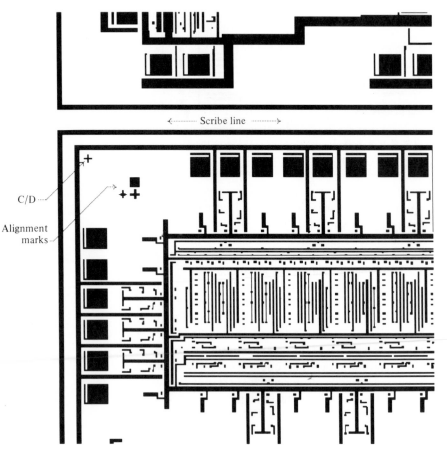

Scribe line

C/D

Alignment marks

Fig. 4.23 Close-up view of the region (Fig. 4.22) containing mask and fab information.

Some other practical details: Participants in the chip set shared some of the commonly used layout items normally required in any project. Examples were input contact pads with attached "lightning arrestor" circuits to protect the input MOSFET gates, and output drivers snaked around and attached to output pads. Even at 1978 device sizes, pads occupy a large fraction of the chip area for large collections of projects, and participants tend to make the pads as small as their bonding skill allows. A square pad $\approx 75 \ \mu$m on a side is a rather small bonding target; $125 \ \mu$m on a side is easier for the novice to hit. Perhaps $\approx 100 \ \mu$m square pads separated by $\approx 75 \ \mu$m is a good compromise, and these should be at least 25μm from any other metal lines to avoid cutting or shorting the lines when bonding. Metal paths (1978) are $\approx 1 \ \mu$m thick and can carry ≈ 1 ma per μm width.

Before submitting final design files for merging into such project chips, participants should be sure to check their projects for problems that might arise at the subsystem level. Metal conductors must be wide enough so that the current density limit is not exceeded (be sure to check major VDD and GND paths). Power densities should be low enough so that thermal problems do not develop. Rough guidelines for power densities can be developed as follows: Even in large systems composed of closely spaced boards fully populated with packaged chips, a power dissipation of ≈ 0.05 watts per cm^2 of board area is easily handled with air cooling. Conventional packaging techniques require about 20 cm^2 of board area for each 1 cm^2 of silicon chip area. Therefore, chips dissipating less than 1 watt per cm^2 of total chip area require no special cooling considerations. The limits of air cooling are reached at about one order of magnitude greater power dissipation. Individual projects can usually ignore these power density considerations, since they cover only a small fraction of the total chip area. It is important however to consider the consequences of enlarging a project to full chip proportions: For example, a 1 cm^2 array of the shift register cells shown in Plate 9(b) would dissipate ≈ 4 to 6 watts. If a large array of such cells were required in an integrated system, we might lengthen the pullups to achieve an acceptable total power dissipation.

The scribe lines on this chip set were laid out as 140 μm wide cuts down to 160 μm wide paths on the diffusion level, to provide lanes free of oxide for scribing or sawing. Metal paths 30 μm wide were then laid out straddling the boundaries of these scribe lines, to provide electrical contact from the substrate to the metal during the etching of the metal layer. Since all the projects on this chip set were prototype designs, and were not intended to be placed in extended use, the chips were not overglassed. Eliminating the overglassing meant that a mask level for defining cuts through overglassing over the contact pads and scribe lines was not needed, reducing maskmaking costs. On the other hand, the chip set included a mask level to pattern the thin gate oxide, to provide buried contacts between diffusion and poly that do not require metal coverage as does the butting contact. Such buried contacts enable more compact layouts, but are subject to a rather complex set of design rules, require an extra mask level, and sometimes reduce yield and reliability.

Deleting the overglassing process step also made it possible to electrically probe interior points on the chips during testing, probing small metal test pads included in the layouts. Such pads must be placed with care, however, because they hang relatively large capacitances onto circuitry and slow it down. Note that test-pad probing requires special jigs and a stereo microscope, and that it is possible to directly probe only the metal layer. Testing uncovered chips may also require reduced light levels. The operation of dynamic circuits (i.e., those which use a pass transistor input into a gate having no other electrical connection) can be severely affected by light. Light induces leakage currents in the p/n junction between source and drain regions and the substrate. At room temperature, charge stored on dynamic nodes can be retained for many milliseconds in the absence of light. However, in normal room light the retention time is reduced to tens of microseconds. Thus care should be taken to avoid high light levels when long clocking periods are used. Dynamic memory chips are packaged in opaque black packages because of this effect.

A software blowback of the metal mask PG file of another project set, organized at Caltech, is shown in Fig. 4.24. The total area of this multiproject chip set is ≈ 1 cm². It is subdivided into four major sections: The lower right quadrant contains the OM2 Data Path Chip described in Chapter 5, laid out using $\lambda = 2.5 \, \mu$m. The upper right quadrant contains a 16 by 16 bit multiplier with on-board accumulator (by Rod Masumoto, Caltech), also using $\lambda = 2.5 \, \mu$m. The lower left quadrant contains a subsystem, laid out using $\lambda = 2.9 \, \mu$m, which converts output from one port of a computer memory into the red, green, and blue analog signals for driving a color TV monitor. The upper left quadrant contains 28 projects, mostly from students in an LSI Systems course at Caltech. Other small projects are located along the left edge of the multiplier, and in the unused area within the TV subsystem project. The source material for this project chip set was generated on three different computer systems, in two different languages. Check plotting and viewing were done on three other systems. In addition to the Caltech projects, this chip set contains projects from Carnegie-Mellon University, Washington University (St. Louis), University of California, Irvine, and the Jet Propulsion Laboratory. Approximately 500,000 pattern generator rectangles were required to pattern the reticles for the five mask levels used in this project set. Conversion from intermediate form to PG files required ≈ 10 CPU hours on the Caltech DECsystem 20.

The masks for the multiproject chip sets shown in Figs. 4.22 and 4.24 were produced by Silicon Valley mask houses from PG tapes, accompanied by PG file software blowbacks showing the locations of auxiliary layout items used during implementation, and by spec sheets containing a list of specs and parameters for mask and fab. These spec sheets contain two types of information:

1. Information the mask house will need for reading the PG tape, generating the reticles, and stepping the master masks. This includes whether dimensions are in metric or English units, whether fiducials and parity marks have been laid out or are to be placed by the mask house, desired reticle magnification (usually 10X,

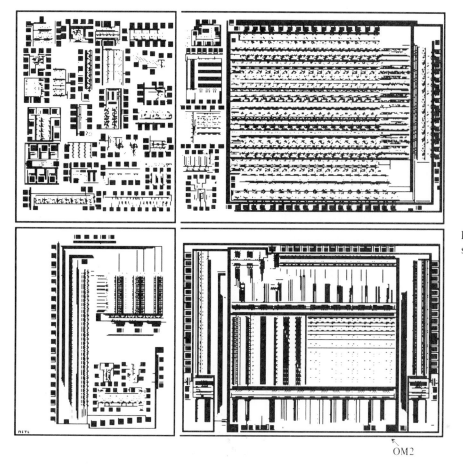

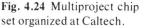
OM2

Fig. 4.24 Multiproject chip set organized at Caltech.

sometimes 5X), the (x,y) step-and-repeat distances, the type and magnification of reticle blowbacks desired, and whether maskmaking beyond reticle generation is to be contingent upon blowback inspection. This information is independent of the chosen fab line.

2. Information that is specific to the fab line, or lines, on which the wafers will be fabricated. Examples here are the number, size, and type of working plates desired, and the photographic polarity of the working plates, i.e., whether they are a positive or negative image of the PG pattern. The polarity of the working plates depends on the process step and on whether positive or negative resist is used. In addition, it is customary to specify how much, if any, the lines in the image will be expanded or contracted to compensate for growth or shrinkage of regions due to the process. This "pulling" of line widths in maskmaking may begin as far back as at pattern generation. Thus, while the patterning and fabrication processes are design and layout independent, they are usually coupled, and masks made for a run on one fab line are not necessarily usable elsewhere.

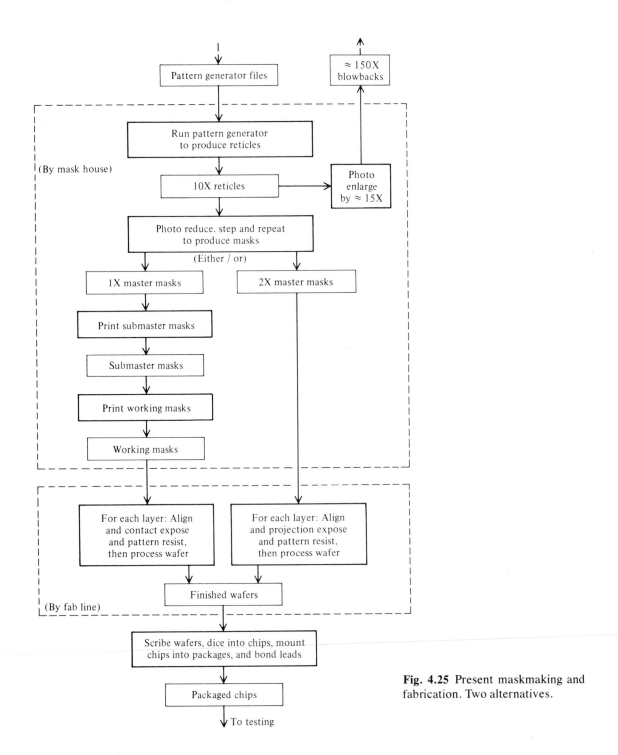

Fig. 4.25 Present maskmaking and fabrication. Two alternatives.

Maskmaking and patterning technology will remain in a state of transition for years to come. The present shift is from contact printing with working plates to projection alignment using original master masks. These two alternatives are illustrated in Fig. 4.25. From the system designer's point of view, at the interface to the mask and fab firms, the alternatives present no essential differences, requiring perhaps slightly different specs and yielding different intermediate artifacts. In the next section we discuss the future evolution of these technologies, presenting several implementation schemes likely to become commonplace over the next decade. These schemes will enable fabrication of systems much denser and faster than present ones. However, the basic concepts of the design methodology will still apply. Remembering our film processing analogy, we will have "finer grain" and "faster" film available as time passes. However, the basic art of photography remains.

4.7 PATTERNING AND FABRICATION IN THE FUTURE

As λ is scaled down toward its minimum value, ultimately limited by the physics of semiconductors to about 0.1 μm, it will become feasible to implement single chip, maximum density VLSI systems of enormous functional power. Patterning and fabrication at such small values of λ require that certain fundamental problems be overcome.[5,6] In this section we will discuss alternative solutions to two of the major problems: At values of λ of $\approx$ 2 μm, a problem of *runout* is encountered, causing successive patterning steps to misalign over large regions of the wafers. This problem is solved by using less than full wafer exposure. At values of λ under 0.5 μm, the wavelength of light used in photolithography is too long to allow sufficient patterning resolution. This problem is solved by using nonoptical lithography, exposing the resist with electron beams or x-rays.

Historically, silicon wafers have been patterned using full wafer exposure, i.e., using masks that covered the entire surface of the wafer. The pattern for one layer of one chip is stepped and repeated during the fabrication of the mask itself, so that the mask contains the patterns for a large array of chips. During the fabrication of each successive layer on the wafer, that layer's mask is aligned at two points with the pattern already on the wafer, and the entire wafer is then exposed through the mask. In the future, as feature sizes are scaled down, full wafer exposure will not likely be possible for reasons developed in this section.

The earliest integrated circuits, circa 1960, were fabricated using wafers of 2.5 cm in diameter, and typical chips were 1 to 2 mm on a side, with a minimum feature size of $\approx$ 25 μm. In 1978, production wafers are 7.5 to 10 cm, typical commercially manufactured LSI chips are 5 mm, and minimum feature size is $\approx$ 5 μ. The concurrent development of ever finer feature sizes and larger wafer sizes has placed an increasingly severe strain on the process of full wafer exposure. The reasons lie in the physics of wafer distortion.

When a wafer is heated to a high temperature, it expands by an amount determined by the thermal coefficient of expansion of silicon. A bare wafer will contract exactly the same amount upon cooling, and will therefore remain exactly the same size. Suppose, however, that a layer of SiO_2 is grown on the wafer when it is at the high temperature. The thermal coefficient of expansion of SiO_2 is approximately 1/5 that of silicon. As the wafer is cooled, the silicon will shrink at a rate much greater than that of the SiO_2. Normally the resulting wafer will not be flat, but convex on the SiO_2 side. If the wafer is cooled slowly enough, it is possible to "relieve" the stress induced by the difference in thermal contraction. Wafers in which such stress relief has been achieved are nearly flat but are, of necessity, a different size than they were originally.[7,8]

It might seem that subsequent masks could be scaled to just match the wafer distortion introduced up to the appropriate point in the process. Unfortunately no such correction can be introduced without a knowledge of the pattern of SiO_2 on the wafer. At high temperature, impurities are incorporated into the silicon surface, making it mechanically, as well as electrically, different from the substrate. During cooling, dislocations are induced in the underlying silicon crystal at the edges of openings in the oxide pattern. Hence, the magnitude and direction of wafer distortion is dependent in complex ways upon the thickness and distribution of SiO_2 on the surface, the amount and type of dopant, and the details of the thermal cycle. While it is in principle possible to compute a geometric correction for each pattern to be produced, it is clearly not possible to apply one correction for all possible patterns. Misalignment between subsequent layers due to distortion of this type is often referred to as *runout*. Runout due to wafer distortion is a major contributor to misalignment between masking steps. Attempts to use finer feature sizes, which require more precise alignment, on larger wafer sizes, which induce larger distortions, seem doomed to failure unless full wafer exposure is abandoned.

Two attractive alternatives to full wafer exposure are now being explored: (1) electron beam exposure, and (2) exposure using "step and repeat" of the chip pattern directly on the wafer.

A scanning electron beam system that can be used to expose resist material is also capable of sensing a previous pattern on the surface of a wafer. The beam can initially scan an area covering the alignment marks of a particular chip. Information gained from this sensing operation can be used to compute the local distortion, and the chip can be exposed in nearly perfect alignment using these computed values. The process can be repeated for each chip on the wafer, until all have been exposed.

This technique has several virtues. No masks are required. A digital description of the chip can be exposed directly onto a silicon wafer. A different chip can be placed at each chip location, and this opens up the possibility of greatly extending the multiproject-chip concept. However, there are also limitations. Data bits are transferred serially. Even at the highest data rates that can be conveniently generated, a long time is required to expose each chip. More fundamentally, the

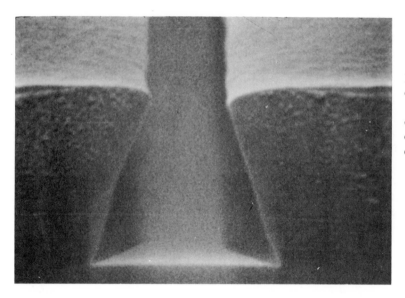

Fig. 4.26 Cross section of a resist-coated Si wafer after exposure with $\approx 1\ \mu$ electron beam and subsequent development. Note widening of opening near the silicon surface due to electron backscattering.

physics of electron beam interactions places severe restrictions on the minimum practical feature size attainable. When a beam of electrons enters a resist-coated wafer, scattering occurs both in the resist and in the wafer. This backscattering contributes a partial exposure at points up to a few microns away from the original point of beam impingement and has a number of implications:

1. The exposure, or spatial distribution of energy dissipation, varies with depth in the resist. Thus resist cross section is not readily controllable. This effect is shown in the resist profile of Fig. 4.26. The resist has been most heavily exposed near the silicon surface. During development, heavily exposed resist is removed faster than lightly exposed resist. Hence the opening is wider at the bottom than at the top.

2. Exposure at any particular point depends on all patterns exposed within a few microns. This is known as the "cooperative exposure" or "proximity" effect and necessitates pattern-dependent exposure corrections.[9]

3. Exposure latitude becomes narrower as the spatial period of a pattern is reduced. This is illustrated in Fig. 4.27, which shows the rise in background level exposure (energy dissipation per unit volume per unit line-charge density) as a function of lateral distance for four different spatial periods: (a) $2\ \mu$m, (b) $1\ \mu$m, (c) $0.5\ \mu$m, (d) $0.3\ \mu$m. The beam diameter is 250 angstrom units, the energy 10 keV, the resist thickness $0.4\ \mu$m. The consequences of this background rise are particularly troublesome for high-speed, low-contrast resists. Experimental results show somewhat greater line broadening than predicted by the model.[10]

For the above reasons, the writing time and the difficulty of exposing desired geometries increase rapidly as linewidths are reduced below about 0.5 micron.[10]

An immediate prospect for achieving feature sizes of 1 μm-2 μm with large wafers is offered by stepping the chip pattern directly on the wafer rather than on a mask. This technique avoids the serial nature of the electron beam writing by exposing an entire chip at once. The use of good optical systems has made it possible for many years to produce patterns with feature sizes in the range 1 μm to 2 μm. Recent progress in the design of optical projection systems may lead to the practical use of ½ to ¾ micron line width patterns over areas several millimeters in diameter.[11] Techniques are known for using light to achieve alignments to a small fraction of a wavelength. Recently, an interferometric optical alignment technique has demonstrated an alignment precision of 0.02 micron and should be capable of a reregistration uncertainty less than 0.01 micron.[12,13] It would seem that devices of ultimately small dimensions (0.25 μm) could be fabricated using

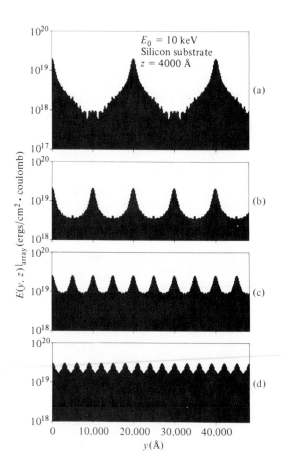

Fig. 4.27 Electron beam exposure of resist on silicon. Monte Carlo calculation of exposure level of a silicon-resist interface, as a function of lateral distance, for four spatial periods. (Contributed by H. I. Smith, Lincoln Laboratory, M.I.T.)

optical alignment. It must be stressed that a realignment to the underlying pattern *must* be done at each chip location to achieve the real potential of the technique.

The step-and-align technique can be extended to ultimately small dimensions by substituting an x-ray source for the optical one, while retaining the automatic optical alignment system. X-rays require a very thin mask support (e.g., polyimide, mylar, silicon, or silicon carbide), upon which a heavy material such as gold or tungsten is used as the opaque pattern. Interactions of x-rays with matter tend to be isolated, local events. Essentially no backscattering of the x-rays occurs, and electrons produced when an x-ray is absorbed are sufficiently low in energy so that their range is limited to a small fraction of a micron. For this reason, patterns formed by x-rays in resist materials on silicon wafers are much cleaner and better defined than those attainable by any other known technique (see Fig. 4.28). Collimated x-ray fluxes of very high intensity can be efficiently obtained from the synchrotron radiation of an electron storage ring. The time required for exposing a chip with such a source is no more than that required at present using optical exposures (a few seconds). Both optical and x-ray exposures (using non-collimated sources) have the property that the total *exposure* time per wafer can be made nearly independent of how much of the wafer is exposed at a step (i.e., the radiation source-to-substrate distance can be adjusted). Therefore, the only

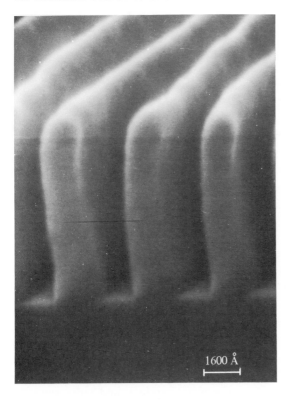

1600 Å

Fig. 4.28 Scanning electron micrograph of the cross section of a 1600-Å linewidth grating pattern exposed in an 8500-Å-thick PMMA film on a SiO_2 /Si substrate using Cu x-radiation at 13.3 Å. The slight curvature of the grating lines was probably caused by overheating during preparation for electron microscopy. (Contributed by H. I. Smith and D. C. Flanders, Lincoln Laboratory, M.I.T.)

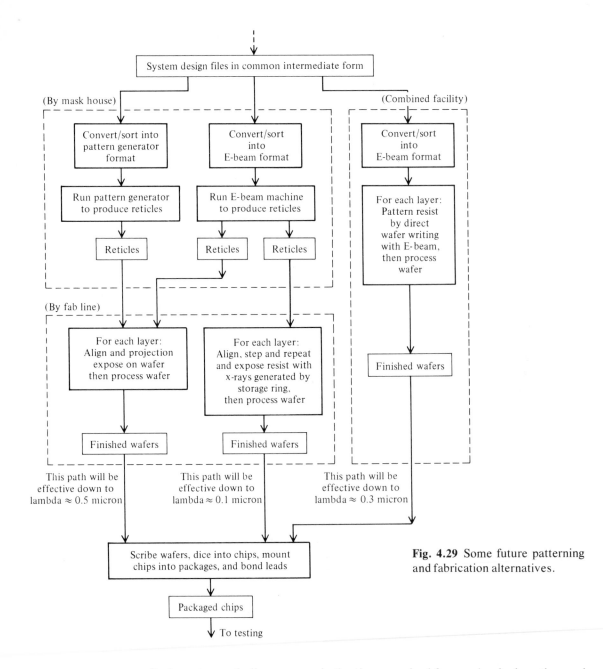

Fig. 4.29 Some future patterning and fabrication alternatives.

penalty in a step-and-align process is the time required for mechanical motion and alignment.

It appears that we have in hand all of the techniques for ultrafine line lithography, even on larger silicon wafers. Electron beam, optical, and x-ray stepping work must, however, focus on local alignment as the crucial element in achieving high-density, high-performance VLSI.

We now describe a production lithography system for ultimately small dimensions. A major component of the system is a 500 to 700 MeV electron storage ring, approximately 5 meters in diameter, shaped in the form of a many-sided polygon, with an exposure station at each vertex. The electron beam within this storage ring is deflected at each vertex by a superconducting magnet. The deflection results in a centripetal acceleration of the electrons, and hence in an intense tangential emission of synchrotron radiation. The most important component of such radiation is soft x-rays in the 280 to 1000 eV quantum energy range (wavelengths of 4.4 to 1.2 nm). Such x-rays are ideal for exposing resist materials with line widths in the 0.1 μm range.[14,15]

Each exposure station has an automatic optical alignment system for individual alignment of each chip. Coarse alignment is controlled by a laser interferometer and the wafer is brought into position by ordinary lead screws moving a conventional stepping stage such as those in current photorepeaters. Automatic alignment to about 0.01 μm is achieved by superposing matching alignment marks on masks and chips using a generalized Moiré technique.[11,12] Piezoelectric transducers driven by the computer system bring the wafer into final alignment under the mask. Each exposure station in such a system is capable of aligning and exposing one layer of one chip every few seconds. Each chip may contain on the order of 10^7 devices, which is the equivalent of several *wafers* at today's scale.

An overview of the possible routes from design files to finished chips with submicron layout geometries is shown in Fig. 4.29. In the immediate future, alignments much better than those achievable today will be possible with the optical step-and-align technique (leftmost path in Fig. 4.29). In addition, this scheme eliminates the step-and-repeat process in maskmaking, enabling considerably shorter turnaround time. The rightmost path, direct electron beam writing on the wafer, promises the ultimate in short turnaround time. It can be viewed as using the fab area as a computer output device. For high-volume manufacturing, at ultimately small dimensions, the center path as described above will most likely become the workhorse of the industry.

REFERENCES

1. GCA / D. W. Mann, "3600 Software Manual, Appendix B: 3600 Pattern Generator Mag Tape Formats," GCA Corporation, IC Systems Group, Santa Clara, Ca.
2. A. C. Kay, "Microelectronics and the Personal Computer," *Scientific American*, September 1977, pp. 210–228.
3. D. G. Fairbairn and J. A. Rowson, "ICARUS: An Interactive Integrated Circuit Layout Program," *Proc. of the 15th Annual Design Automation Conf., IEEE*, June 1978, pp. 188–192.
4. N. Wirth, "What Can We Do about the Unnecessary Diversity of Notations for Syntactic Definitions?," *Communications of the ACM*, November 1977.
5. I. E. Sutherland; C. A. Mead; and T. E. Everhart, "Basic Limitations in Microcircuit Fabrication Technology," ARPA Report R-1956-ARPA, November 1976.

6. C. A. Mead, "Ultra Fine Line Lithography," Display File #1179, December 2, 1977, Department of Computer Science, California Institute of Technology.
7. I. A. Blech and E. S. Meieran, "Enhanced X-ray Diffraction from Substrate Crystals Containing Discontinuous Surface Films," *J. App. Phys.*, vol. 38, June 1967, pp. 2913–2919.
8. E. S. Meieran and I. A. Blech, "High Intensity Transmission X-ray Topography of Homogeneously Bent Crystals," *J. App. Phys.*, vol. 43, February 1972, pp. 265–269.
9. M. Parikh, "Self Consistent Proximity Effect Correction Technique for Resist Exposure," *J. Vac. Sci. Tech.*, vol. 15, May/June 1978, pp. 931–933.
10. R. J. Hawryluk; H. I. Smith; A. Soares; and A. M. Hawryluk, "Energy Dissipation in a Thin Polymer Film by Electron Beam Scattering: Experiment," *J. App. Phys.*, vol. 46, June 1975, pp. 2528–2537.
11. J. S. Wilczynski, "A Step and Repeat Camera for Direct Device Fabrication," Semicon East, Boston, September 1977; M. Huques and M. Babolet, "Lenses for Microelectronics," and P. Tigreat, "Use in a Photodemagnifier," International Conference on Microlithography, Paris, June 1977.
12. D. C. Flanders; H. I. Smith; and S. Austin, "A New Interferometric Alignment Technique," *App. Phys. Lett.*, vol. 31, October 1977, pp. 426–428.
13. S. Austin; H. I. Smith; and D. C. Flanders, "Alignment of X-Ray Lithography Masks Using a New Interferometric Technique—Experimental Results," *J. Vac.Sci.Tech.*, vol. 15, May/June 1978, pp. 984–986.
14. B. Fay et al., 'X-Ray Replication of Masks Using the Synchrotron Radiation Produced by the ACO Storage Ring," *App. Phys. Lett*, vol. 29, September 1976, pp. 370–372.
15. E. Spiller et al., "Application of Synchrotron Radiation to X-Ray Lithography," *J. App. Phys.*, vol. 47, December 1976, pp. 5450–5459.

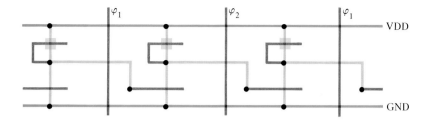

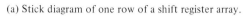

(a) Stick diagram of one row of a shift register array.

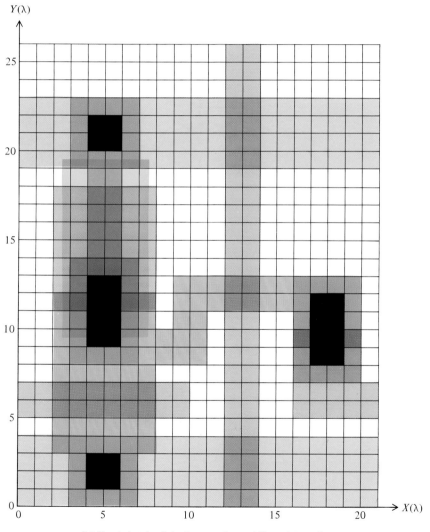

(b) Hand sketch of the layout of one shift register cell.

PLATE 9 Hand layout from stick diagram

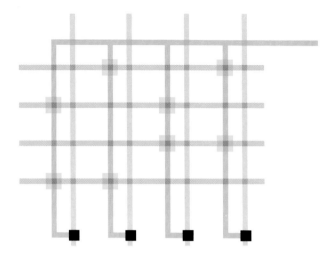

(a) Stick diagram of the function block.

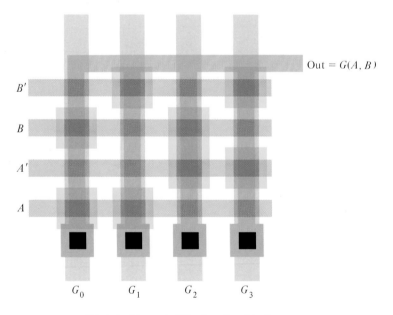

Out = $G(A, B)$

B'

B

A'

A

G_0 G_1 G_2 G_3

(b) Actual layout of the function block.

PLATE 10   The function block

Bus A Out *P P'* Cin Cin' Bus B

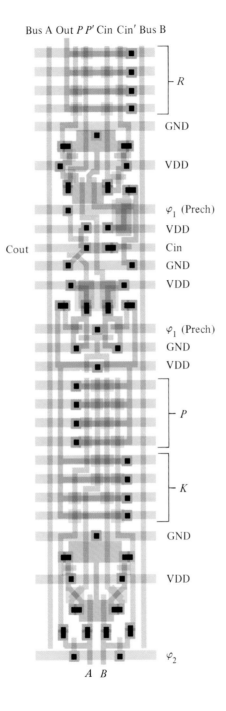

R

GND

VDD

φ_1 (Prech)

VDD

Cout   Cin

GND

VDD

φ_1 (Prech)

GND

VDD

P

K

GND

VDD

φ_2

A B

PLATE 11   Layout of ALU bit slice and input registers

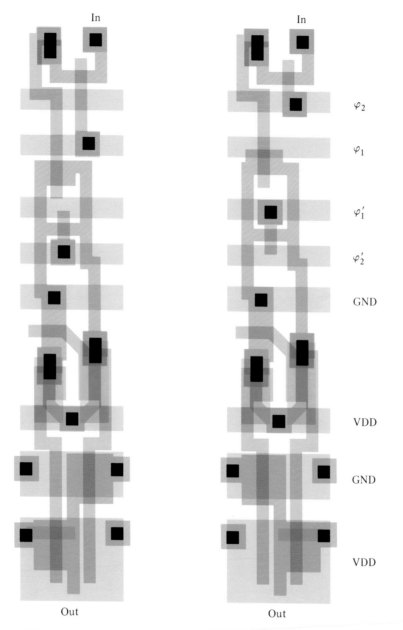

(a) ALU control driver layout.

(b) Select control driver layout

PLATE 12 Driver layouts

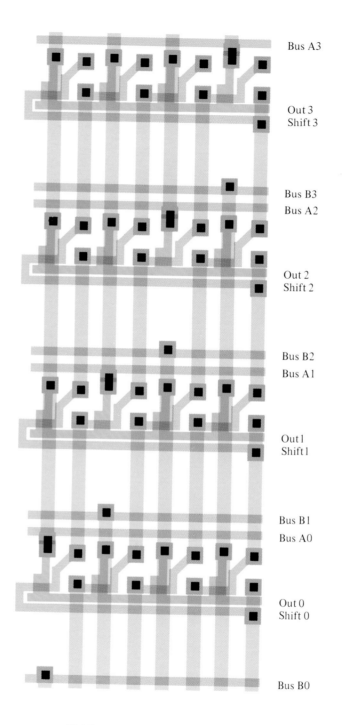

Bus A3

Out 3
Shift 3

Bus B3
Bus A2

Out 2
Shift 2

Bus B2
Bus A1

Out 1
Shift 1

Bus B1
Bus A0

Out 0
Shift 0

Bus B0

PLATE 13 Layout of a 4-bit barrel shifter

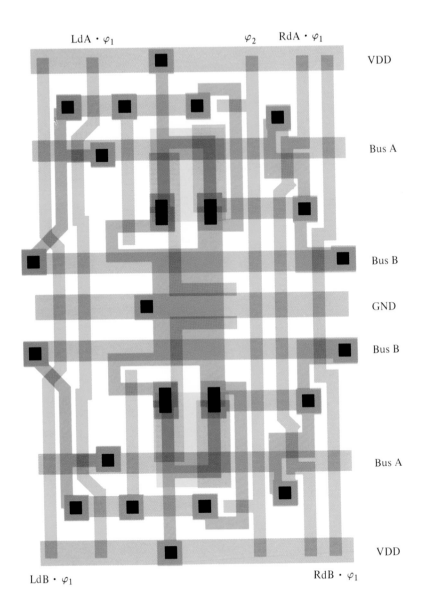

PLATE 14 Layout of two dual-port register cells

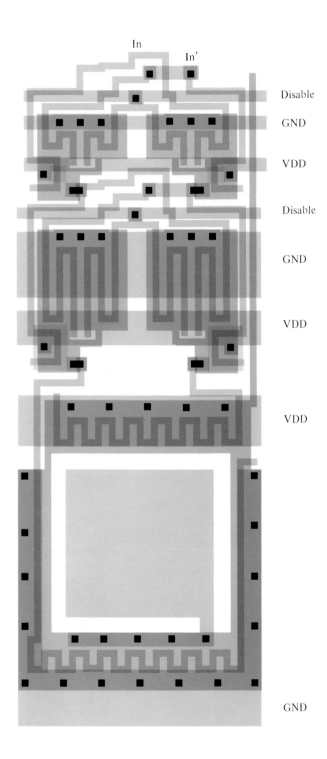

PLATE 15 Pad driver layout

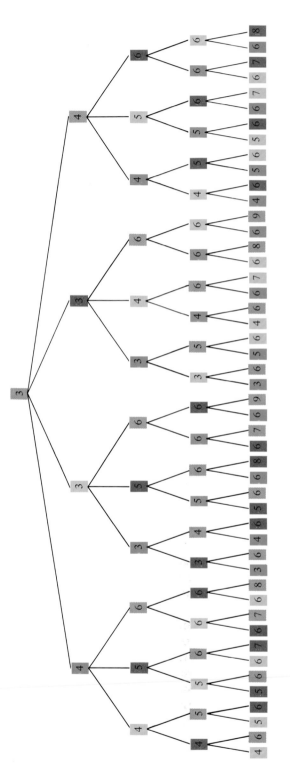

PLATE 16 Example of the color–cost tree

OVERVIEW
OF AN LSI
COMPUTER SYSTEM,
AND THE DESIGN
OF THE OM2
DATA PATH CHIP*

5.1 INTRODUCTION

Up to this point, we have used simple examples to illustrate the fundamental properties of integrated systems, and to illustrate a design methodology for creating hierarchically organized, complex systems. To more fully clarify these concepts, we now present examples drawn from the design of an LSI computer system. In this chapter, we provide a brief overview of this computer system and then describe in detail one of its major components, the *data path chip*. Much of the detail in this chapter is intended to provide the reader with a source of examples of the implementation of digital logic subsystems into LSI circuit layout structures, under the constraints imposed both by the design methodology and by the architectural requirements of a real computer system. Chapter 6 similarly describes the *controller chip* of this computer system, and provides additional information on the sequencing of the overall system.

In this chapter we assume that the reader is familiar with the structure and function of the classical stored-program digital computer, and with the concept, and computer design implications, of microprogrammed control. An informal review of these basic concepts is given in the introductory portions of Chapter 6, so that the mapping of the required controller subsystems into silicon can be examined. The less-experienced reader may benefit from a study of that material in parallel with reading this chapter.

It is important to note that the computer system discussed in Chapters 5 and 6, while composed of structured LSI subsystems, is nevertheless of classical von Neumann form. The architectural possibilities of VLSI are just now beginning to be explored. Future lower cost, higher density, higher speed devices, combined with major reductions in integrated system implementation time, will make completely new forms of computing machines, and new notions of programming, not only feasible but practical. (Some of these issues will be discussed in Chapter 8.)

---

*Much of the material in this chapter was contributed by Dave Johannsen, Caltech, who played a leading role in the architecture and design of the OM system.

5.2   THE OM PROJECT AT CALTECH

The design of this computer system was undertaken as a university project in experimental computer architecture. The "Our Machine" (OM) project, as it has come to be known, was started by Carver Mead in 1976, as part of the LSI Systems course at Caltech. The project involved the design of a number of LSI chips, as described in Section 5.3.

The initial focus of the project was the architecture and design of the system's primary data processing module, the *data path chip*. Early contributions to this effort were made by Mike Tolle (Litton Industries) while attending the LSI systems course. Other participants were Caltech students Dave Johannsen and Chris Carroll, who received much inspiration from Ivan Sutherland. By December 1976, the first design (OM0) of the data path chip was nearly completed. The participants decided at that time that the design had become "baroque" and "ugly," and it was scrapped. A new data path design (OM1) was completed by March 1977 by Dave Johannsen, Chris Carroll, and Rod Masumoto. Fabricated chips were received in June 1977. It was this chip that appeared in the article by Sutherland and Mead in *Scientific American* (September 1977). The chip was fully functional except for a timing bug in the dynamic register array, which had been designed in departure from the structured design methodology developed in this text.

A complete redesign of the data path chip was undertaken in June 1977, by Dave Johannsen. By September 1977, a complete set of new cells had been constructed. The design was completed by December, and chips were fabricated by April 1978. The redesign included improvements in the encoding of the microcode control word and rigorously applied the structured design methodology. (Certain cells from the OM2 data path chip, and from its companion controller chip, were used as examples in Chapter 3.)

During 1977, the controller chip was designed as one of four class projects in the Caltech LSI Systems course. It was finished in the summer of 1977, and fabricated chips were received in early 1978.

During 1978, the architecture of an overall system was planned. Design has begun of the three remaining chips in the OM computer system: the system bus interface chip, the memory manager chip, and the clock chip.

All of the detailed LSI design on the OM project has been done by students. Throughout most of the project's history only rather limited design aids were available, notably a simple symbolic layout language and graphic plotters for checkplotting. The efforts of students to quickly create large integrated systems, using only primitive designs aids, helped to motivate the development and refinement of the structured design techniques described in this text.

The OM project has also required the *implementation* of many prototype designs and complete chip designs. Since early in the project, the Caltech group collaborated with researchers in industry, who were similarly completing many prototype LSI system designs, on the development of practical methods for simplifying and speeding up prototype project implementation. This led to the formulation and debugging of the standard starting frame for conveying multiproject chips through maskmaking and wafer fabrication, as described in Chapter 4.

5.3 SYSTEM OVERVIEW

An informal block diagram of one OM system is shown in Fig. 5.1. Such a system is a complete stored-program, general-purpose computer. Input/output devices are usually interfaced via the external data bus and control lines, located to the left in Fig. 5.1. Several such systems may be interconnected via the system bus to augment one user's overall system. Tasks may then be distributed among the OM systems, for example, using different ones to independently control different input/output devices, thereby improving overall system performance. Groups of different user systems may also share the system bus.

Each OM system is composed of five LSI chips, along with some standard memory chips and a few MSI chips. A brief description follows of the five LSI chips being designed as part of this project.

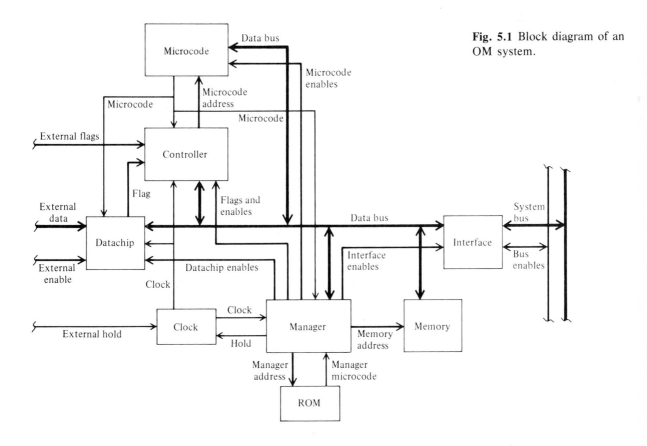

Fig. 5.1 Block diagram of an OM system.

The *data path chip* performs most of the data manipulation functions for the system. The operations are performed as directed by sequences of control micro-instructions, which are fetched from a microcode memory using addresses generated by the controller chip. The main subsystems of the data path chip are a register array, a shifter, and an arithmetic logic unit (ALU). Two buses connect these subsystems. This chip's internal structure is described in detail later in this chapter.

The *controller chip* contains the microprogram counter (μPC) that stores the microcode memory address, and a counter for the control of microprogram loops. The chip also contains stacks for both the microprogram counter and loop control counter values. The concepts of controller structure and function are fundamental in computer architecture. (Chapter 6 provides an introduction to these ideas and then describes the organization and layout of the controller chip.)

The *memory manager chip* provides addresses for the data memory and directs the communication between chips on the data bus. It also implements some simple data structures in the data memory. The manager can divide the memory into separate partitions and implement a different data structure in each partition. Four basic data structures are implemented: stacks, queues, linked lists, and arrays. When accessing a stack partition, for example, the microcode need only ask the manager to push or pop data on or off the stack. The manager maintains the stack pointers, performs bounds checking to see if the stack is full or empty, and transfers the data.

The *system bus interface chip* provides asynchronous communication with other OM systems via the system bus. There are a whole host of subtleties associated with interfacing asynchronous buses. These issues are among those discussed in Chapter 7.

The *clock chip* generates the two-phase clock signals needed by the system. The clock can be stopped to allow for the synchronization of asynchronous signals. Some chips in the system have only a single clock input; those chips generate the two clock phases on-chip.

A few words about timing may be helpful: In general, during φ_1 data bits are transferred from one subsystem to another on the same chip, while during φ_2 data bits are transferred from one chip to another. The data chip's ALU, and other data modification units, operate during φ_2. Microcode is available on both phases and is pipelined by one phase. Thus, the opcodes that control the ALU enter the data chip during φ_1. The microprogram address is generated by the controller chip during φ_2, gets driven off-chip into the data chip's microcode memory's latches during φ_1, and is used to look up the next opcode on the following φ_2. Because of these timing requirements, all jumps in the microcode are pipelined by one clock cycle.

The remainder of this chapter describes the data path chip and is presented in two distinct parts. The first part outlines the architectural requirements for the data path chip and then illustrates, via the detailed design and layout of the chip's subsystems and cells, how the design methodology was applied to satisfy these requirements. The second part is an external functional description of the data path chip, intended as a user manual for those who microprogram the computer system, and for reference while studying the OM2 controller chip in Chapter 6.

5.4 THE OVERALL STRUCTURE OF THE DATA PATH

The basic requirements initially established for the data path chip were (1) that it be gracefully interconnectable into multiprocessor configurations, (2) that it effectively support a microprogrammed control structure, thus enabling machine instruction sets to be configured to the application at hand, (3) that it be able to do variable field operations for emulation instruction-decoding, assembly of bit-maps for graphics, etc., and (4) that its performance be as fast as possible.

In order to satisfy the first requirement, the data path chip was designed with two ports: one port to be used for a system interconnection, and the other for connection to local memory, input/output devices, etc. The requirement for gracefully handling variable-length words required a shifter at least sixteen bits long. The performance requirement dictated an arithmetic logic unit having considerable flexibility without sacrificing speed. In many systems time is lost in assembling the two operands required for most operations. Therefore, the data path has two internal buses, and all registers on the chip are two-port registers. In order to avoid extensive random wiring for connecting the major subsystems on the chip, the following strategy was adopted at the outset: two internal buses would run through the entire processing array, from one end of the chip to the other. One port was to be located at the left end of the chip, and the other port at the right end.

The three main functional blocks on the chip are the arithmetic logic unit, the shifter, and the register array. These blocks are placed next to each other in the center of the chip, between the two ports. The arrangement of the major subsystems is shown in Fig. 5.2. The system buses run horizontally, on the polysilicon level, through these functional blocks. The major control lines run vertically across these blocks, on the metal level. The power, ground, and clock lines are run parallel to the control signal lines. The details of these functional blocks will be

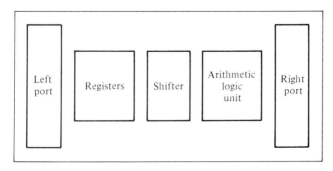

Fig. 5.2 General floor plan of the data path chip.

described in subsequent sections of this chapter. Included are descriptions of peripheral circuits needed to interface subsystems with each other and to the out-side world. Detailed layouts of certain cells in the system are also included. The overall layout of the data chip is shown in the frontispiece.

5.5   THE ARITHMETIC LOGIC UNIT

The carry chain of the ALU, and its associated logic, was the first functional block to be designed in detail, since it was believed that the carry chain would limit the performance of the system. Simulations of several look-ahead carry circuits indi-cated that they would add a great deal of complexity to the system without much gain in performance. For this reason a decision was made early in the project to implement the fastest possible Manchester-type carry chain, [Chap. 1: 5,6] having a carry propagation circuit similar to that shown in Fig. 1.28. The carry chain and its associated logic were allowed to dictate the repeat distance of the cells in the vertical direction. In *n*MOS technology, a Manchester carry chain is particularly limited in its ability to propagate a *high* carry signal. However, it can quite rapidly propagate a *low* carry signal.

In the arithmetic logic unit there will be a null period when the OP code for the next operation is being brought in. Advantage can be taken of this null period to precharge the carry chain and other sections of the data path where timing is particularly crucial. In this way, it is not necessary to propagate high signals through pass transistors where the rise transient would be particularly slow. This strategy was applied in OM's ALU; the resulting carry chain is shown in Fig. 5.3.

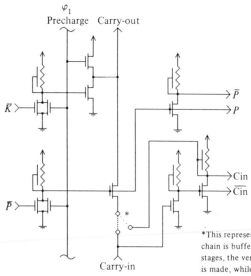

Fig. 5.3 Carry-chain circuit for the arithmetic logic unit.

*This represents how the carry chain is buffered. In most of the stages, the vertical connection is made, while in the stages with the amplification the diagonal connection is made.

The main carry chain runs through a pass transistor from carry-in to carry-out. The carry-in signal is detected by the gate of an inverter that feeds the signal into the subsequent logic of the ALU. Three transistors are used to control the state of the carry-out of each stage. The first one merely precharges the node associated with carry-out during the null period of the ALU. The second is the carry-kill signal that is derived from the inputs to the ALU, and it simply grounds the carry-out through a single transistor. The third is the pass transistor that causes carry-out to be equal to carry-in. These last two signals associated with the carry chain in each stage, carry-kill and carry-propagate, are generated by two NOR gates that have kill-bar and propagate-bar as one input and precharge as the second input. Hence, it is assured that the kill signal and propagate signal are disabled during the null period when the precharging takes place.

After some analysis, we found that nearly all interesting combinations of carry-in and the input signals could be generated using propagate and carry-in from each stage. Thus, as in Fig. 5.4, the carry chain may be seen as a logic block with two inputs, carry-kill and carry-propagate; the outputs, propagate and carry-in; vertical signals, carry-in and carry-out; and one control wire, precharge.

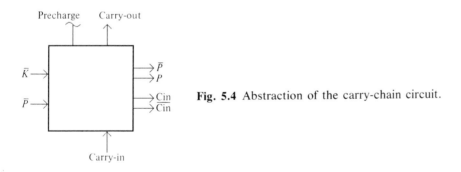

Fig. 5.4 Abstraction of the carry-chain circuit.

The task of designing the balance of the ALU is now reduced to that of designing functional blocks to (a) combine the two input variables to form a propagate-bar and kill-bar, and (b) combine carry-in and propagate to form the output signal, and then designing drivers for controlling the logic function blocks and deriving a timing for precharge.

A number of random logic implementations of function blocks for deriving kill, propagate, and the output were attempted. All seemed to be at variance with the horizontally microprogrammed architecture of the data path and required a large amount of area and power. For this reason it was decided to use the general logic function block illustrated in color Plate 7(a). Recall that the depletion mode transistors, i.e., those covered by ion implanted regions represented by yellow, are always on. Such logic function blocks are used to generate kill-bar and propagate-bar, and for combining carry-in and propagate to form the output.

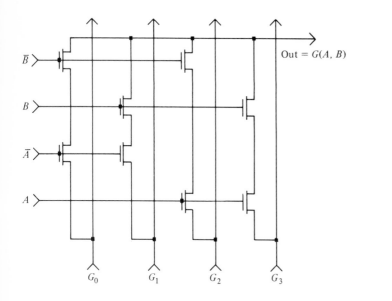

$Out = G(A, B)$

G_0 G_1 G_2 G_3

Fig. 5.5 Circuit diagram for a general logic
function block.

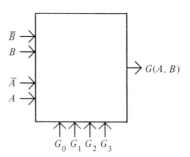

$\bar{B}$
B
$\bar{A}$
A

$G(A, B)$

G_0 G_1 G_2 G_3

Fig. 5.6 Functional abstraction of the
general logic function block.

The circuit, shown in Fig. 5.5, implements the sixteen logic functions of two input
variables. It consists of a set of transistors that fully decode the input combination
of A and B. The set connects one and only one of the vertical control lines to the
output, depending on this input combination. For example, when A and B inputs
are both low, the vertical control wire labeled G_0 is connected to the output. The
truth table entries for the desired logic function are placed on the G vertical control wires, and the output is then the desired logic function of the two input variables. For example, if the Exclusive-OR of A and B is desired, a logic-0 will be
applied to the control wires G_0 and G_3, and logic-1 will be applied to control wires
G_1 and G_2. Since it is desired to implement the same logic function on all bits of the
word, the control variables G_0 through G_3 need not be generated in every bit slice,
but may be generated once at either the top or bottom of the array. The functional
abstraction of the circuit of Fig. 5.5 is shown in Fig. 5.6. For a color-coded stick
diagram of the function block, see Plate 10(a). Plate 10(b) shows a color-coded
actual layout of the function block.

The block diagram for a complete arithmetic logic unit is shown in Fig. 5.7.
The functional dependence of the output on the two inputs and the state of the
carry is determined by a 12-bit number: P_0 through P_3, K_0 through K_3, and R_0
through R_3, together with the carry-in to the least significant bit of the ALU. The
ALU is quite general, and its detailed operation set may be left unbound until the
control structure of the computer system is designed.

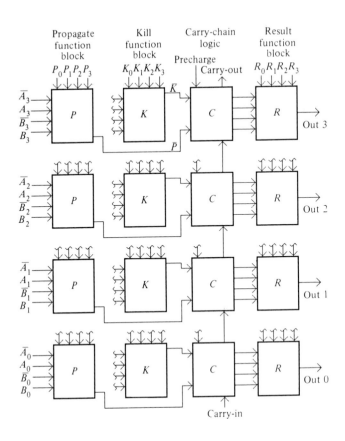

Propagate function block — $P_0 P_1 P_2 P_3$

Kill function block — $K_0 K_1 K_2 K_3$

Carry-chain logic — Precharge — Carry-out

Result function block — $R_0 R_1 R_2 R_3$

Fig. 5.7 Block diagram of a 4-bit ALU.

There are two general principles illustrated by this design. (1) It is often less expensive in area, time, and power to implement a general function than to implement a specific one. (2) If a general function can be implemented, the details of its operation can be left unbound until later, and hence provide a much cleaner interface to the next level of design. The detailed selection of which functional entities to leave unbound and which to bind early requires a considerable amount of judgment.

Two details must be dealt with before the arithmetic logic unit subsystem is complete. Drivers are needed for the $P_0 \ldots P_3$, $K_0 \ldots K_3$, and $R_0 \ldots R_3$ control lines that will generate signals with the appropriate timing. In addition, inverters must be interposed in the carry chain occasionally to minimize the propagation delay through the entire carry chain. The way we have chosen to implement the interposition of inverters is to recognize that each carry chain function block contains two inverters that produce at their output the carry-in, having been twice inverted from the actual carry-in signal. If we merely substitute this signal for the carry-in signal to the pass transistor, we have doubly inverted our carry-in and buffered it to minimize the propagation delay. This approach avoids putting spaces for inverters between the carry function blocks. It is illustrated by the dotted connection lines in Fig. 5.3. In the actual implementation, the connection through the inverters was made in every fourth stage (see Section 1.11).

Drivers for the P, K, and R control lines have the following function: At some time during the null period of the ALU (which occurs during φ_1), an OP code specifying the state of each control line arrives at the drivers. It must be latched while the ALU itself is being precharged, and then it must be applied to the P, K, and R control lines as soon as the ALU is activated. The P, K, and R function blocks are themselves composed of pass transistors, and their outputs are more effectively driven low than high. For this reason, we will precharge the outputs of the P, K, and R function blocks as well as the carry chain itself. This is most conveniently done by requiring that all of the P, K, and R control signals be high during the null period of the ALU. Then, independent of the states of A and B inputs, the outputs will be charged high by the time the ALU active period commences. The control driver implementing this function is shown in Fig. 5.8.

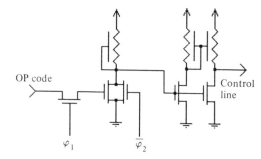

Fig. 5.8 ALU control driver. All outputs high during $\bar{\varphi}_2$; selected terms low during φ_2; OP code valid during φ_1.

The OP code is latched through a pass transistor whose gate is connected to φ_1, and the OP code runs into a NOR gate, the other input of which is $\bar{\varphi}_2$. Thus, the output of the NOR gate is guaranteed to be low during the φ_1 period. The NOR gate output is then run through an inverting super buffer, so that during φ_1 the output is guaranteed to be high. During φ_2, the OP code is driven onto the P, K, and R control lines. The only interface specification for the ALU that must be passed to the next level of system design is that the inputs to the P, K, and R control drivers be valid before the end of φ_1, and that the A and B inputs likewise be valid by the end of φ_1 and be stable throughout φ_2, the active period of the ALU. We are then guaranteed that after enough time has passed to allow the carry to propagate, the output of the R function block will accurately reflect the specified function of the ALU and may be latched at the end of φ_2.

A color-coded plot of the layout of a 1-bit slice through this ALU is shown in color Plate 11. This ALU went through another design iteration before inclusion in the OM2. A plot of the ALU control driver is shown in color Plate 12.

5.6 ALU REGISTERS

In order for the arithmetic logic unit described in the last section to be useful, it must be equipped with a set of registers both for its input variables and for its output. Let us consider the input registers first. Inputs to the ALU may be derived from either the shifter, the buses, or other sources. They may be latched and left unchanged during any φ_1-φ_2 machine cycle or set of machine cycles. This is one of the situations in which combining the multiplexing function with the latching function simplifies the design and achieves better performance. A register operating in this manner is shown in Fig. 5.9.

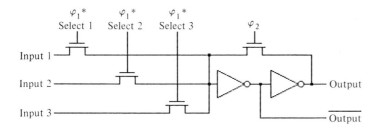

Fig. 5.9 ALU input register and multiplexer.

The input to the first inverter can be derived from four sources: three external sources such as shifter output, bus, etc., and a fourth, the output of the second inverter. When it is desired to latch a new signal into the register, one of the source pass transistors is driven on during φ_1. The feedback transistor around the two inverters is always activated during φ_2. Thus, with three vertical control wires plus the φ_2 timing signal, it is possible to select one of three sources into the register, or none of the three sources, thereby leaving the previous value of the register stored on the gate of the first inverter during the φ_1 period. Since it is necessary to have two inverters to form the stable pair when the feedback transistor is on, both the input and its complement are available as required by the P and K function blocks of the arithmetic logic unit. The OP code signal that selects the source that will be applied to the ALU input register during φ_1 must enter the chip during the previous φ_2. Each of the select signals must be low during φ_2, and at most one of them may come high during the following φ_1. A driver appropriate for these control signals is shown in Fig. 5.10, with the corresponding layout shown in color Plate 12. The control OP code is latched during φ_2, during which time the NOR gate shown disables the output driver. Since the output driver in this case is noninverting, the output select line is held low during all of φ_2. At the end of φ_2, the OP code signal is latched and at the beginning of φ_1 the particular select line to be enabled that cycle is allowed to go high.

Note that this timing allows two incoming OP code bits per external wire per machine cycle. In particular, if it were desirable to share a microcode bit between the ALU function and the ALU selector inputs, this could be done by bringing the ALU OP code in during φ_1 and the ALU input selection code in during φ_2. This technique was suggested by Ivan Sutherland.

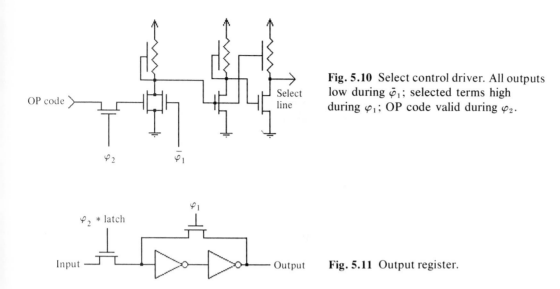

Fig. 5.10 Select control driver. All outputs low during $\bar{\varphi}_1$; selected terms high during φ_1; OP code valid during φ_2.

Fig. 5.11 Output register.

The ALU output register is similar to the ALU input register, except that the timing is reversed. The result of the ALU operation is available at the end of φ_2. An OP code bit will, if desired, enable the latch signal to go high during φ_2. The feedback transistor is always enabled during φ_1, and thus the latch is effectively static even though in the absence of a latching signal the data is stored dynamically on the gate of the first inverter through the φ_2 period. Once again, both the output and its complement are available if desired.

5.7 BUSES

An early design decision was made to have data flow through the data path chip on two buses that communicate with all of the major blocks of the system. We have already seen that the ALU performs its operation during the φ_2 period and does not have valid data to place into its output register until the end of φ_2. If data are to be transferred from the output register of the ALU to its input register, this must be done during the φ_1 period. If we adopt a standard timing scheme in which all transfers on the buses occur during φ_1, we can make use of the φ_2 period when the ALU is performing its operation to precharge the buses in the same manner that the carry chain was precharged during the φ_1 period. In this way we solve one of the knotty problems associated with a technology designed for ratio logic. If we had insisted that the tri-state drivers associated with various sources of data for a bus be able to drive up as well as down, we would have required both a sourcing and sinking transistor, together with a method for disabling both transistors. While it is perfectly possible to build such a driver (we shall undertake the exercise as part of the design of the output ports), it is a space-consuming matter to use such a driver at every point where we wish to source data onto an internal bus.

By using the bus precharge scheme, our "tri-state drivers" become simply two series transistors as shown in Fig. 5.12. Here the data from one source, for example, the ALU output register, is placed on the gate of one of the series transistors. An enable signal, which may come high during φ_1, is placed on the other series transistor. If one and only one of the enable signals is allowed to come high during any one φ_1 period, the bus can be driven from as many sources as necessary. The performance of such a bus is limited by the pull-down capability of the two series transistors. We attach such a driver to each of the output registers for the ALU.

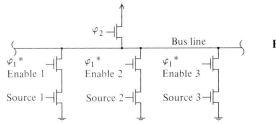

Fig. 5.12 Precharged bus circuit.

5.8 BARREL SHIFTER

Since shifting is basically a simple multiplexing function, one might think that a shifter could be combined with the input multiplexer to the ALU. A simple 1-bit, right-left shifter implemented in this manner is shown in Fig. 5.13. It is identical with the three-input ALU register, with the three inputs used to select among the bus, the bus shifted left by one, and the bus shifted right by one. To support the

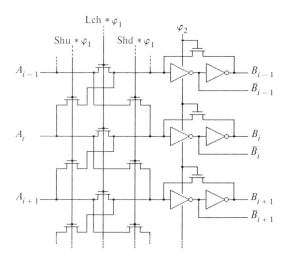

Fig. 5.13 A simple 1-bit, right-left shifter.

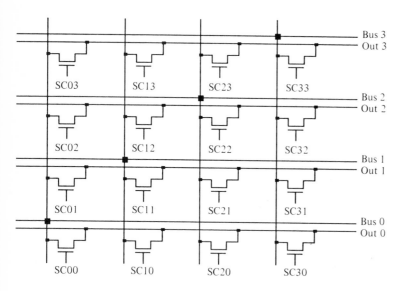

Fig. 5.14 A 4-by-4 crossbar switch.

multibit shifts necessary for field extraction and building up odd bit arrays, something more is required. One is tempted initially to build up a multibit shift out of a number of single shifts. However, for word lengths of practical interest, the n^2 delay problem (see Section 1.11) makes such an approach unworkable.

The basic topology of a multibit shift dictates that any bus bit be available at any output position. Therefore, data paths must run vertically at right angles to the normal bus data flow. Once this simple fact is squarely faced, a multibit shifter is seen as no more difficult than a single bit shifter. A circuit enabling any bit to be connected to any output position is shown in Fig. 5.14. It is basically a crossbar switch with individual MOS transistors acting as the crossbar points, the basic idea being that each switch SC_{ij} connects bus i to output j. In principle this structure can be set to interchange bits as well as to shift them, and it is completely general in the way in which it can scramble output bits from any input position. In order to maintain this generality, the control of the crossbar switch requires n^2 control bits. In some applications, the n^2 bits may not be excessive, but for most applications a simple shift would be adequate. The gate connections necessary to perform a simple barrel shift are shown in Fig. 5.15. The shift constant is presented on n wires, one and only one of which is high during the period the shift is occurring. If the shifter's output lines are precharged in the same manner as the bus, the pass transistors forming the shift array are required to pull down the shifter's outputs only when the appropriate bus is pulled low. Thus, the delay through the entire shift network is minimized and effective use is made of the technology.

A second topological observation is that in every computing machine, it is necessary to introduce literals from the control path into the data path. However,

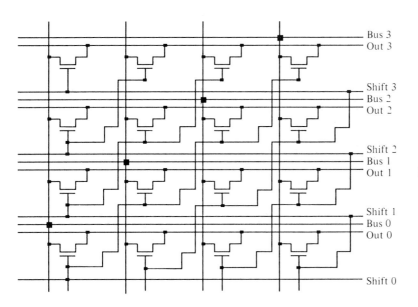

Bus 3
Out 3

Shift 3
Bus 2
Out 2

Shift 2
Bus 1
Out 1

Shift 1
Bus 0
Out 0

Shift 0

Fig. 5.15 A 4-by-4 barrel shifter.

our data path has been designed in such a way that the data bits flow horizontally while the control bits from the program store flow vertically. In order to introduce literals, some connection between the horizontal and vertical flow must occur. It is immediately obvious in Fig. 5.15 that the bus is available running vertically through the shift array. That is then the obvious place to introduce literals into the data path or to return values from the data path to the controller.

At the next higher level of system architecture, the shift array bit slice may be viewed as a system element with horizontal paths consisting of the bus, the shifter output, and if necessary, the shift constant since it appears at both edges of the array. The literal port is available into or out of the top edge of the bit slice, and the shift constant is available at the bottom of the bit slice. These slices, of course, are stacked to form a shift array as wide as the word of the machine being built.

One more observation concerning the multibit shifter is in order. We stated earlier that our data path was to have two buses. Therefore, in our data path, any bit slice of a shifter such as the one shown in Fig. 5.15 will of necessity have two buses running through it rather than one. We chose to show only one for the sake of simplicity. There remains the question of how the two buses are to be integrated with the shifter. Since we are constructing a two-bus data path, we have two full words available, and a good field extraction shifter would allow us to gracefully extract a word that crosses the boundary between two data path words. The arrangement shown in Fig. 5.15 performs a barrel shift on the word formed by one bus. Using the same number of control lines and pass transistors, and adding only the bus lines that are required for the balance of the data path anyway, we may construct a shifter that places the words formed by the two buses end to end and extracts a full-width word that is continuous across the word boundary between

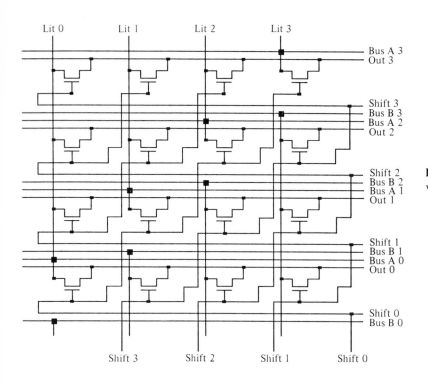

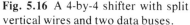

Fig. 5.16 A 4-by-4 shifter with split vertical wires and two data buses.

the A and B buses. This function is accomplished, in as compact a form as just described, with a circuit shown in Fig. 5.16. Notice that the vertical wires have a split in them. The portion of the wire above the corresponding shift output is connected to bus A, and that below the corresponding shift output to bus B. The layout of the barrel shifter is shown in color Plate 13.

It can be seen by inspection that this circuit performs the function shown in Fig. 5.17, which is just what is required for doing field extractions and variable word length manipulations. The literal port is connected directly to bus A and may be run backward in order to discharge the bus when a literal is brought in from the control port. A block diagram representing the shifter at the next level of abstraction is shown in Fig. 5.18.

In order to complete the shifter functional block, it is necessary to define the drivers on the top and bottom that interface with the system at the next higher level. Let us assume that the literal bus from outside the chip will contain data valid on the opposite phase of the clock from that of the internal buses. For that case, a very simple interface between the two buses that will operate in either direction is shown in Fig. 5.19.

The internal shifter output is precharged during φ_2, and active during φ_1. It may be sourced from the shifted combination of either the A and B buses or the literal bus and B bus, as shown in Fig. 5.16. The external literal bus itself may be

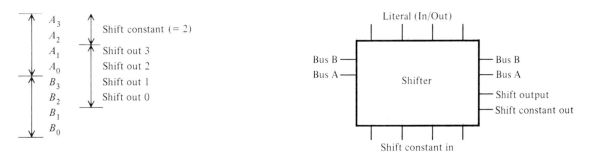

Fig. 5.17 Conceptual picture of the shifter's operation.

Fig. 5.18 Block diagram of the shifter.

sourced either from the opposite end (the external paths from the program source) or from the end attached to bus A in the shift array shown.

The bus to the external literal path is precharged during φ_1, and data bits from the literal port of the shifter are enabled onto it by a signal active during φ_2, as shown in Fig. 5.19. The two signals, $\varphi_1 * $ In, and $\varphi_2 * $ Out, are derived from buffers identical to those shown earlier. The shift constant itself is represented by one line out of n, which is high, the others remaining low. Buffers for these lines are identical to those shown in Fig. 5.10.

There is one more observation concerning the n-bit shift constant. It is represented most compactly by a log n bit binary number. However, in order to generate from such a form a signal that can be used in the actual data path, a decoder is required for converting the binary number into a 1-of-n signal suitable for feeding the buffers. There are a number of ways of making decoders in nMOS technology. The most common form is the NOR form, which is the fully decoded equivalent of the AND-plane in the programmable logic array (Chapter 3). It is shown in Fig. 5.20. Notice that the output is a high-going 1-of-n pattern.

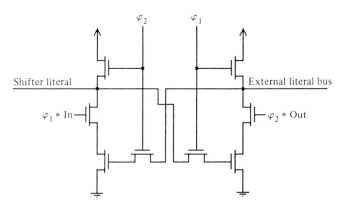

Fig. 5.19 Literal interface.

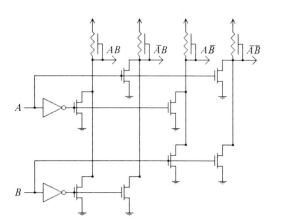

Fig. 5.20 A NOR form 1-of-n decoder.

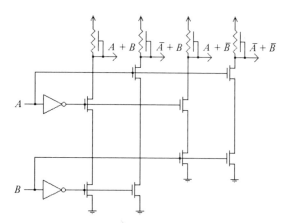

Fig. 5.21 A NAND form 1-of-n decoder.

Decoders can also be made in other forms. For small values of n, the NAND form shown in Fig. 5.21 is often convenient. We used a variant of this form for the ALU function block described earlier. Notice that the output of this form, when used as a decoder, is a low-going 1-of-n pattern. There is also a complementary form of decoder that can be built with this technology, as suggested by Ivan Sutherland. It takes advantage of the fact that in any decoder both the input term and its complement must be present. In this case, the input term can be used to activate pull-up transistors in series, while the complement can be used to activate pull-down transistors in parallel. This logic form is similar in principle to that used with complementary technologies and has similar benefits. It can generate either a high-going or a low-going 1-of-n number and dissipates no static power. A decoder of this sort is shown in Fig. 5.22. Once we have added the appropriate buffers and decoders to our shift array, we have a fully synchronized subsystem ready to be integrated with the system at the next level up. The block diagram of this subsystem is shown in Fig. 5.23.

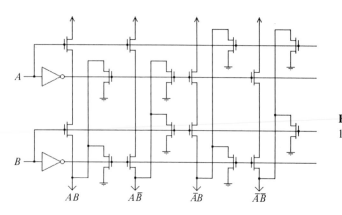

Fig. 5.22 A complementary form 1-of-n decoder.

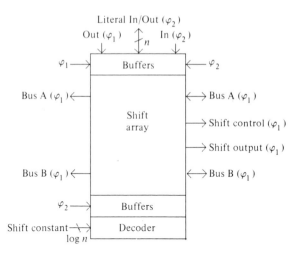

Fig. 5.23 A fully synchronized shifter.

5.9 REGISTER ARRAY

In any microprogrammed processor designed for emulating a higher level instruction set, it is convenient to have a number of miscellaneous registers available, both for working storage during computations and for storing pointers (stack pointers, base registers, program counters, etc.) of specific significance in the machine being emulated. Since the data path has two buses and the ALU is a two-operand subsystem, it is convenient if the registers in the data path are two-port registers. The circuit design of a typical two-port register cell is shown in Fig. 5.24. The layout of a pair of these cells is shown in color Plate 14. This register is a simple combination of the input multiplexer described earlier, the φ_2 feedback transistor, and two output drivers, one for each bus. The registers can be combined into an array m bits long and n bits wide. Each cell of the array can be viewed at the next level up as shown in Fig. 5.25. Drivers for the load inputs and the read outputs are

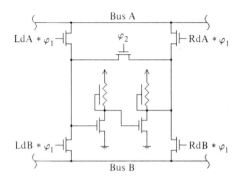

Fig. 5.24 A two-port register cell.

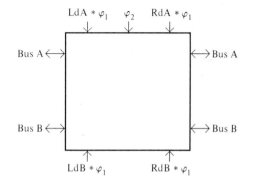

Fig. 5.25 Block diagram definition of the two-port register cell.

identical to those shown in Fig. 5.10. While we could immediately encode the load and read inputs to the registers into log *n* bits, we shall delay doing so until the next level of system design. There are a number of sources for each bus besides the registers, and we will conserve microcode bits by encoding them together.

Before we proceed, there is one important matter that must be taken care of in the overall topological strategy. Routing of VDD and ground paths must generally be done in metal, except for the very last runs within the cells themselves. Often the metal must be quite wide, since metal migration tends to shorten the life of conductors if they operate at current densities much in excess of 1 milliampere per square micron cross section. Thus, it is important to have a strategy for routing ground and VDD to all the cells in the chip before doing the detailed layout of any of the major subsystems. Otherwise, one is apt to be faced with topological impossibilities because certain conductors placed for other reasons interfere with the routing of VDD and ground. A possible strategy for the overall routing of VDD and ground paths is shown in Fig. 5.26. Notice that the VDD and ground paths form a set of interdigitated combs, so that both conductors can be run to any cell in the chip. Any such strategy will do, but it must be consistent, thoroughly thought through at the beginning, and rigidly adhered to during the execution of the design.

5.10 COMMUNICATION WITH THE OUTSIDE WORLD

Although in particular applications the interface from a port of the data path to the outside world may be a point-to-point communication, the ports will often connect

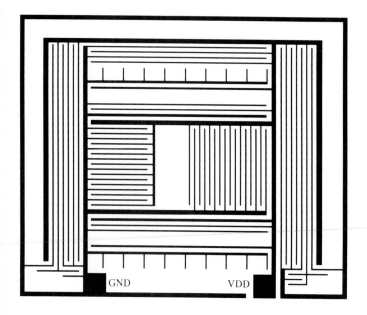

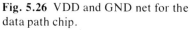

Fig. 5.26 VDD and GND net for the data path chip.

to a bus. Thus it is desirable to use port drivers that may be set in a high impedance state. Drivers that can either drive the output high, drive the output low, or
appear as a high impedance to the output are known as *tri-state* drivers. Such
drivers allow as many potential senders on the bus as necessary. Figure 5.27
shows the circuit for a tri-state interface to a bonding pad. Here, either bus A or
bus B can be latched into the input of a tri-state driver during φ_1. Likewise the pad
may be latched into an incoming register at any time independent of the clocking
of the chip. Standard bus drivers are enabled on bus A and bus B. The only remaining chore is the design of the tri-state driver that drives the pad directly.
Details of the tri-state driver are shown in Fig. 5.28. The layout of an output pad
and its associated driver circuitry is shown in color Plate 15.

The terms *out* and *outbar* are fed to a series of buffer stages that provide both
true and complement signals as their outputs and are disabled by a *disable* signal.
Note that this Disable signal does not cause all current to cease flowing in the
drivers, since the pull-up transistors are depletion type. In general there will be a
number of super buffer stages of this sort. The very last stage of the driver is
shown in Fig. 5.28(b). It is not a super buffer but employs enhancement mode
transistors for both pull-up and pull-down. These transistors are very large in
order to drive the large external capacitance associated with the wiring attached to
the pad. They are disabled in the same manner as the super buffers, except that
when the gates of both transistors are low, the output pad is truly tri-stated. The

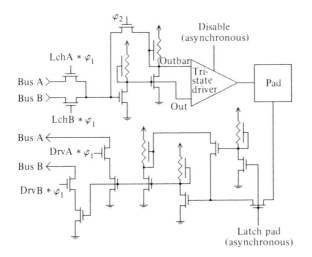

Fig. 5.27 Data port tri-state pad circuit.

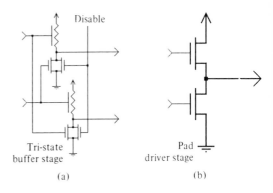

Fig. 5.28 The tri-state driver consists of
any number of tri-state buffer stages followed
by a pad driver stage. The current design
used two tri-state buffer stages.

two output transistors are a factor of approximately e larger than the last super buffer in the buffer string.

As we have seen, a rather large inverter string is required to transform the impedance from that of the internal circuits on-chip to that sufficient for driving a pad attached to wiring in the outside world; the large size imposes a delay, of some factor times a logarithm of this impedance ratio, upon communications between the chip and the outside world. Any help that can be obtained in making this transformation is of great value. For example, the latch and buffers associated with the input bus circuit to the pad drivers can themselves be graded in impedance level, so that by the time the out and outbar signals are derived, they are at a considerably higher current drive capability than the buses. Note that the buses are of a considerably larger capacitance than minimum nodes on the chip, and thus the initial latch buffers can be larger than typical inverters on the chip. All such tricks help to minimize the number of stages between the bus and the outside pad and consequently the total delay in going off-chip.

5.11 ENCODING THE CONTROL OPERATION OF THE DATA PATH

By now we have defined a complete functional data path with ports on each end and functional blocks through the center, as shown in Fig. 5.29. The data path operation code bits required to control the data path are shown, as is the phase of the clock on which they are latched. There are forty-nine such bits together with

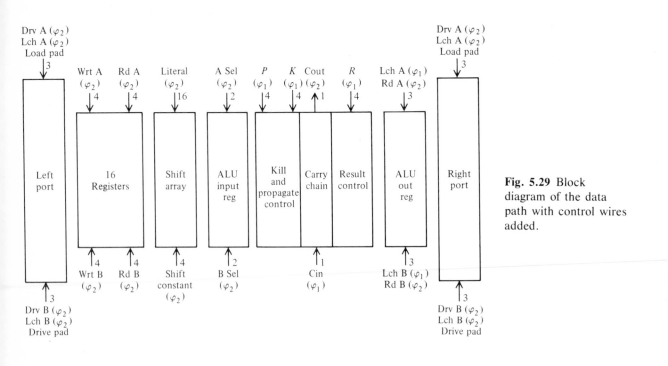

Fig. 5.29 Block diagram of the data path with control wires added.

the four asynchronous bits for latching and driving the pad to the external world. In addition, there are the carry-out wire and the sixteen literal wires. These sixty-six wires together with the thirty-two from the left and right ports must go to and come from somewhere. (Schemes for encoding internal data path operations into microinstructions of various lengths are discussed in Chapter 6.) At one extreme all the data path control wires can be brought out to a microcode memory driven by a microprogram counter and controller, in which case all operations implementable by the data path may be done in parallel. The opposite extreme is to tightly encode the operations of the data path into a predefined microinstruction set. In the present system, this encoding would be most conveniently done by placing a programmable logic array or set of programmable logic arrays along the top and the bottom of the data path. A condensed microinstruction could then be fed to the programmable logic arrays that would then decode the compact microinstruction into the data path operation code bits.

An important point of the design strategy used here is that we can orthogonalize the design of the data path and the design of the microinstruction set in such a way that the interface between the two designs is not only very well defined and very clean, but it can be described precisely, in a way that system designers at the next higher level can understand and work with comfortably. The data path can then be viewed as a component in the next level system design.

Using the approximate capacitance values given in Table 2.1, we can estimate the minimum clock period for sequencing the data path. We would expect a φ_1 time for the data path of $\approx 50\,\tau$ (same as the general estimate given in Section 1.13 on transit times and clock periods). However, the φ_2 time of the data path is limited by the carry chain, as discussed earlier in this chapter. The relative areas of metal, diffusion, and gate can be estimated from the ALU layout shown in Plate 11. The metal and diffusion occupy ≈ 15 and ≈ 8 times the area of the propagate pass transistor gate, respectively. Metal is ≈ 0.1 and diffusion is typically 0.25 times the gate capacitance per unit area. Thus the total capacitance of each stage of the carry chain is ≈ 4.5 times that of the pass transistor gate. The effective delay time is correspondingly longer than the transit time τ of the transistor itself. The effective delay through n stages of such pass transistor logic is $\approx \tau n^2$. In the OM2, $n = 4$ and the effective delay for 4 bits of carry chain is $\approx 4.5 * 16\tau = 72\tau$. To this must be added the delay of the doubly inverting buffers at the end of every 4 bits of straight Manchester logic. This delay is $(1 + k)$ times the transit time of the inverter pull-down, properly corrected for stray capacitance in the inverter. Here the inverter ratio k is ≈ 8, since its input is driven through the pass transistors. Conservatively, strays in such a circuit are always several times greater than the basic gate capacitance, and we may estimate the inverter delays at $\approx 30\,\tau$. Our estimate for the total carry time is thus ≈ 100 times the transit time for each block of 4 ALU stages. The total φ_2 time should then be $\approx 400\,\tau$. In 1978, the fastest commercial n MOS processes yield a transit time τ of ≈ 0.3 ns, and we would expect a minimum total clock period of $\approx 450\,\tau$, or ≈ 135 ns.

5.12 FUNCTIONAL SPECIFICATION OF THE OM2 DATA PATH CHIP*

5.12.1 Introduction

This specification describes a 16-bit data path chip referred to as OM2 [#986]. The OM2 contains 16 registers, an ALU, and a 16-bit shifter, and is designed as part of a microprogrammed writable-control-store digital computer. The companion chip is the controller chip, which contains the program counter, stacks, and so on. The controller is described in Chapter 6. The entire system is designed to run on a single 5-volt supply.

The OM2 data chip has two data ports for communication with the external system and a communication path to the controller chip. The data ports are tri-state with either internal or external control. Communication with the controller consists of a 16-bit literal port and a single flag bit. Seven control bits come directly from the microcode memory.

The system runs on a single clock, generating φ_1 and φ_2 internally. When the clock is high, the internal buses transfer data; when the clock is low, the ALU is performing its operation. Microcode bits enter the data chip the phase before that code is to be executed. Therefore, the bus transfer code enters the data chip when the clock is low, and the ALU code enters when the clock is high. Figure 5.30 sketches a possible OM system.

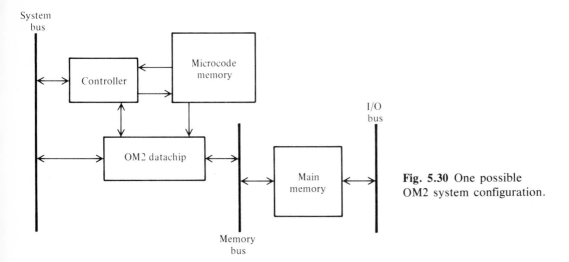

Fig. 5.30 One possible OM2 system configuration.

---

*Section 5.12 contains a functional specification of the OM2 data path chip, contributed by Dave Johannsen of California Institute of Technology. This specification was originally documented in Display File #1111, by Dave Johannsen and Carver Mead of the Caltech Computer Science Department, and copyrighted by Caltech. The specification is reprinted here with the permission of the California Institute of Technology. See also the later document, "Our Machine: A Microcoded LSI Processor," Display File #1826, by Dave Johannsen, Caltech Computer Science Department, for a general description of the OM System.

Throughout this section a positive logic convention is used. A "1" refers to a high voltage level, while a "0" refers to a low voltage level.

5.12.2 Data paths

A block diagram of OM2 is shown in Fig. 5.31. There are two buses that connect the various elements of the chip. The buses transfer data while the clock is high, the period referred to as φ_1. During φ_2, when the clock is low, the buses are precharged. Each bus can get data from only one source and give data to only one destination during any one cycle.

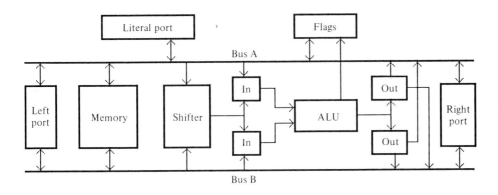

Fig. **5.31** Block diagram of the OM2.

The left and right ports communicate between the data chip and the outside world. The right port has been traditionally known as the memory bus port while the left port has been the system bus port, but since the two ports are identical, this is an arbitrary convention. Each port has both an input latch and an output latch to provide facilities for synchronizing the data chip to the outside buses. Under program control either of the two buses can load the output latch during φ_1. There are three modes of driving data from the output latch to the pins, two of which are under program control and one of which is under hardware control. The first method is to output the data as soon as it comes from the bus, during the same φ_1. The second method is to latch the data from the bus during φ_1 and drive it out during the following φ_2. The final method is to latch the data from the bus during φ_1, but output the data when an enable pin is pulled low. The enable pin would be controlled by a bus manager, and can be asynchronous with respect to the data chip. Inputting from the port is similar. By pulling down on another enable pin, data bits from the external bus are loaded into the input latch, which can be read later under program control. Alternatively, the microcode can force the data currently on the external bus into the internal bus during the current φ_1. With this scheme, many types of synchronous and asynchronous buses may be interfaced to OM2's. For internal control only, the external enable pins can be left floating.

5.12.3 Registers

The registers are static and dual port. Any one of the 16 registers may source either or both of the buses, while any one of the 16 may be the destination for either bus, but not both. There are only two restrictions on the use of the registers:

1. One register may not be the destination for both buses on the same cycle.

2. One register may not be both the source for one bus and the destination for the other bus on the same cycle.

5.12.4 Shifter

The shifter concatenates the two buses, resulting in a 32-bit word, with bus A being the more significant half. The shift constant then selects the bit position where the 16-bit output window starts. The shift constant specifies the number of bits from bus B present in the output; i.e., a shift constant of 0 returns bus A, while a shift constant of 15 returns the LSB of bus A in the MSB of the output, followed by all but the LSB of bus B in the rest of the word. A conceptual picture of the shifter is shown in Fig. 5.32. The ALU can select as inputs either the bus, the shift output, or shift control. If shift control is selected, the entire word is 0 except where the LSB of bus A appears in the shift output. The shifter operates on φ_1; it may be viewed as an extension of the buses.

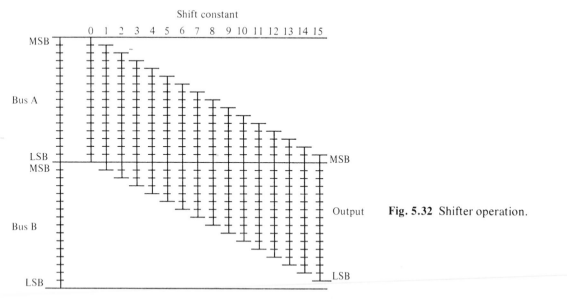

Fig. 5.32 Shifter operation.

5.12.5 ALU

A block diagram of a single bit of the ALU is shown in Fig. 5.33. The ALU operates on the data that is contained in its two input latches. Input latch A may be

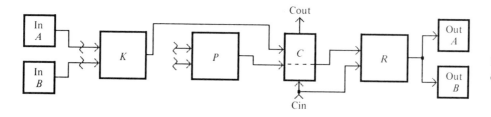

Fig. 5.33 Block diagram of one bit of the ALU.

loaded from bus A, the shifter output, or the shift control, while the input latch *B* may be loaded from bus B, the shifter output, or the shift control.

The outputs of the two latches become the inputs to two function blocks that determine what will happen on the carry chain. Function block *P* determines whether the carry chain propagates, while *K* decides if it is to kill the carry. If neither is true, the carry chain generates a carry. Each function block has four control inputs, which, for the propagate function block, are referred to as *PFF*, *PFT*, *PTF*, and *PTT*. If *PFF* is enabled, the *P* block output is high if both input latches are false (contain 0). Enabling *PFT* activates the output if input *A* is false and input *B* is true, and so on. If, for example, both *PFF* and *PFT* are enabled, the output is active if input *A* is false, regardless of the state of input *B*. To further illustrate the operation of the function blocks, consider addition. If both inputs contain a 1, the carry is to be generated, while if both inputs are 0, the carry is killed. If the two inputs are different, the carry is to be propagated (carry-out←carry-in). To do this operation, the kill output should be active if both inputs are false, so *KFF* is enabled. Both *PFT* and *PTF* should be enabled to propagate properly. Therefore, $K = (KFF, KFT, KTF, KTT) = (1,0,0,0)$, and $P = (PFF, PFT, PTF, PTT) = (0,1,1,0)$.

The result of the ALU is produced by the *R* function block, which has as inputs *P*-block out and carry-in. For the addition example above, the output should be the exclusive-OR of *P* and Cin, so $R = (0,1,1,0)$. *P*, *K*, and *R* values for common ALU operations are listed in Table 5.2.

Two ALU output latches (*A* and *B*) can be loaded from the *R* block output; either one may later be used to source either bus.

5.12.6 Flags

The carry input to the LSB of the ALU is a logical combination of a flag bit and two control inputs. The two control inputs can force the carry-in to be either 1 or 0, or they can select either flag or flag bar as the input.

There is also a method for doing conditional ALU operations under the control of a 2-bit conditional OP field. A conditional operation performed by the ALU is not only a function of the control inputs but also of the flag bit. The conditional operation control forces some of the control inputs low, regardless of what the *P*, *K*, and *R* microcode says. The coding for conditional operations allows the use of operations like multiply step and divide step without the necessity for branching in the microcode.

There is a 16-bit flag register that can also be a source or destination of bus A. This register can also be loaded with the ALU flags during φ_2. The ALU flags include *carry-out, overflow, carry-in to the MSB, zero, MSB, LSB, less than, less than or equal to,* and *higher* (in unsigned value). The last three flags are comparison flags used after a subtraction. For example, after subtracting ALU input latch *B* from latch *A*, the "less than" flag is true if the value in ALU input latch *B* is larger than the value in ALU input latch *A*. The MSB of the flag register is called the flag bit, and this bit may be modified every φ_1 by loading it with the value of one of the other bits of the flag register. The flag bit is used in the calculation of carry-in and modification of conditional ALU OPs. This bit is also sent to the controller chip to be used for conditional branching, etc.

5.12.7 Literal

The one remaining data path is the literal port. It is used to send data from the data chip to the controller, and vice versa. It is a source or destination for bus A. When the literal port is being used, standard bus operations are suspended for that cycle.

5.12.8 Programming

The data chip requires 23 bits of microcode on each phase of the clock. This section of the memo specifies the encoding of the fields within that microcode. Figure 5.34 shows the arrangement of the microcode word.

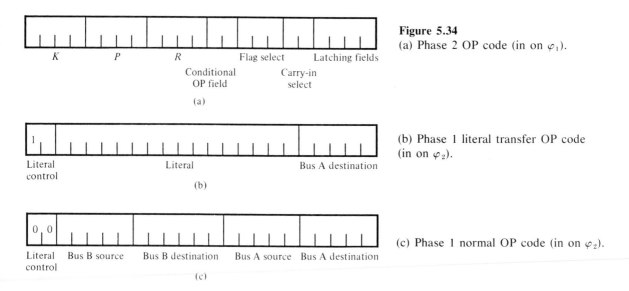

Figure 5.34
(a) Phase 2 OP code (in on φ_1).

(b) Phase 1 literal transfer OP code (in on φ_2).

(c) Phase 1 normal OP code (in on φ_2).

Bus Transfer

The bus transfer control bits enter the data chip during φ_2 and are used during the following φ_1. There are two buses, bus A and bus B, that interconnect the modules

Table 5.1

	Bus A Source		Bus A Destination
0*nnnn*	Register *n*	0*nnnn*	Register *n*
10000	Right port pins	10000	Left port, drive now
10001	Right port latch	10001	Left port, drive φ_2
10010	Left port pins	1001*x*	Left port, no drive
10011	Left port latch	10100	Right port, drive now
10100	ALU output latch *A*	10101	Right port, drive φ_2
10101	ALU output latch *B*	1011*x*	Right port, no drive
10110	Flag register	11000	ALU input latch *A*
		11001	ALU input latch *A* gets shift out
		11010	ALU input latch *A* gets shift control
		11011	Flag register

	Bus B Source		Bus B Destination
0*nnnn*	Register *n*	00*nnnn*	Register *n*
10000	Right port pins	010000	Left port, drive now
10001	Right port latch	010001	Left port, drive φ_2
10010	Left port pins	01001*x*	Left port, no drive
10011	Left port latch	010100	Right port, drive now
10100	ALU output latch *A*	010101	Right port, drive φ_2
10101	ALU output latch *B*	01011*x*	Right port, no drive
		0110*xx*	ALU input latch *B*
		10*nnnn*	ALU input latch *B* gets shift output, shift constant = *n*
		11*nnnn*	ALU input latch *B* gets shift control, shift constant = *n*

of the data chip. These two buses are similar in many respects; however, there are a few asymmetries as to sources and destinations. Also, when a literal is being transferred, the only bus transfer field that is active is the bus A destination, which stores the literal entered on bus A. A listing of bus sources and destinations appears in Table 5.1.

ALU Input Selection

The two ALU input latches are destinations for the two buses, as shown above. In addition to being loaded directly from the buses, these two latches can be loaded from the outputs of the shift array. The shift constant always comes from the four least significant bits of the bus B destination field, even though the destination of bus B is not the ALU input latch *B*. For example, bus B may be transferring the contents of register 3 into register 5 while bus A is transferring the contents of register 4 to the ALU input latch *A* through the shifter. In this case, the shift constant would be "5" because the four least significant bits of the bus B destination field contain "0101".

Table 5.2

	K	P	R	Cin	Cond	
$A + B$	1	6	6	0	0	Add
$A + B + \text{Cin}$	1	6	6	1	0	Add with carry
$A - B$	2	9	6	2	0	Subtract
$B - A$	4	9	6	2	0	Subtract reverse
$A - B - \text{Cin}$	2	9	6	1	0	Subtract with borrow
$B - A - \text{Cin}$	4	9	6	1	0	Subtract reverse with borrow
$- A$	12	3	6	2	0	Negative A
$- B$	10	5	6	2	0	Negative B
$A + 1$	3	12	6	2	0	Increment A
$B + 1$	5	10	6	2	0	Increment B
$A - 1$	12	3	9	2	0	Decrement A
$B - 1$	10	5	9	2	0	Decrement B
$A \wedge B$	0	8	12	0	0	Logical AND
$A \vee B$	0	14	12	0	0	Logical OR
$A \oplus B$	0	6	12	0	0	Logical EXOR
$\neg A$	0	3	12	0	0	Not A
$\neg B$	0	5	12	0	0	Not B
A	0	12	12	0	0	A
B	0	10	12	0	0	B
Mul	1	14	14	0	1	Multiply step
Div	3	15	15	0	2	Divide step
A/O	0	14	12	0	3	Conditional AND/OR
Mask	10	5	8	2	0	Generate mask
SHL A	3	0	10	0	0	Shift A left
Zero	0	0	0	0	0	Zero

ALU Operations

Table 5.2 shows coding for ALU operations that are commonly found useful. The user is encouraged to encode other operations if these are not suitable. The numbers given are the decimal representation of the 4-bit control word. For P and K, $A'B' = 1$, $A'B = 2$, $AB' = 4$, $AB = 8$. For R, $P'C' = 1$, $P'C = 2$, $PC' = 4$, $PC = 8$. Cin is the carry-in select, and Cond is the conditional OP select.

Carry-In Select

The carry-in select field determines what the carry into the LSB of the ALU will be, according to the following table:

00	0
01	Flag bit
10	1
11	Flag bit complemented

Conditional OP Select

The conditional OP select field is used to generate three basic conditional type operations: multiply, divide, AND/OR step. In a great many cases, the conditional OP allows functions dependent on a flag to be performed in one cycle, rather than sending the flag to the controller and branching to two separate instructions depending upon that flag. When a conditional OP is selected, certain ALU control bits are forced to zero. Which bits are zeroed depends on the conditional OP select and the flag bit, as follows:

Select	Flag bit	K	P	R	
0	x	----	----	----	Unconditional
1	0	---0	--0-	--0-	Multiply step
	1	----	0---	0---	
2	0	0--0	-00-	-00-	Divide step
	1	-00-	0--0	0--0	
3	0	----	----	----	AND/OR
	1	----	-00-	----	

For example, consider multiplication: If the flag bit is high, P_3 and R_3 are grounded, so the ALU OP $(1,14,14)$ becomes $(1,6,6)$, which is the code for "ADD". If the flag bit is low, K_0, P_1, and R_1 are pulled low, transforming $(1,14,14)$ into $(0,12,12)$, i.e., the code for "input A".

Flags

The flag select field determines which of the ALU flags becomes the new flag bit. The following table lists the selection options.

Select	New flag bit
0	Old flag bit
1	Carry-out
2	MSB
3	Zero
4	Less than
5	Less than or equal
6	Higher (in absolute value)
7	Overflow

The ALU flags are loaded into the flag register under the control of the latching field, bit 3. They are loaded into the following positions.

Bit	Flag
0	Not changed
1	Not changed
2	Not changed
3	Not changed
4	Not changed
5	Previous value of flag bit
6	¬ Carry into MSB stage
7	Less than or equal
8	¬ Higher (in absolute value)
9	¬ Less than
10	LSB
11	¬ Zero
12	MSB
13	Overflow
14	¬ Carry-out
15	Current flag bit

Latching Field

The latching field specifies which of four registers should be loaded, as shown in the following table:

Latching field	Register loaded
1xxx	Flag register loaded with current ALU flags
x1xx	ALU output latch A loaded with the ALU output
xx1x	ALU output latch B loaded with the ALU output
xxx1	The literal field during the next φ_2 is loaded with the contents of bus A during the last φ_2
0000	None of these registers are affected

Literals

The 2-bit literal field specifies when a literal is to be used and which direction it goes. If both bits are 0, no literal transaction will occur. If the first bit is 1, a literal will be transferred. If the second bit is 1, the literal goes off-chip, while if the bit is 0, the literal comes on-chip.

5.12.9 Programming Examples

Here we present three programming examples that should provide a better understanding of the various data paths within OM2.

The first example is 16-bit integer multiplication. The two inputs, X and Y, are multiplied to produce the result, Z. In the multiply loop, the number X is shifted left and the MSB is stripped off. Z is shifted left, then Y is added to the new Z if the MSB of X was a 1. The sequence of instructions is repeated 16 times, using the counter in the controller to signal when the 16 iterations have been performed.

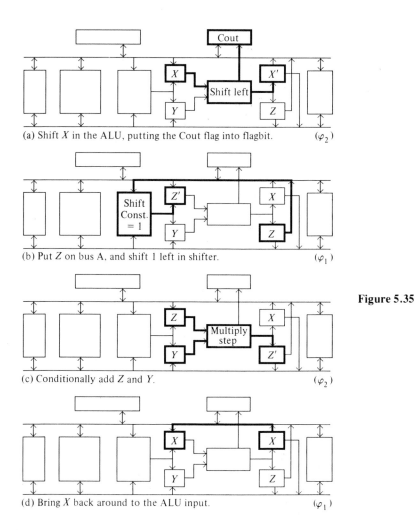

(a) Shift X in the ALU, putting the Cout flag into flagbit. (φ_2)

(b) Put Z on bus A, and shift 1 left in shifter. (φ_1)

Figure 5.35

(c) Conditionally add Z and Y. (φ_2)

(d) Bring X back around to the ALU input. (φ_1)

Assume contents of R[0] = 0. Figure 5.35 illustrates each step of the loop listed here:

φ_2:   ALU.Out.A←ALU(Shift A left)←ALU.In.A;
         Latch Flags;
φ_1:   ALU.In.A←Shift.out, Bus.A←ALU.Out.B;
         R[1]← Bus.B← R[0];   This gives a shift constant of 1.
φ_2:   ALU.Out.B←ALU(Multiply Step);   *conditionally add*.
         Flag←Cout;
φ_1:   ALU.In.A←Bus.A←ALU.Out.A

The second example will be to generate a parity flag, which is not directly available from the ALU. Parity is generated by exclusive-oring all of the bits of the data together. If the data are loaded into both ALU inputs, with input B rotated by one, performing an exclusive-or operation will give an output that is the exclusive-or of adjacent bits; bit i of the output will be bit i of the input $\oplus$ bit $i-1$ of the same input. If this same operation is performed, this time rotating input B by two, bit i becomes $i \oplus i-1 \oplus i-2 \oplus i-3$. By doing this two more times, rotating B first by four and then by eight, every bit of the output is equal to the parity, that is, the EXOR of all of the bits. The MSB flag is the parity odd flag, while the zero flag is the parity even flag. The program is listed below and illustrated in Fig. 5.36.

φ_1: ALU.In.A←Bus.A←R[0];   *generate the parity of register 0.*
       ALU.In.B←Shift.out(1); Bus.B←R[0];
φ_2: ALU.Out.A←ALU(Exor);
φ_1: ALU.In.A←Bus.A←ALU.Out.A;
       ALU.In.B←Shift.out(2); Bus.B←ALU.Out.A;
φ_2: ALU.Out.A←ALU(Exor);
φ_1: ALU.In.A←Bus.A←ALU.Out.A;
       ALU.In.B←Shift.out(4); Bus.B←ALU.Out.A;
φ_2: ALU.Out.A←ALU(Exor);
φ_1: ALU.In.A←Bus.A←ALU.Out.A;
       ALU.In.B←Shift.out(8); Bus.B←ALU.Out.A;
φ_2: ALU(Exor);

The third example illustrates how the data path can compute its own instruction. When driving a literal off-chip, the literal values appear in 16 of the microcode bits. If we have the literal port drive the data off chip, but don't set the disable bits in the instruction decoder, the data path will "execute" the literal. In the code below we sum all the registers (as further illustrated in Fig. 5.37). The basic literal transfers R[1] to the ALU to be added to R[0]; if we increment and execute the literal, then R[2] is transferred; etc.

φ_1: ALU.In.A←Literal "Bus.A←R[1]; ALU.In.B←Bus.B←ALU.Out.B";
φ_2: ALU.Out.B←ALU(A);
φ_1: ALU.In.A←Bus.A←R[0];
φ_2: ALU.Out.A←ALU(0);   *This is just setup, now the loop!*
φ_1: Bus.A←ALU.Out.B;
       ALU.In.B←Bus.B←ALU.Out.A;
φ_2: ALU.Out.A←ALU(add);
       Execute Literal;
φ_1: ALU.In.A←Bus.A;   *The rest of this instruction is the literal!*
φ_2: ALU.Out.B←ALU(increment B)←ALU.In.B;   *point to next register.*

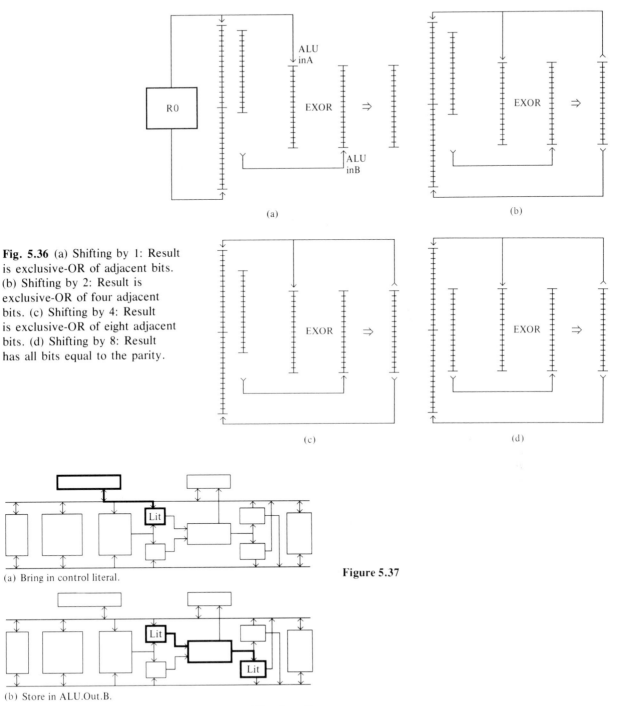

Fig. 5.36 (a) Shifting by 1: Result is exclusive-OR of adjacent bits. (b) Shifting by 2: Result is exclusive-OR of four adjacent bits. (c) Shifting by 4: Result is exclusive-OR of eight adjacent bits. (d) Shifting by 8: Result has all bits equal to the parity.

(a)

(b)

(c)

(d)

(a) Bring in control literal.

(b) Store in ALU.Out.B.

Figure 5.37

(*Continued*)

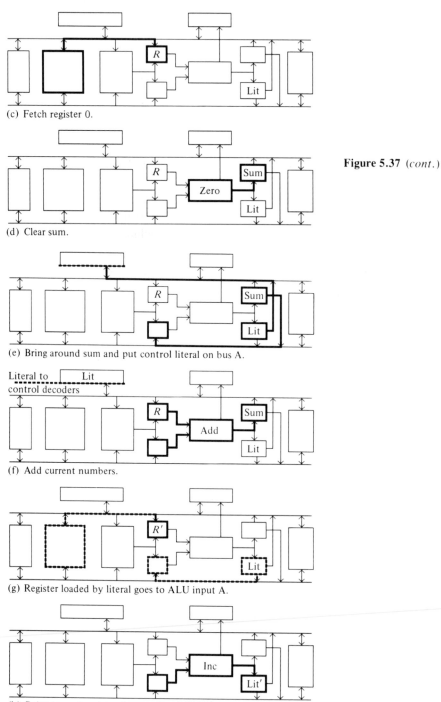

(c) Fetch register 0.

(d) Clear sum.

(e) Bring around sum and put control literal on bus A.

(f) Add current numbers.

(g) Register loaded by literal goes to ALU input A.

(h) Point to next register, loop to Fig. 5.37(e).

Figure 5.37 (*cont.*)

APPENDIX

This appendix contains additional detailed specifications for the OM2 Data Chip:

A description of the format and function of OM2 microinstructions is given, using ISP notation. The ISP (Instruction-set Processor) notation is defined, and examples of its use given, in Chapter 2 of C. G. Bell and A. Newell, *Computer Structures: Readings and Examples,* New York: McGraw-Hill, 1971.

A concise microinstruction command summary is included for reference.

A microcode "coding form" template is included, which can be used to plan the flow of data movement during microinstruction execution, as shown in the earlier examples (Fig. 5.37 a–h).

Finally, the pinout (identity of the 64 individual pins) of the packaged OM2 chip is illustrated.

ISP DESCRIPTION OF THE OM2 DATA CHIP

Pin States
    lp < 0:17 >   *left port*
    rp < 0:17 >   *right port*
    new.code < 0:22 >   *microcode*
    flag.pin < 0 >   *flag to controller*
    power < 0:3 >   *power, ground, clock, substrate*

Pin Formats
    left.port.data < 0:15 >   := lp < 0:15 >
    left.out.async < 0 >   := lp < 16 >
    left.in.async < 0 >   := lp < 17 >
    right.port.data < 0:15 >   := rp < 0:15 >
    right.out.async < 0 >   := rp < 16 >
    right.in.async < 0 >   := rp < 17 >
    literal < 0:15 >   := new.code < 5:20 >
    clock < 0 >   := power < 3 >

Mp State
    reg[0:15] < 0:15 >   *registers*
    a.bus < 0:15 >   *bus a*
    a.bus.old < 0:15 >   *bus a latched for a literal*
    b.bus < 0:15 >   *bus b*
    left.out < 0:15 >   *left pad output latch*
    left.in < 0:15 >   *left pad input latch*
    right.out < 0:15 >   *right pad output latch*
    right.in < 0:15 >   *right pad input latch*
    left.out.later < 0 >   *for output during φ2 operations*
    right.out.later < 0 >   *for output during φ2 for right port*
    alu.in.a < 0:15 >   *alu input latch a*
    alu.in.b < 0:15 >   *alu input latch b*
    alu.out.a < 0:15 >   *alu output latch a*
    alu.out.b < 0:15 >   *alu output latch b*
    old.code < 0:22 >   *microcode that came in last phase*
    flags < 0:15 >   *flag register*

Instruction format
    a.source < 0:4 >   := old.code < 5:9 >
    b.source < 0:4 >   := old.code < 16:20 >
    a.destination < 0:4 >   := old.code < 0:4 >
    b.destination < 0:5 >   := old.code < 10:15 >
    literal.in < 0 >   := old.code < 22 >
    old.literal < 0:15 >   := old.code < 5:20 >
    alu.p.op < 0:3 >   := old.code < 19:22 >
    alu.k.op < 0:3 >   := old.code < 15:18 >
    alu.r.op < 0:3 >   := old.code < 11:14 >
    alu.conditional < 0:1 >   := old.code < 9:10 >
    flag.select < 0:2 >   := new.code < 6:8 >
    carry.in.select < 0:1 >   := old.code < 4:5 >
    latch.flags < 0 >   := old.code < 3 >
    latch.alu.out.a < 0 >   := old.code < 2 >
    latch.alu.out.b < 0 >   := old.code < 1 >
    literal.control < 0 >   := old.code < 0 >
    reg.select.1 < 0:3 >   := a.source < 0:3 >

```
reg.select.2 < 0:3 >                    : = a.destination < 0:3 >
reg.select.3 < 0:3 >                    : = b.source < 0:3 >
reg.select.4 < 0:3 >                    : = b.destination < 0:3 >
select.1 < 0 >                          : = a.source < 4 >
select.2 < 0 >                          : = a.destination < 4 >
select.3 < 0 >                          : = b.source < 4 >
select.4 < 0:1 >                        : = b.destination < 4:5 >
shift.constant < 0:3 >                  : = b.destination < 0:3 >
sharay < 0:31 >                         : = b.bus < 0:15 > □a.bus < 0:15 >
```

Temporary State
```
    kill.control < 0:3 >
    propagate.control < 0:3 >
    result.control < 0:3 >
    kill < 0:15 >
    propagate < 0:15 >
    carry < 0:16 >
    alu.out < 0:15 >
```

Instruction Execution
```
Instruction.execution: = (
    left.out.async = 0⟹(left.port.data←left.out);next
    left.in.async = 0⟹(left.in←left.port.data);next
    right.out.async = 0⟹(right.port.data←right.out);next
    right.in.async = 0⟹(right.in←right.port.data);next
    phi1(: = clock = 1)⟹(
        left.out.later←0;next
        right.out.later←0;next
        literal.in = 1⟹(a.bus←old.literal);next
        literal.in = 0⟹(
            select.1 = 0⟹(a.bus←reg[reg.select.1]);
            select.1 = 1⟹(
                reg.select.1 = 0⟹(a.bus←right.in←right.port.data);
                reg.select.1 = 1⟹(a.bus←right.in);
                reg.select.1 = 2⟹(a.bus←left.in←left.port.data);
                reg.select.1 = 3⟹(a.bus←left.in);
                reg.select.1 = 4⟹(a.bus←alu.out.a);
                reg.select.1 = 5⟹(a.bus←alu.out.b);
                reg.select.1 = 6⟹(a.bus←flags);next);next
            select.3 = 0⟹(b.bus←reg[reg.select.3]);
            select.3 = 1⟹(
                reg.select.3 = 0⟹(b.bus←right.in←right.port.data);
                reg.select.3 = 1⟹(b.bus←right.in);
                reg.select.3 = 2⟹(b.bus←left.in←left.port.data);
                reg.select.3 = 3⟹(b.bus←left.in);
                reg.select.3 = 4⟹(b.bus←alu.out.a);
                reg.select.3 = 5⟹(b.bus←alu.out.b);next);next
            select.4 = 0⟹(reg[reg.select.4]←b.bus);
            select.4 = 1⟹(
                reg.select.4 = 0⟹(left.port.data←left.out←b.bus);
                reg.select.4 = 1⟹(
```

(Continued)

```
                    left.out←b.bus;next
                    left.out.later←1;next);
            reg.select.4 = 2⟹(left.out←b.bus);
            reg.select.4 = 3⟹(left.out←b.bus);
            reg.select.4 = 4⟹(right.port.data←right.out←b.bus);
            reg.select.4 = 5⟹(
                    right.out←b.bus;next
                    right.out.later←1;next);
            reg.select.4 = 6⟹(right.out←b.bus);
            reg.select.4 = 7⟹(right.out←b.bus);
            reg.select.4∈{8,9,10,11}⟹(alu.in.b←b.bus);next);
        select.4 = 2⟹(alu.in.b⟨0:15⟩←sharay⟨16-shift.constant:31-shift.constant⟩);
        select.4 = 3⟹(alu.in.b←2↑shift.constant);next);next
    select.2 = 0⟹(reg[reg.select.2]←a.bus);
    select.2 = 1⟹(
        reg.select.2 = 0⟹(left.port.data←left.out←a.bus);
        reg.select.2 = 1⟹(
            left.out←a.bus;next
            left.out.later←1;next);
        reg.select.2 = 2⟹(left.out←a.bus);
        reg.select.2 = 3⟹(left.out←a.bus);
        reg.select.2 = 4⟹(right.port.data←right.out←a.bus);
        reg.select.2 = 5⟹(
            right.out←a.bus;next
            right.out.later←1;next);
        reg.select.2 = 6⟹(right.out←a.bus);
        reg.select.2 = 7⟹(right.out←a.bus);
        reg.select.2 = 8⟹(alu.in.a←a.bus);
        reg.select.2 = 9⟹(alu.in.a⟨0:15⟩←sharay⟨16-shift.constant:31-shift.constant⟩);
        reg.select.2 = 10⟹(alu.in.a←2↑shift.constant);
        reg.select.2 = 11⟹(flags←a.bus);next);next
    flag.select = 1⟹(flags⟨15⟩←flags⟨14⟩);
    flag.select = 2⟹(flags⟨15⟩←flags⟨12⟩);
    flag.select = 3⟹(flags⟨15⟩←flags⟨11⟩);
    flag.select = 4⟹(flags⟨15⟩←flags⟨9⟩);
    flag.select = 5⟹(flags⟨15⟩←flags⟨7⟩);
    flag.select = 6⟹(flags⟨15⟩←flags⟨8⟩);
    flag.select = 7⟹(flags⟨15⟩←flags⟨13⟩);next

phi2(: = clock = 0)⟹(
    left.out.later = 1⟹(left.port.data←left.out);next
    right.out.later = 1⟹(right.port.data←right.out);next
    kill.control←alu.k.op;next
    propagate.control←alu.p.op;next
    result.control←alu.r.op;next
    alu.conditional = 1⟹(
        flags⟨15⟩ = 1⟹(
            propagate.control⟨0⟩←0;next
            result.control⟨0⟩←0;next);
        flags⟨15⟩ = 0⟹(
            kill.control⟨3⟩←0;next
            propagate.control⟨2⟩←0;next
            result.control⟨2⟩←0;next);next);
```

184

```
alu.conditional = 2⟹(
    flags ⟨ 15 ⟩ = 1⟹(
        kill.control ⟨ 2 ⟩ ←0;next
        kill.control ⟨ 1 ⟩ ←0;next
        propagate.control ⟨ 3 ⟩ ←0;next
        propagate.control ⟨ 0 ⟩ ←0;next
        result.control ⟨ 3 ⟩ ←0;next
        result.control ⟨ 0 ⟩ ←0;next);
    flags ⟨ 15 ⟩ = 0⟹(
        kill.control ⟨ 3 ⟩ ←0;next
        kill.control ⟨ 0 ⟩ ←0;next
        propagate.control ⟨ 2 ⟩ ←0;next
        propagate.control ⟨ 1 ⟩ ←0;next
        result.control ⟨ 2 ⟩ ←0;next
        result.control ⟨ 1 ⟩ ←0;next);next);
alu.conditional = 3⟹(
    flags ⟨ 15 ⟩ = 1⟹(
        propagate.control ⟨ 2 ⟩ ←0;next
        propagate.control ⟨ 1 ⟩ ←0;next);next);next
kill ⟨ 0:15 ⟩ ←(
    kill.control ⟨ 3 ⟩ ∧(¬alu.in.a ⟨ 0:15 ⟩ )∧(¬alu.in.b ⟨ 0:15 ⟩ )∨
    kill.control ⟨ 2 ⟩ ∧(¬alu.in.a ⟨ 0:15 ⟩ )∧alu.in.b ⟨ 0:15 ⟩ ∨
    kill.control ⟨ 1 ⟩ ∧alu.in.a ⟨ 0:15 ⟩ ∧(¬alu.in.b ⟨ 0:15 ⟩ )∨
    kill.control ⟨ 0 ⟩ ∧alu.in.a ⟨ 0:15 ⟩ ∧alu.in.b ⟨ 0:15 ⟩ );next
propagate ⟨ 0:15 ⟩ ←(
    propagate.control ⟨ 3 ⟩ ∧(¬alu.in.a ⟨ 0:15 ⟩ )∧(¬alu.in.b ⟨ 0:15 ⟩ )∨
    propagate.control ⟨ 2 ⟩ ∧(¬alu.in.a ⟨ 0:15 ⟩ )∧alu.in.b ⟨ 0:15 ⟩ ∨
    propagate.control ⟨ 1 ⟩ ∧alu.in.a ⟨ 0:15 ⟩ ∧(¬alu.in.b ⟨ 0:15 ⟩ )∨
    propagate.control ⟨ 0 ⟩ ∧alu.in.a ⟨ 0:15 ⟩ ∧alu.in.b ⟨ 0:15 ⟩ );next
carry ⟨ 0 ⟩ ←carry.in.select ⟨ 1 ⟩ ⊕(carry.in.select ⟨ 0 ⟩ ∧flags ⟨ 15 ⟩ );next
for k = 1 step 1 until 16 do:
    (carry ⟨ k ⟩ ← ¬(kill ⟨ k-1 ⟩ + propagate ⟨ k-1 ⟩ * ¬carry ⟨ k-1 ⟩ ) + kill ⟨ k-1 ⟩ *
            propagate ⟨ k-1 ⟩ *x);next                 in OM2, x is undefined
    If kill(i) and propagate(i) are both high, the carry chain does funny things.
    We represent that here by use of the "x" in the carry function.
alu.out ⟨ 0:15 ⟩ ←(
    result.control ⟨ 3 ⟩ ∧(¬propagate ⟨ 0:15 ⟩ )∧(¬carry ⟨ 0:15 ⟩ )∨
    result.control ⟨ 2 ⟩ ∧(¬propagate ⟨ 0:15 ⟩ )∧carry ⟨ 0:15 ⟩ ∨
    result.control ⟨ 1 ⟩ ∧propagate ⟨ 0:15 ⟩ ∧(¬carry ⟨ 0:15 ⟩ )∨
    result.control ⟨ 0 ⟩ ∧propagate ⟨ 0:15 ⟩ ∧carry ⟨ 0:15 ⟩ );next
latch.alu.out.a = 1⟹(alu.out.a←alu.out);next
latch.alu.out.b = 1⟹(alu.out.b←alu.out);next
literal.control = 1⟹(literal←bus.a.old);next
latch.flags = 1⟹(
    flags ⟨ 5 ⟩ ←flags ⟨ 15 ⟩ ;next
    flags ⟨ 6 ⟩ ←carry ⟨ 15 ⟩ ;next
    flags ⟨ 10 ⟩ ←alu.out ⟨ 0 ⟩ ;next
    flags ⟨ 11 ⟩ ←0;next
    alu.out = 0⟹(flags ⟨ 11 ⟩ ←1);next
    flags ⟨ 12 ⟩ ←alu.out ⟨ 15 ⟩ ;next
    flags ⟨ 14 ⟩ ←carry ⟨ 16 ⟩ ;next
    flags ⟨ 13 ⟩ ←flags ⟨ 14 ⟩ ⊕flags ⟨ 6 ⟩ ;next
    flags ⟨ 9 ⟩ ←flags ⟨ 12 ⟩ ⊕flags ⟨ 13 ⟩ ;next
    flags ⟨ 7 ⟩ ←flags ⟨ 11 ⟩ ∨flags ⟨ 9 ⟩ ;next
    flags ⟨ 8 ⟩ ← ¬(flags ⟨ 14 ⟩ ∨flags ⟨ 11 ⟩ );next);next);next
                        )                    end of instruction execution
```

SUMMARY OF COMMANDS FOR THE OM2

Transfer phase, φ_1

Literal control	Bus B source	Bus B destination	Bus A source	Bus A destination

Bus A Source

0nnnn	Register n
10000	Right Port Pins
10001	Right Port Latch
10010	Left Port Pins
10011	Left Port Latch
10100	ALU Output Latch A
10101	ALU Output Latch B
10110	Flag Register
-------	Literal (see Literal Control)
other	No Source

Literal Control

000	Microcode In
001	Illegal
010	Literal In
011	Illegal
100	Execute old A Bus
101	Illegal
110	A Bus gets old A Bus
111	Literal Out

LSB of the Latching Field during last PHI 2.

Bus B Source

0nnnn	Register n
10000	Right Port Pins
10001	Right Port Latch
10010	Left Port Pins
10011	Left Port Latch
10100	ALU Output Latch A
10101	ALU Output Latch B
other	No Source

Bus A Destination

0nnnn	Register n
10000	Left Port, drive now
10001	Left Port, drive PHI 2
1001x	Left Port, no drive
10100	Right Port, drive now
10101	Right Port, drive PHI 2
1011x	Right Port, no drive
11000	ALU Input Latch A
11001	ALU Input Latch A gets Shift Out
11010	ALU Input Latch A gets Shift Control
11011	Flag Register
other	No Destination

Bus B Destination

00nnnn	Register n
010000	Left Port, drive now
010001	Left Port, drive PHI 2
01001x	Left Port, no drive
010100	Right Port, drive now
010101	Right Port, drive PHI 2
01011x	Right Port, no drive
0110xx	ALU Input Latch B
0111xx	No Destination
10nnnn	ALU Input Latch B gets shift output, shift constant=n
11nnnn	ALU Input Latch B gets shift control, shift constant=n

Operation phase, φ_2

ALU operation	Flag select	Carry-in select	Latching field

ALU Operation

1000	0110	0110	00	00	Add
1000	0110	0110	00	01	Add with Carry
0100	1001	0110	00	10	Subtract
0010	1001	0110	00	10	Subtract Reversed
0100	1001	0110	00	01	Subtract with Borrow
0010	1001	0110	00	01	Subtract Reversed with Borrow
0011	1100	0110	00	10	Negative A
0101	1010	0110	00	10	Negative B
1100	0011	0110	00	10	Increment A
1010	0101	0110	00	10	Increment B
0011	1100	1001	00	10	Decrement A
0101	1010	1001	00	10	Decrement B
0000	0001	0011	00	00	Logical AND
0000	0111	0011	00	00	Logical OR
0000	0110	0011	00	00	Logical Exclusive Or
0000	1100	0011	00	00	Not A
0000	1010	0011	00	00	Not B
0000	0011	0011	00	00	A
0000	0101	0011	00	00	B
1000	0111	0111	01	00	Multiply Step
1100	1111	1111	10	00	Divide Step
0000	0111	0011	11	00	Conditional AND/OR
0101	1010	0001	00	10	Generate Mask
uuuu	uuuu	uuuu	uu	uu	User Defined Op

Carry In Select Field

Carry In Select

00	0
01	Flagbit
10	1
11	Flagbit Complemented

Flag Select

000	Old Flagbit
001	Carry Out
010	MSB
011	Zero
100	Less than flag
101	Less than or equal flag
110	Higher flag
111	Overflow

Latching Field

1xxx	Latch Flags
x1xx	Load ALU Output Latch A
xx1x	Load ALU Output Latch B
xxx1	Literal bits get old A Bus next PHI 1
0000	Nop

SAMPLE CODING FORM FOR THE OM2

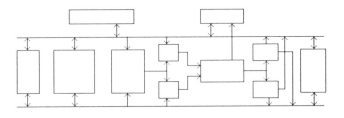

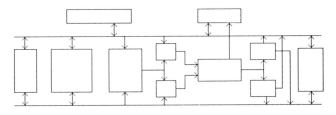

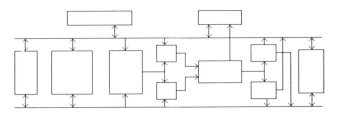

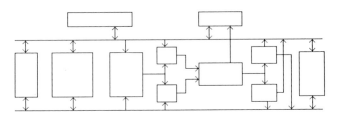

PINOUT OF THE OM2 DATA CHIP

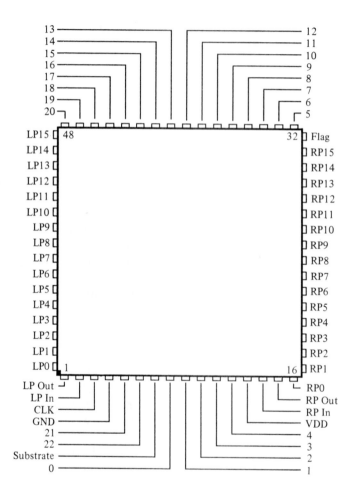

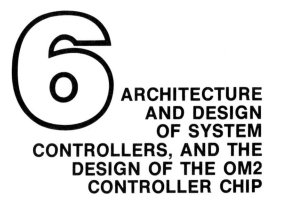

6 ARCHITECTURE AND DESIGN OF SYSTEM CONTROLLERS, AND THE DESIGN OF THE OM2 CONTROLLER CHIP

6.1 INTRODUCTION

In this chapter we present alternative structures for controlling a data path of the type described in Chapter 5. We review the basic concepts of the stored-program computer and how such computers are constructed from a combination of (1) a data processing path, (2) a controller, and (3) a memory to hold programs and data. We describe some of the ideas behind the architecture of a specific controller chip, designed at Caltech for use with the OM2 data chip, and we provide several examples of controller operations.

We have previously used the OM2 data path chip as a source of illustrative examples, primarily at the *circuit layout* level, to help the reader span the range of concepts from devices, to circuit layout, to LSI subsystems. In this chapter, the controller chip is used as a source of examples one level higher, at the *subsystem* level, to help the reader span the range from digital logic circuits, to LSI subsystems, to arrangements of subsystems for constructing LSI computer systems. The computer system one can construct using the OM2 data chip, the OM2 controller chip, and some memory chips, contains rather simple, regular layout structures. Yet the system is functionally quite powerful, comparing well with other classical, general-purpose, stored-program computers.

All present general-purpose computers are designed starting with the stored-program, sequential-instruction, fetch-execute concepts described in this chapter. These concepts are important not only for understanding present machines, but also for understanding their limitations.

As we look into the future and anticipate the dimensional scaling of the technology, we must recognize that it will ultimately be possible to place large numbers of simple machines on a single chip. When mapped onto silicon, classical stored-program machines make heavy use of a scarce resource: communication bandwidth. They make little use of the most plentiful resource: multiple, concurrent, local processing elements. What might be the alternatives? We will reflect on some of these issues at the end of this chapter and examine them in detail in Chapter 8.

6.2 ALTERNATIVE CONTROL STRUCTURES

In this section we will clarify the distinction between the data processing functions and control functions in a digital computer system; then we will examine several alternative forms of control structures.

The data processing path described in Chapter 5 is capable of performing a rich set of operations on a stream of data supplied from its internal registers or from its input/output ports. How is it that a structure having such a static and regular appearance as the OM data path can mechanize such a rich set of operations? An analogy may help in visualizing the data path in operation. Imagine the data path as being like a piano, with the interior regions of the chip visualized as the array of piano wires, and the control inputs along the edge of the chip as the keys. Under the external control of the controller chip, now visualized as the piano player, a sequence of keys is struck. During some cycles, many keys are struck simultaneously, forming a chord. A complex function may thus be performed over a period of time by the data path, just as the static-appearing array of piano wires may produce a complex and abstract piece of music when a series of notes and chords are struck in a particular order.

We see from this analogy, however, that the data path in itself is not a complete system. A mechanism is required to supply, during each machine cycle, the control bits that determine the function of the path during that cycle. The overall operations performed on data within the data path are determined by sequences of control bit patterns supplied by the system controller.

Mechanisms for supplying these sequences of control inputs to a data path can be either very simple or highly complex. There are many alternative sorts of control structures. The detailed nature of the controller has many important effects on the structure, programming, and performance of the computer system. Let us begin with the description of the simplest form of finite-state machine controller. Then, through a sequence of augmentations of this controller, we will build up to the concepts of the stored-program computer and microprogramming.

Simple block diagrams, such as Figs. 6.1, 6.2, and 6.3, are used here to convey the essential distinctions between various classes of controllers, without requiring the diagramming of the internal details of any particular controller. Although the detailed internal logic of any particular controller may be rather complex, there is only a small set of key ideas involved in the hierarchy of controller structures presented by the sequence of block diagrams.

If you closely examine the controllers of typical computers, you will find that every one either is, or contains within it, a finite-state machine such as those described in Chapter 3. The very simplest form of controller for the data path is a

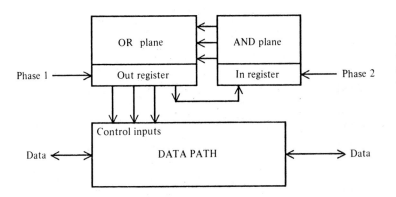

Fig. 6.1 Finite-state machine controlling the data path. In this case there is periodic cycling through a fixed sequence of states.

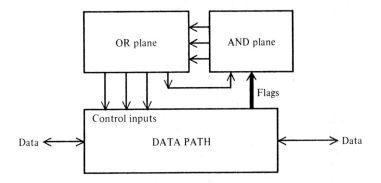

Fig. 6.2 Finite-state machine controlling the data path. In this case the next state can be a function of the previous operation's outcome.

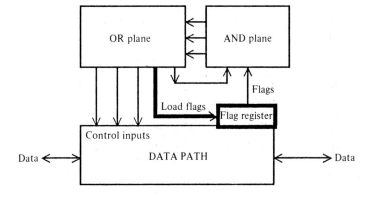

Fig. 6.3 Finite-state machine controlling the data path. In this case a data path operation result may control machine sequencing for a number of later cycles.

finite-state machine having no inputs other than state feedback lines, as shown in Fig. 6.1. The operations performed by the data path are determined by the sequencing of the state machine. During each clock cycle, the output of the OR plane is fed back into the AND plane and determines the next state of the state machine, which periodically cycles through a fixed sequence of states. The data path is clocked in synchronism with the controller, although for simplicity we haven't shown clock inputs to the data path in the figures. Thus a fixed algorithm implemented in the code of the state machine operates on the data in the data path.

Such a control structure could be used with the data path to implement a function such as a digital filter, in which data bits are taken in from the left port of the data path, a fixed set of operations is performed on the data, and a result is output at the right port of the data path. However, this elementary control structure provides no way to perform operations that depend on the outcome of a previous operation or upon the data itself.

A simple augmentation, shown in Fig. 6.2, enables the control sequencing to be a function of the outcome of the previous operation. Figure 6.2 shows that some of the data, or some logical functions of the data, called *flags*, are fed into the AND plane inputs of the state machine along with the next state information. Some typical flags are (1) *ALU output* = 0; (2) *ALU output* > 0; (3) *ALU input A* = *ALU input B*. The next state can thus be a function of flags generated during the preceding operation. To simplify Fig. 6.2, we have not shown the clock inputs to the PLA. However, assume that all subsystem structures shown in the figure, and throughout this chapter, are appropriately operated in a synchronous manner using our normal two-phase clock scheme and proper design methodology.

While in principle the structure pictured is quite general, improvements are possible that allow greater flexibility and compactness of representation of the algorithm in the state machine. One of these improvements is shown in Fig. 6.3. Here an additional output from the OR plane of the state machine is used to control the loading of the flag outputs of the data path into a flag register. The flag register is used as an input into the AND plane of the state machine. This enables flags generated by a particular operation to be used as control inputs for the state machine for a number of later operations. The stored flag values are replaced by a new set only when the flag load signal is raised. One difficulty inherent in this structure is the limited amount of information provided by the few flags generated by the data path's ALU.

6.3 THE STORED-PROGRAM MACHINE

A very general and powerful arrangement is shown in Fig. 6.4. This structure is similar to the one discussed in the previous section. In this case the state machine sequencing is controlled not only by the last state and flags, but also by the data coming from some memory attached to the machine. The memory contains the data upon which the data path is operating, in addition to encoded information for influencing the sequencing of the state machine.

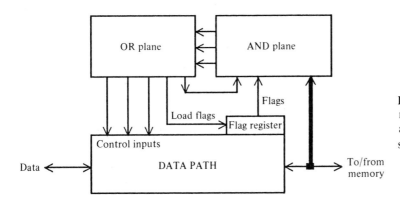

Fig. 6.4 A simple stored-program machine, where data read from a memory can affect machine sequencing.

This scheme gets around the limitation of the structure shown in Fig. 6.3 and also provides a complete new dimension of possibilities. The gist of the idea is to design the state machine controller so that it may perform *any of a set* of different predefined operations, called the *machine instruction set*, rather than perform just one dedicated, predefined operation. The machine instruction set is carefully defined so as to enable the system composed of the data path, controller, and memory to mechanize any of a number of different algorithms of interest to a number of different users. These algorithms are implemented as *programs* composed as *sequences of machine instructions* loaded into the memory. These programs operate upon data also contained in the memory.

It is possible to show that this arrangement is perfectly general and can implement any digital data processing function. John von Neumann[1] is generally credited with originating this idea of a stored-program machine, and such machines are often called von Neumann machines. The abstract notion of the most basic form of stored-program machine was proposed by Turing[2] in 1936 for application in the development of the theory of algorithms. The abstract *Turing Machine* is important not only for historical reasons, but also because of its present use in the development of the theory of computational complexity of sequential algorithms.

The way in which the stored-program machine operates is as follows. One of the internal registers of the data path is selected to hold a pointer into the program stored in the memory. This register is commonly called the *program counter* (PC), or alternatively, the *instruction address register*. In one particular state of the controlling state machine, which we will call the *fetch next instruction* (FNI) state, the program counter is caused by the state machine to output its data as an address to the memory, and the state machine initiates a memory read from this address. The data bits from this memory read operation are taken into the AND plane of the state machine, placing the state machine into a state that is the first of the sequence of states that mechanize the machine instruction corresponding to the code just read from the memory. The state machine then sequences the data path through a number of specific operations sufficient to perform the function defined by that instruction. At some point during instruction execution the next PC value is calculated, usually by simply incrementing the current PC value.

When the state machine has completed the interpretation, or execution, of the machine instruction, it returns to the FNI state. The instruction fetch is then repeated, sending a new program counter value to the memory as an address, reading the next instruction from the memory, and beginning its interpretation. The system can thus perform any set of required operations on data stored in memory, as specified by encoded instructions stored in memory.

There is a problem with the organization of the controller shown in Fig. 6.4. Most of the steps of an instruction execution sequence need as input the encoding of the instruction that initiated the sequence. In the machine outlined in Fig. 6.4, this information must be duplicated each cycle by the next state information. The number of bits in the feedback path for this information can be reduced by the arrangement shown in Fig. 6.5. Here the incoming instruction is stored in a register, called the *instruction register* (IR), which is loaded under the control of an output from the state machine. It stays in the instruction register and is available for state machine input during the entire period that particular instruction is being interpreted by the machine. This new arrangement is not fundamentally different from the preceding one, but it is more efficient in its use of the PLA's.

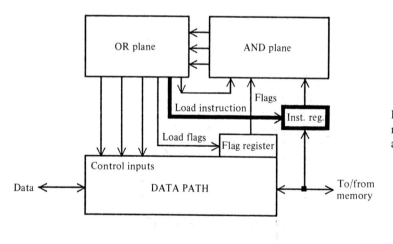
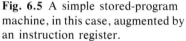

Fig. 6.5 A simple stored-program machine, in this case, augmented by an instruction register.

The separation and naming of the instruction register also enables us to take another step in the structuring of the state machine controller's operations: the conception and naming of stages of the interpretation of instructions fetched and held in the IR.

Suppose we have defined a machine instruction set that, for example, includes arithmetic-logic instructions, memory instructions, and branch instructions. Suppose we also have a data path such as the OM data chip, or any other typical data path, containing registers, an ALU, buses for moving data around, and inputs for control signals to control the movement of data and the ALU operations. What functions must a control unit, such as that shown in Fig. 6.5, perform in order to fetch and execute machine instructions? We find that in most stored program machines the execution or interpretation of each machine instruction is typically broken down into the following *six basic stages*. Note that some instruction types may skip one or more of the stages, and that each of the stages may require sequencing through several controller states:

1. *Fetch next instruction.* This is the starting point of the fetch-execute sequence. The machine instruction at the address contained in the PC is fetched from the memory into the IR.

2. *Decode Instruction.* As a function of the fetched machine instruction's type, encoded in its OP code field, the controller must "branch" to the proper next control state to begin execution of the operations specific to that particular instruction type.

3. *Fetch instruction operands.* Instructions may specify operands such as the contents of registers or of memory locations. During this execution stage, the controller cycles through a sequence of states outputting control sequences to fetch the specified operands into specified locations; for example, into the input registers of the ALU.

4. *Perform Operation.* The operation specified by the OP code is performed upon the operands.

5. *Store Result(s).* The results of the operation are stored in destinations, such as in registers; memory locations, flags.

6. *Set up next address, and return to FNI.* Most instructions increment the PC by one and return to the FNI state (1). Branch instructions may modify the PC, perhaps as a function of flags, by replacing its contents with a literal value, fetched value, or computed value.

Now, how would we go about designing such a controller? We can construct the state diagram for the controller just as we did for the traffic light controller example in Chapter 3. Then we proceed to build up the detailed state transition table and finally derive the AND and OR plane codes for the PLA. However, in this case the state diagram will be rather more complex than that in our earlier example. One hundred or more states may be required to implement the controller for a simple machine instruction set. How do we even begin to construct the state diagram? The above list of stages of instruction execution provides a simple means of structuring the diagram. Figure 6.6 contains part of the controller state

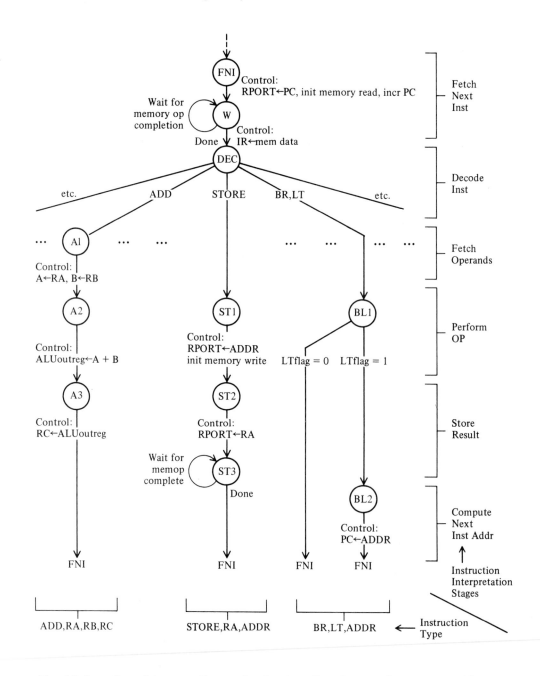

Fig. 6.6 A portion of the state diagram for the controller of a stored-program machine, illustrating some typical instruction interpretation state sequences and their associated control outputs.

diagram for a typical stored-program machine. The diagram is structured as a matrix of regions, where the instruction execution stages proceed from top to bottom, and the columns contain specific state sequences for each instruction type. The FNI state is placed at the top of the diagram, followed by the states leading to the decode. The decode results in a many-way branch, each path leading to a sequence for executing a particular instruction type. The figure contains some (informal) details indicating the sorts of specific control operations performed at each stage of the instruction execution or interpretation. One will encounter many variations on the simple state diagram structure shown, usually easily understood elaborations. For example, groups of machine instructions may share common sub-sequences of control operations. To reduce the number of states, we might have another level of decoding, first decoding to groups of instructions and performing operations common to a particular group, then decoding to individual instruction types. In any event, the generation of the state diagram and eventually the PLA code is just a matter of grinding out the details. The generation of these details is another activity that is made more tractable by following a structured approach.

Some examples follow that will clarify how a machine instruction's execution can be divided into parts, and how the parts interact with each other. Instead of using the graphical notation of Fig. 6.6, an informal tabular form is used, containing a list of statements that are normally performed sequentially, as encountered. In these examples, the unbracketed statements under "control [& state] sequence" indicate control actions. However, the bracketed statements, [ ], explicitly set the next state Y'; they indicate a more complex state transition than simple state-to-state progression (shown in Fig. 6.6 by a single arrow between circles).

ALU Example

Suppose that an arithmetic-logic instruction in our machine instruction set has the general form { ALUOP, REGA, REGB, REGC }, specifying that ALUOP be performed on operands REGA and REGB, and the result stored in REGC. Then the instruction { ADD, R7, R2, R5 } might be executed by the following control sequence. Note that certain of the individual control steps may occur in the same machine cycle (for example, $A \leftarrow R7$, $B \leftarrow R2$) as a function of the capabilities of the data path; the more the data path can do in parallel, the fewer machine cycles it will require to complete an instruction.

The example assumes there is some sort of shared access to the memory, and thus the time for completion of memory accesses is not predictable. That is why we wait, testing for the presence of a completion signal before proceeding. In some computer systems, such memory accesses might proceed in lockstep with the controller sequencing, and then the data might be taken from, or placed on, the memory bus at some fixed number of cycles following initiation of the memory

Function of sub-sequence	Control [& State] sequence	Comments
Fetch Next Inst:	RPORT ← PC	Place next instr. address in right port.
	read memory	Raise control line to initiate memory read.
	PC ← PC + 1	Increment PC, overlapping incr. with fetch.
	[Y' = fcn(memop complete)]	Loop here till memory read completes.
	IR ← mem data	Load IR with inst, when read completed.
Decode Instruction:	[Y' = fcn(IR)]	Set machine state as fcn of instruction.
Fetch Operands:	A ← R7	Load ALU input registers with operands.
	B ← R2	
Perform Operation:	ALUoutreg ← A + B	Add A and B, store in output register.
Store Result:	R5 ← ALUoutreg	Send result address to R5 .
	[Y' = FNI]	Inst. not a branch, so simply return to FNI state.

operation. Normally, most machine instructions are not branches, so we usually just have to increment the PC sometime during instruction execution. This incrementing can often be overlapped with other operations. In the example, the incrementing of the PC is done during the FNI stage, while waiting for the completion of the instruction-fetch memory operation.

Memory Example

A memory instruction in our set might have the general form { MEMOP, REGA, ADDRESS }, specifying the loading or storing, according to MEMOP, of the contents of register REGA to or from the memory address ADDRESS. The instruction { STORE, R3, ADDRESS } might then be executed by the following control sequence:

Function of sub-sequence	Control [& State] sequence	Comments
Fetch Next Inst:	RPORT ← PC	Place next instr. address in right port.
	read memory	Raise control line to initiate memory read.
	PC ← PC + 1	Increment PC.
	[Y' = fcn(memop complete)]	Loop here till memory read completes.
	IR ← mem data	Load IR with inst, when read completed.
Decode Instruction:	[Y' = fcn(IR)]	Set machine state as fcn of instruction.
Perform Operation:	RPORT ← IR(ADDRESS)	Send the contents of IR address field to memory.
	write memory	Raise write control line to init. memory write.
Store Result:	RPORT ← R3	Place data in right output port.
	[Y' = fcn(memop complete)]	Loop here till memory write completes.
	[Y' = FNI]	Inst. not a branch, so simply return to FNI state.

Branch Example

Suppose that branch instructions have the form { BR, COND, ADDRESS }, specifying that if the condition COND is true according to the flags, then the PC is to be loaded with memory address ADDRESS. The branch instruction { BR. LT. ADDRESS } might then be executed by the following control sequence:

Function of sub-sequence	Control [& State] sequence	Comments
Fetch Next Inst:	RPORT ← PC	Place next instr. address in right port.
	read memory	Raise control line to initiate memory read.
	PC ← PC + 1	Increment PC.
	[Y′ = fcn(memory complete)]	Loop here till memory read completes.
	IR ← mem data	Load IR with inst, when read completed.
Decode Instruction:	[Y′ = fcn(IR)]	Set machine state as fcn of instruction.
Perform Operation:	[Y′ = fcn(LT flag)]	Set machine state as fcn ALU LTflag. Set to FNI if notLT. Else continue and generate new address.
Next Address:	PC ← IR(ADDRESS)	Extract new address field from IR.
	[Y′ = FNI]	Return to the FNI state.

Now, how are the next higher-level system software control functions mapped onto this basic machine structure? Higher-level functions common to all machine instructions are often performed within the FNI stage of instruction execution. After return to the FNI state, but prior to the decode state, one machine instruction has been completely executed but no action has yet been taken to execute the next instruction. Therefore, that is a natural place to check for interrupts from I/O devices, to test the priorities for task switching in a multi-programming environment, and so forth. The testing of these logical signals, which are input to the state machine, can often be overlapped with other FNI activity. Multiple tasks may then be implemented by having the controller manipulate multiple registers as program-counters.

In summary, once both a machine instruction set and a data path have been defined, then the control sequences required to interpret the machine instructions can be "programmed," the overall controller state diagram can be constructed, the "code" for the AND and OR sections of the state machine can be generated, and software systems can be built upon the resulting stored-program machine. Interestingly, the control sequences in the above examples look somewhat like "programs" written in a very primitive machine language. This observation anticipates the concept of microprogrammed control, which is described in the next section.

For more information on this material, including the various trade-offs involved in the definition and encoding of instructions, see the many examples in Bell and Newell.[3] Dietmeyer[4] works out an example all the way from state diagram through the design of the controls of an elementary digital computer. Dietmeyer also gives formal methods for describing state machine algorithms. An interesting alternative method, based on ideas of T. E. Osborne, is presented along with practical examples in Clare.[5]

The abstract concepts behind the arrangement shown in Fig. 6.5 are used in almost all stored-program digital computers manufactured today. A computer having any sort of machine instruction set can be implemented with the arrangement shown. In many cases, the state machine is implemented in random logic and therefore is not easily recognizable as one of the forms shown. However, the operations performed are equivalent to those described here. Note that, in any case, the number of machine cycles required to mechanize particular algorithms trades off against the functional capability of the data path.

6.4 MICROPROGRAMMED CONTROL

Sometimes the complete machine instruction set is not definable at the time a computer is being designed. This contingency often arises when certain operations, defined by some later user, must be executed at very high speed. Perhaps the data path is inherently capable of satisfying the required performance constraints, but not when operated under the control of any sequence composed of standard machine instructions. In such cases, special new machine instructions would have to be defined and then implemented in the state machine control logic.

Another common situation is the need to execute the instruction set of another computer system for which the user has existing programs. While such instructions could be executed by simulation, that is, by interpreting them via a program written in the original machine instruction set, such simulations usually pay a high performance penalty. It would be much better if the machine could execute them directly. However, a substantial augmentation and/or modification of the controller's logic would have to be made, for such direct execution to be possible.

In both of these situations it would be desirable if the state machine were implemented in some writable medium, rather than in the fixed code of a standard programmable logic array and thus patterned permanently in the silicon. While it is quite possible to build writable programmable logic arrays, none are currently in use. Instead, machine designers have invented many clever ways of using standard writable memories to hold the feedback logic of the state machine.

The simplest such arrangement is shown in Fig. 6.7. Here the state machine is implemented using a set of memory chips. Collectively, this set of memory chips functions externally exactly as the programmable logic array shown earlier. However, this very elementary structure has a problem in supporting wide machine instruction words, since the decoder must exhaustively decode all combinations of the input variables. Thus, if f is the number of flag bits and n is the number of next state lines, then the memory must have $2^{(i + f + n)}$ words to be of sufficient size

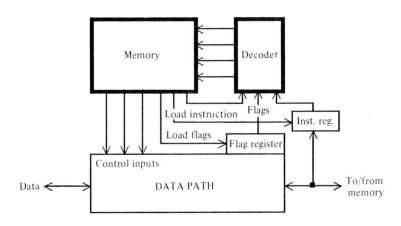

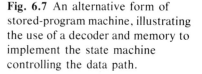

Fig. 6.7 An alternative form of stored-program machine, illustrating the use of a decoder and memory to implement the state machine controlling the data path.

to allow emulation of any machine having instructions i bits wide. For this reason designers have taken to inserting more complex logic than just a simple instruction register into the path between the data source and the memory decoder section of state machines of this form.

A system using a logic path between the memory bus, or source of instructions, and the memory decoder section of the state machine is shown in Fig. 6.8. Here a logic block we have termed the *microprogram counter path* is inserted between the source of machine instructions and the inputs to the decoder. This type of control, using either writable or read-only memories, is generally referred to as *microprogrammed control*. Notice in Fig. 6.8 that the flags and the machine instruction fetched from the memory both act as input data to the small microprogram counter data path, and the outputs of this data path are the microcode memory address lines. The arrangement shown is very powerful and general, and capable of emulating any set of instructions for which there is sufficient microcode memory.

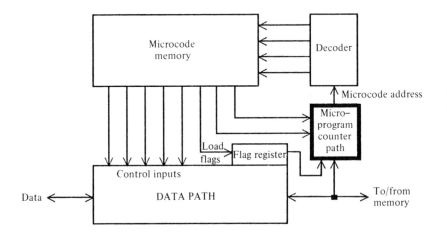

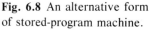

Fig. 6.8 An alternative form of stored-program machine.

In a microprogrammed controller, the design of the control logic is reduced to encoding sequences of control bit patterns to be stored, along with control memory address sequencing information, in the microcode memory. The encoded control bit patterns for each clock cycle or machine cycle are visualized, as in the examples in the past section, as a primitive form of "instruction" and are called microinstructions. Rather than create a "circles and arrows" state diagram and then "assembling" PLA code, we write a symbolic microprogram and assemble it in the same manner as we would a symbolic machine language program.

The microprogram counter (μPC) data path is similar to the main data path: it is controlled by a number of outputs from the microcode memory section of the state machine. Its main purpose is to decrease the amount of microcode memory required to emulate the particular machine instruction set being implemented. This is done in two ways: (1) the path maps the $f + n$ bits of state into a smaller number of bits that are then decoded to address the microcode memory, and (2) it reduces n by allowing complex operations within the path to be specified with only a few bits of control information. The controller chip described in the later sections of this chapter is the microprogram counter data path portion of a microprogrammed controller for OM2.

The concept of microprogramming was originated by M. V. Wilkes[6,7] in 1951. In those days when controller logic functions were implemented using gates constructed out of vacuum tubes, switching hardware was very expensive compared to wires, and great efforts were expended toward gate minimization. This inevitably led to rather intertwined connections in the controller logic, and any change in function might require a complete redesign. Wilkes presented the notion of microprogrammed control using a read-only memory to hold the control sequences, as a means of bringing regularity and structure to the design of system controllers and thus simplifying their design and redesign. There is a large body of knowledge associated with the architectural implications of microprogrammed control, and the serious reader will benefit from a study of the literature.[3,4,8]

Today, although we can easily implement control logic in a structured way using a PLA, we still often use microprogrammed control in order to obtain the advantages offered by writable control logic. An additional present advantage of microprogrammed control is that the detailed design/redesign of control logic is extended into the wide arena of those familiar with linear sequential programming concepts. In the future as the "programming" of structures into silicon becomes easier, as the time to implement designs becomes much shorter, and as state machine "coding" becomes more widely understood, we may find that these activities will be viewed as a natural extension of microprogramming.

There is an alternative way of viewing the machine shown in Fig. 6.8. Examine carefully the loop formed by the microprogram counter data path, the decoder section of the microcode memory, and the outputs of the microcode memory that are used to control the microprogram counter. We can view the microcode memory address as an instruction address and the wires coming from the microcode

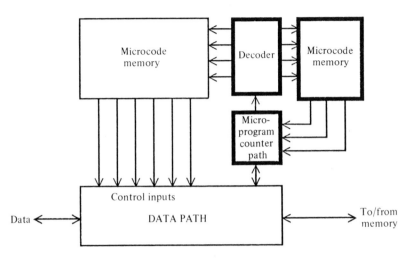

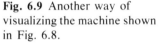

Fig. 6.9 Another way of visualizing the machine shown in Fig. 6.8.

memory to control the microprogram counter path as an instruction. This alternative view is illustrated in Fig. 6.9. Observe that we have constructed another stored programmed machine of the same form as that shown in Fig. 6.5. We have come full circle in our machine design: in our zeal to put as much capability as possible in the path between the machine instruction and the decoder of the state machine, we have in fact created a stored programmed machine within a stored programmed machine. This phenomenon is referred to by Ivan Sutherland as the "great wheel of reincarnation." Computers often have many such levels of machine within them, each a general-purpose stored-program machine in its own right. We thus find that elaborate computing machines are often only simple machines, nested and connected in complex ways.

6.5 DESIGN OF THE OM2 CONTROLLER CHIP

We now describe some of the ideas behind the design of one particular microprogram counter path used for controlling the OM data chip in the system configuration[9] described in Chapter 5. The design of the controller chip will be examined at several stages in its development. This material illustrates the mapping into LSI, and the topological/geometrical planning in LSI of various subsystems such as stacks, incrementers/decrementers, and multiplexers that are useful in constructing controllers.

Even at the 1978 value of $\lambda = 3$ microns, the OM2 data path and certain forms of controller can be integrated onto a single chip. The separation of these modules onto two chips was primarily for research and tutorial purposes in the university environment, so that different controllers could be used with the OM2 data chip and vice versa. The fact that data path and controller are on separate chips does, however, lead to detailed system partitioning decisions aimed at minimizing interchip communication. These decisions might be made differently were data path and controller integrated onto the same chip. Nevertheless, the issue of minimization of interchip communication would still be involved at the next system level and is worthy of study.

The basic function of the *microprogram counter path*, which we call the *controller* for short, is to provide microprogram memory addresses. The microprogram memory addresses are stored in a latch called the *microprogram counter, or* μPC. The μPC should be distinguished from the *program counter*, or PC, which stores the main memory addresses of higher level machine instructions. The most common address calculation is to increment the address by one; so in addition to the μPC latch, the controller should contain an incrementer. The second most important address calculation is the jump or branch; so there should be some means of forcing values into the μPC latch. With the hardware mentioned so far, we have progressed one step beyond the controller type shown in Fig. 6.7: our instruction register also increments, so we don't need the feedback terms that originate in the microcode memory and drive the memory decoder.

A great deal of microcode memory space can be saved if *subroutines* are available at the microcode level. These subroutines can be shared among microcode sequences emulating instructions at a higher level. For example, many different machine instruction types may have the same set of operand fetch sequences. If the machine instruction set encodes a variety of indexing or relative addressing schemes, these operand fetch sequences may be quite lengthy, and repeating these sequences for every instruction type would waste a great deal of microcode memory. To provide such microcode subroutine capabilities, provisions must be made for saving μPC values, which is most easily done with a stack. Stacks are easily constructed in LSI. An example of stack cell and subsystem design, and stack control driver design, is given in Chapter 3.

The microcoder may also wish to use *relative* jumps or subroutine calls so that relocatable microcode can be written. To provide for relative operations, an adder must be included that can add displacements to the μPC contents. The displacements can either be fixed displacements and come from the microcode or be calculated displacements and come from the data path. Calculated displacements enable many-way branching, or *dispatching*, in the microcode, which is an almost essential operation for emulating instructions at a higher level. (An example of dispatching will be given in a later section.) Therefore, provisions should be made for accepting displacements from either the microcode or the data path.

Another microcode address operation that could be considered is a form of loop operation, which is useful when sections of microcode are to be executed n times, where n can either be a constant and come from the microcode or be the result of a calculation done in the data path. One way to implement this instruction is to dedicate one register in the data path to be the loop counter and to do conditional branches in the controller based on the result of decrementing the value in that register. This is simple to do, because the hardware of the controller and data

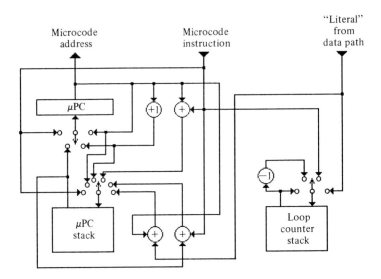

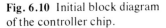

Fig. 6.10 Initial block diagram of the controller chip.

path discussed so far will allow the execution of this instruction. Unfortunately, there is a time penalty when doing interchip communication: the loop counter must be decremented during one cycle, the result of the decrement must be sent to the controller during the following cycle, and a conditional branch must be performed in the controller on the third cycle. If the loop counter were in the controller chip, this operation would only take one cycle and would not require the use of the ALU for one cycle in the data path.

　　With only one loop counter, loops could not be nested, and loops could not be used inside of subroutines. If a stack were provided for the counter values, however, nested loops and loops within subroutines could both be accommodated.

　　The first OM2 controller proposal was based primarily on the arguments presented above. Figure 6.10 shows a block diagram of the proposed controller. Table 6.1 lists the operations possible for each of the three sections of the controller chip: the μPC source selection, the μPC stack operation, and the loop stack operation. In each cycle, the controller executes one operation in *each* of the three sections. For most operations, all three sections work together to perform the programmed operation. There are cases, however, when only one or two of the sections are needed to perform the controller's instruction, so the other section(s) are free to perform other tasks. For example, the loop stack may be loading a count from the data path, while the μPC sections are performing a subroutine call. This concurrency saves having to load the count later and may save microcode space. Because the controller's instruction is broken into three fields, more than one thing can be happening in parallel in the controller. This is why the instruction was not kept as one field, but was decoded into the three sections on chip.

Table 6.1. Operations of the Initial Controller Proposal.

μPC sources	μPC stack operations
μPC + 1	Push μPC + 1
microcode	Push microcode
μPC Stack Top	Push μPC + microcode
True: μPC Stack Top; False: μPC + 1	Push μPC Stack Top + microcode
	Pop
Loop stack operations	Push μPC + literal
Push microcode	True: Pop; False: NOP
Push literal	True: NOP; False: Pop
Push count	NOP
Pop	
Decrement Count	
NOP	

The controller shown could handle all of the microcode address operations listed above, and a few new operations were discovered and added to the list. However, there are a few problems with this design. It is a "brute force" design: rather than view the whole chip at one time and look for generalizations, we looked at each section of the chip and at the chip's operation individually, and filled the chip with specialized hardware for performing specialized operations. We found that by adding one circuit "here," a new operation could be performed; and that by adding another circuit "there," a different operation could be added to the repertoire. Many designs suffer from "creeping features" of this sort. While it may be easy to draw circles and arrows on paper, it can be more difficult to draw adders and multiplexers on silicon. It would be very difficult to route all the wires needed to interconnect the devices shown in the proposal.

So let's make a few generalizations about the circuits in the design. First, there are too many adders on the chip. A close look at the proposal shows that almost all operations used only one adder for any one cycle, and the few operations that used more than one adder are not critical operations. Incrementing the μPC can also be done in the adder, by clearing one of the data inputs to the adder and forcing a carry into the first stage. Thus, all three of the adders and the incrementer can be combined into one adder, and multiplexers can be put on the inputs to that adder. Another simplification would be to always load the μPC latch from the output of the adder, which would allow the removal of the multiplexer on the input to the latch. The only operations that were sacrificed in making the simplifications involved loading the μPC stack with the output of an adder. Figure 6.11 shows the block diagram of our simplified controller, and Table 6.2 lists the operations it performs. Notice that the controller's instruction is now broken into five fields: controlling the μPC sources, the μPC stack sources, the counter operation, the condition selection, and the μPC stack operation.

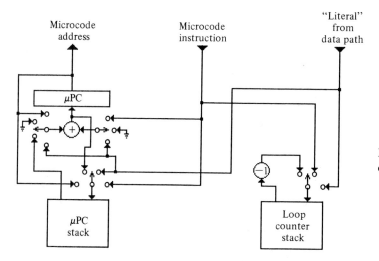

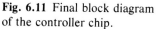

Fig. 6.11 Final block diagram of the controller chip.

Table 6.2. Operations of the Final Controller Design.

μPC sources	μPC stack sources
μPC + 1	Adder output
μPC + microcode + 1	μPC
microcode	microcode
Stacktop + 1	literal
Stacktop + microcode + 1	**Counter operations**
Stacktop + literal + 1	No Operation
μPC + literal + 1	Push microcode
literal + microcode	Push literal
Condition selection	Pop to literal bus
False	true: decrement; false: pop
True	true: decrement; false: NOP
Data path flag	**μPC stack operations**
Complement of Data path flag	Push if condition is true
Count = 0	No push
Count < > 0	Pop if condition is true
Data path flag AND Count = 0	No pop
Data path flag OR Count = 0	

Now we will develop the geometrical and topological arrangement of the controller's subsystems. Such arrangements are often called *floor plans*. A translation of the preceding ideas into the starting floor plan of the controller is shown in Fig. 6.12. The plan is composed of subsystems built of horizontal bit slices that are then stacked vertically. The number of bit slices is equal to the microcode address width for the machine, which in this case is 12 bits.

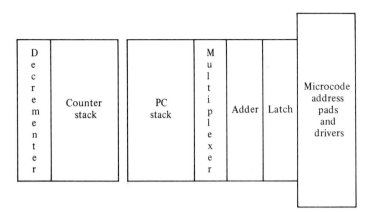

Fig. 6.12 Starting floor plan for the controller.

The following points were considered when deciding upon the basic framework of the floor plan. First, the μPC latch is placed adjacent to the microcode memory address pins. This is done to minimize the delay when driving addresses to the memory, as this operation is in the critical timing path for the entire machine. The input of the latch comes only from the output of the adder, so the adder should logically be placed next to the μPC latch. The adder is considerably simpler than the full arithmetic logic unit used in the data path. However, it employs the same principles as the ALU: the Manchester carry chain, the insertion of double inverters every four bits to minimize the delay in the carry chain, and the logic block to implement the desired functions with the minimum delay and power. The multiplexer is placed adjacent to the left side of the adder. This multiplexer operates in the same manner as the input multiplexer to the ALU in Chapter 5. The μPC stack is then placed to the left of the multiplexer.

The only problem with this arrangement of the floor plan is that the microcode bus and the data path bus must also connect to the multiplexer. A large amount of area is wasted if these two buses are connected to the multiplexer from the side. Instead, if the buses are placed where the μPC stack is located, they then connect to the loop counter circuits directly. But then there is the problem of where to place the μPC stack. The solution is to run the buses *through* the μPC stack. Each cell of the stack thus has the two buses designed right in. The two buses run right on through the loop counter stack to the loop counter decrementer and the pads.

Having placed the major blocks of the chip into the floor plan, we can examine the layout of the control circuits and work out a detailed floor plan. Each of the stacks require push and pop drivers, as discussed in Chapter 3. As in the Chapter 3 example, one set of drivers is placed along the top, and the other set along the bottom of the stack. The control drivers for the latch, adder, multiplexer, and counter are identical to those discussed in Chapter 5. The control bits for these control drivers could all be derived directly from the outputs of the microcode memory, but this technique would result in an exceedingly wide microinstruction. By encoding the operations to be performed by the adder and its input multiplexers, we can dramatically decrease the width of the microinstruction. With proper encoding of these operations, the functional capability of the chip is not impaired. (A number of possible control signal combinations are in fact illegal and thus redundant. For instance, if more than one control line for the multiplexer is enabled, the outputs of two or more sources would be shorted together, and the resulting multiplexer output would contain erroneous data.) The placements of the control circuits and encoding PLA's are shown in Fig. 6.13, which also shows additional details of the final floor plan. Notice that the counter stack is *higher* than the 12-bit high μPC stack, so that it can contain entire 16-bit data path words for parameters passed to subroutines in the microcode. The stacks are aligned on their least significant bit position, and the additional length of the counter stack allows space for the control PLA's for the adder and μPC stack.

The programmable logic arrays employed in instruction decoding do not have feedback from their outputs back into their inputs. Their only function is to serve as combinational logic for condensing the number of control wires and thus saving microcode memory bits. The *finite-state machine* for the control of this path is made up of the microcode memory address feedback through the adder and stack PLA's and also the microcode literal path feedback into the input of the adder. If there were feedback terms in any of the PLA's, provisions would have to be made for access to the state of the feedback terms from off-chip. Without such access,

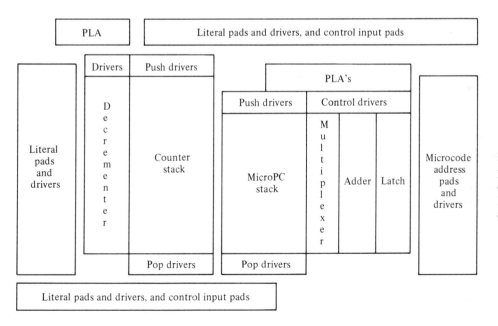

Fig. 6.13 Final floor plan of the controller chip. (For a detailed layout, see the back endpaper.)

the *untestable state* information on the chip would make the testing of the completed chip next to impossible: the current operation of the chip would be a function not only of the control signals and data that we supply to the chip at a particular moment, but also of the past control signals and data. In the absence of a practical way to directly probe all the signal lines on the chip, it is imperative that all of the chip's state be accessible somehow from off-chip.

One of the problems encountered in many multichip microprogrammed machines is that a great deal of interchip communication is required in their operation. Although the effective computing bandwidth of the machine can be made large by *pipelining* the operations, any operation that requires the full circle through the feedback loop of the state machine will require a great deal of time for its execution. In the OM2 system, we have included hardware features in both the data path and the controller to reduce the chip-to-chip communication as much as possible. As already mentioned, we included the loop counter circuitry on the controller chip, thus reducing the loop operation time from 3 cycles to 1 cycle. Chapter 5 mentions the *conditional ALU operations* in the data path that can modify the actual ALU operation as a function of the *flag bit*. An example of the utility of this capability is provided by the multiply operation. When performing a multiply, the ALU should either add two numbers or just pass one of the numbers straight through, depending upon the state of a flag. One way to do this operation would be to send the flag to the controller chip and execute a conditional branch to one of two locations. One of the two appropriate ALU operations would be at each of the two microcode locations. However, it would take several cycles to perform each step of a multiply were this method used. Since in OM the ALU on the data chip has the capability of modifying its instruction as a function of the flag, a single cycle will perform this part of the multiply step.

There are times when it would be convenient to communicate many bits between the controller and the data path in one cycle. For instance, when emulating the instruction set of a higher level machine, the data path can examine various fields in the instruction currently being emulated and calculate microcode branch locations. It is then necessary to load the μPC latch with the calculated branch location. To facilitate this loading, a 16-bit bus, referred to as the "literal bus," connects the two chips. To economize on the data path's pin count, this bus is used to load microcode into the data path chip when it is not transferring literal data between the two chips. A large number of pads is required for the microcode and data path literal interconnections. There was insufficient space along the left edge of the chip for all of the pairs of pads required for this communication. Hence, we placed some pads along the top of the chip and others along the bottom, and we made connections between these pads and the buses by running vertical wires to the appropriate bus lines where they run between the two stacks.

The layout of the completed controller chip is shown on the endpaper. Examples of the use of several of the controller's operations are given in the following section.

6.6 EXAMPLES OF CONTROLLER OPERATION*

This section will illustrate the operation and programming of the controller presented in the last section, through the use of four programming examples: subroutine linkage, For-loops, Do-loops, and field dispatches. Refer to Table 6.2 for a
tabulation of the controller's operations. It should be noted that the μPC operations
are pipelined by one cycle so that if one particular microinstruction contains a
controller jump opcode, the following microinstruction will also be executed before
the jump actually occurs.

To call a subroutine, we would like to save the current value of the μPC on the
μPC stack and load the μPC latch with the microcode address of the subroutine.
When we have finished executing the subroutine code and wish to return, we just
pop the return address off the μPC stack and load it into the μPC latch. To save
the μPC value on the stack, "μPC" should be selected as the stack source and
"Push" should be selected as the stack operation. As Table 6.2 shows, the *condition* must be true in order for the stack to push a value. Therefore, the condition
selection should be "True" to guarantee that the stack will save the return address. While we are saving the current μPC value on the stack, we must also load
the μPC latch with the subroutine address. To do this, we select "Microcode" as
the μPC source and put the subroutine address in the literal field of the microcode.
Since we are not using the counter, the counter operation should be "Nop." For
the return, we load the μPC latch with the return address by selecting
"Stacktop+1" as the μPC source and pop the stack by selecting "Pop" as the
μPC stack operation. In order to guarantee that the stack pops the old value off
the stack, we must make sure the condition is true by selecting "True" as the
condition selection.

Figure 6.14 illustrates the execution of subroutine linkages. Four "snapshots"
of the microcode and μPC contents are shown at the various steps as the execution
proceeds. Snapshot (a) gives us a background for what is happening: The μPC is
stepping through a segment of microcode and is about to execute a Call operation.
The Call operation contains a pointer to a subprogram located somewhere in the
microcode memory. Snapshot (b) shows the state of the machine just after the Call
operation is executed. The μPC now points to microcode addresses inside the
subroutine, while the return address to the main "program" is saved on the stack.
Snapshot (c) shows that the μPC has advanced to the end of the subroutine, and
the Return operation is about to be executed. The return address is popped off the
stack and loaded into the μPC latch, and program execution resumes where it left
off in the main program, as shown in the last snapshot.

A For loop should execute the same section of code many times. We can use
the loop counter to store the number of times we have executed the code so that
we know when we have finished the specified number of executions. Thus, when
starting a For loop, we should push the repetition number onto the loop counter

---

*This section is contributed by Dave Johannsen, California Institute of Technology.

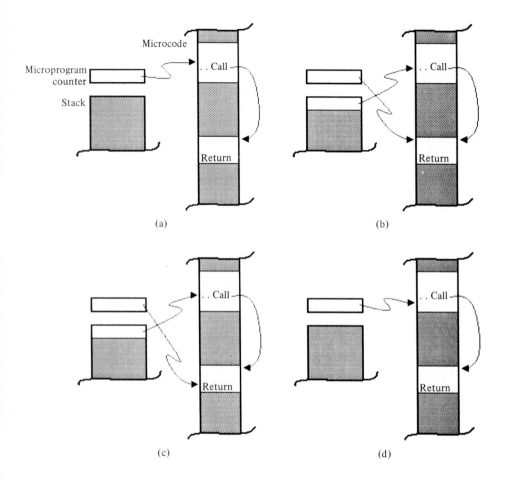

Fig. 6.14 Illustration of subroutine linkage. (a) Before execution of the Call instruction. (b) Just after execution of the Call instruction. (c) Just before execution of the Return instruction. (d) After execution of the Return instruction.

stack. At the end of the loop we decrement the count and, if the result is not zero, we should jump back to the start of the loop. If the decremented result is zero, we have finished execution of the For loop, and we should pop the count off the loop counter stack. Execution of the For loop in this manner requires that the end-of-loop command contain the address of the start of the loop. How then can we construct relocatable code containing For loops? We can eliminate the need for the end-of-loop command to contain the loop's start address, by saving the start address on the μPC stack. The μPC latch then merely has to be loaded with the value contained at the top of the stack. When we use this method of saving the loop address, the start-of-loop command becomes

μPC Source $\leftarrow \mu$PC + 1
μPC Stack Source $\leftarrow \mu$PC
μPC Stack Operation $\leftarrow$ Push
Condition $\leftarrow$ True
Counter Operation $\leftarrow$ either Push Microcode or Push Literal

The end-of-loop command becomes

μPC Source $\leftarrow$ True: Stacktop + 1; False: μPC + 1
μPC Stack Operation $\leftarrow$ Pop
Condition $\leftarrow$ Count NOT EQ 0
Counter Operation $\leftarrow$ True: decrement; False: Pop

The operation of For loops is illustrated in Fig. 6.15. Again, four snapshots are shown which represent the state of the controller and microcode at various points in the execution of the loop. Snapshot (a) shows the state of the machine just prior to the execution of the For operation. When the For operation is executed, the value in the μPC latch is pushed onto the μPC stack, and the number of iterations specified by the For command is pushed onto the counter stack. The μPC continues advancing through the microcode. Snapshot (b) shows the state of the controller and microcode at some point in the middle of the For loop execution. When the end of the loop is reached, the value on the top of the counter stack is decremented. If the result is not zero, the new value is pushed onto the stack and the μPC latch is loaded with the value on the top of the μPC stack, as shown in snapshot (c). Notice that the value is not popped off the top of the μPC stack because we will need the loop address again if the loop is not completed after executing one more time. When the result is zero, data is popped off the top of both stacks (to remove the loop address and the old count, which is now = 0) while the μPC value is just incremented, causing the controller to exit from the For loop, as shown in the last snapshot.

The Do loop is similar to the For loop, except that the code is repeatedly executed until a condition becomes True. That condition may be, for instance, when the data path flag becomes True. In this case, the condition selection in the end-of-loop command becomes ''Data path flag'' instead of ''Count not EQ 0''. Also, since the counter is not being used, the counter operation in both the start-of-loop and the end-of-loop commands becomes ''Nop''.

Figure 6.16 shows some snapshots associated with the execution of a Do loop. By comparing Figs. 6.15 and 6.16, we can observe the similarities between For loops and Do loops. Basically, the only difference between these two types of loops is the decision of when to exit the loop. In a For loop a counter decides when the loop should be exited, while in the Do loop a flag, such as the flag from the

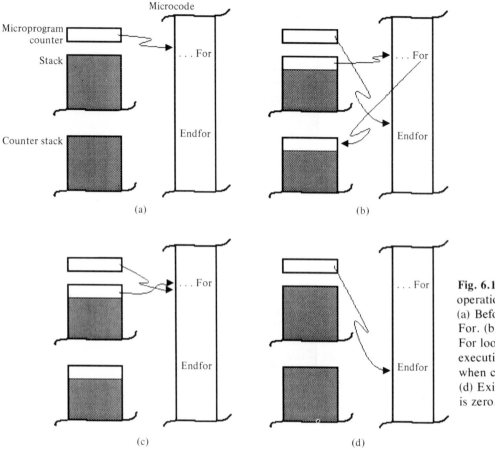

Fig. 6.15 Illustration of the operation of For loops. (a) Before execution of the For. (b) Execution of the For loop. (c) Repeat execution of the For loop when count is not zero. (d) Exit loop when count is zero.

data path, decides when the loop should be exited. Since the Do loop does not use the counter, the counter is not shown in the snapshots of Fig. 6.16.

When emulating the instruction set of a higher-level machine, we find it is often convenient to do a multi-way branch. Suppose, for example, that the machine we are emulating has a 16-bit instruction word that contains a 4-bit opcode field and a 12-bit address field. In this case, we would have 16 code segments in the microcode, one for emulating each of the 16 possible opcodes of the higher-level machine. We would like to be able to perform a 16-way branch, depending on the contents of the 4-bit opcode field, that would take us directly to the correct microcode segment, thus implementing the *decode stage* of instruction interpretation. We could use the ALU in the data path for calculating the microcode address for the proper segment, and load the μPC latch with the result of this calculation. This works especially well if the starting addresses of the segments are evenly spaced, because to calculate the branch address we merely multiply the 4-bit opcode by the segment length and add the displacement of the first segment. The

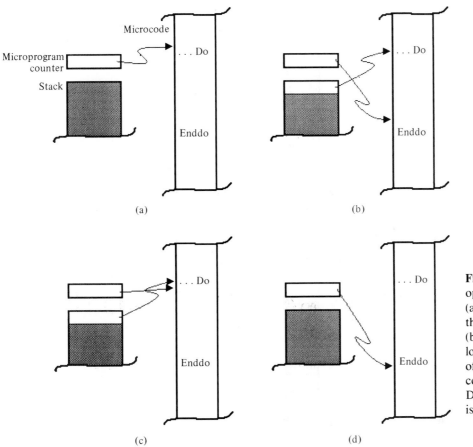

Fig. 6.16 Illustration of the operation of the Do loop. (a) Before execution of the Do instruction. (b) Execution of the Do loop. (c) Repeat execution of the Do loop while condition is false. (d) Exit Do loop when condition is true.

multiplication is particularly easy to perform if the segment length is a power of 2, because then we have only to shift the 4-bit opcode value of the appropriate number of places to the left.

A problem with the above method of field dispatching is that the microcode segments have to be evenly spaced in the microcode, preferably by a power of 2. In practice, segments are seldom of the same length. Even if they were of the same length, if one of the segments had to be modified, extensive corrections might have to be made all though the microcode. As an alternative, a *dispatch table* can be inserted into the microcode, which just contains a series of jump instructions to the appropriate microcode segments. If this is done, the 4-bit opcode value need only be shifted left once (because jump instructions are two microcode words long, due to pipelining), added to the dispatch table displacement, and loaded into the μPC latch. To load the value into the μPC latch, the data path sends the result of the above calculation across the literal bus to the controller, and the controller selects a μPC source of "Literal".

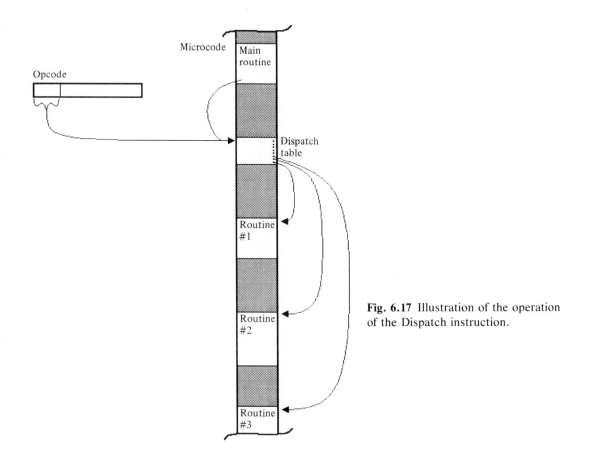

Fig. 6.17 Illustration of the operation of the Dispatch instruction.

Figure 6.17 illustrates the operation of the dispatch instruction. The controller jumps to a location in the dispatch table that is a function of one of the fields in the opcode. The dispatch table contains Jump instructions to the various routines that perform the microinstructions necessary to emulate each of the possible opcodes. The selection of the proper field in the opcode and the calculation of the dispatch table address are performed in the data path prior to the dispatch.

6.7 SOME REFLECTIONS ON THE CLASSICAL STORED-PROGRAM MACHINE

In the future, very large quantities of computing machinery may be placed on a single chip. Such chips will be easily and quickly designed, and rapidly implemented. This capability will present both a great opportunity and a great challenge. How are we to organize and program such a wealth of hardware? Certainly not the way we do now.

Scaling of the technology to higher densities is producing effects that may be clarified by analogy with events in civil architecture. Decades ago, standard bricks, "two-by-fours," and standard plumbing were used as common basic building blocks. Nevertheless, architects and builders still explored a great range of architectural variation at the top level of the time: the building of an individual home. Today, due to the enormous complexities of large cities, many architects and planners have moved on to tackle the larger issues of city and regional planning. The basic building blocks have become the housing tract, the business district, and the transportation network. While we may regret the passing of an older style and its traditions, there is no turning back the forces of change.

In present LSI, where we can put many circuits on a chip, we are like the earlier builder. While we no longer tend to explore and locally optimize at the circuit level (the level of bricks and two-by-fours), we still explore a great range of variation at the level of the individual computer system. In future VLSI, where we may put many processors on a chip, architects will, like the city planner, be more interested in how to interconnect, program, and control the flow of information and energy among the components of the overall system. They will move on to explore a wider range of issues and alternatives at that level, rather than occupy themselves with the detailed internal structure, design, and coding of each individual stored program machine within a system. If systems are to work at all, they must at the least be understood at their highest level. These are some of the issues explored in Chapter 8.

REFERENCES

1. A. W. Burks; H. H. Goldstine; and J. von Neumann, "Preliminary Discussion of the Logical Design of an Electronic Computing Instrument," Institute for Advanced Study report to the Army Ordnance Dept., 1946, contains the original description of a stored program electronic digital computer. Reprinted as Chapter 4 in Bell and Newell.[3]
2. A. M. Turing, "On Computable Numbers, with an Application to the Entscheidungsproblem," *Proc. of the London Mathematical Society*, Series 2, vol. 42, 1936–1937, pp. 230–265.
3. C. G. Bell and A. Newell, *Computer Structures: Readings and Examples*, New York: McGraw-Hill, 1971, presents some of the history of computer architecture and design and contains many interesting examples of computer structures.
4. D. L. Dietmeyer, *Logic Design of Digital Systems*, Boston: Allyn and Bacon, 1971. Chapter 6 contains an example of the design of an elementary digital computer, starting with machine instruction set and controller state diagram.
5. C. R. Clare, *Designing Logic Systems Using State Machines*, New York: McGraw-Hill, 1973.
6. M. V. Wilkes, "The Best Way to Design an Automatic Calculating Machine," address to the Manchester University Computer Inaugural Conference, July 1951. The basic principles of microprogramming were first stated by Wilkes in this address.
7. M. V. Wilkes and J. B. Stringer, "Microprogramming and the Design of the Control Circuits in an Electronic Digital Computer," *Proc. Cambridge Phil. Soc.*, pt. 2, vol. 49, April 1953, pp. 230–238. An extension of the earlier work by Wilkes, this paper includes a worked out example. Reprinted as Chapter 28 in Bell and Newell.[3]
8. P. M. Davies, "Readings in Microprogramming," *IBM Systems Journal*, no. 1, 1972, pp. 16–40. Provides a good primer and guide to the literature on microprogramming.
9. D. L. Johannsen, "Our Machine: A Microcoded LSI Processor," Display File #1826, July 1978, Dept. of Computer Science, California Institute of Technology.

7

SYSTEM TIMING

CHARLES L. SEITZ

Department of Computer Science
California Institute of Technology

7.1 THE THIRD DIMENSION

The successful design of large scale integrated systems requires careful management not only of the two-dimensional silicon area but also of the operation of the system in the time dimension. Although time is physically different than the spatial dimensions, the general strategies already introduced for carrying the spatial design from conception to layout apply to system timing as well. These are the usual strategies for containing complexity: use of abstraction and structured design.

Much of the functional design of the spatial aspect of a system is done with the help of block diagrams, logic diagrams, circuit diagrams, and stick diagrams, in a metric-free topological domain. These representations are helpful because they allow designers to suppress detail, so that they can think about system behavior at a level of abstraction that is effective for the task at hand. One specific abstraction employed in these diagrammatic representations is the suppression of geometrical detail, while focusing on the topological structure of the circuit or system. Topology is sufficient to specify information flow between functional parts, so diagrammatic representations are a useful abstraction to the functional or logical structure of a system.

The third dimension, time, may also be regarded as having features analogous both to geometry and topology. The definition of a sequential process—whether represented by a program, flowchart, state diagram, or in plain English—specifies only the ordering, or partial ordering, of the individual steps that compose it. Thus it is the metric-free "topological" concept of *sequence*, rather than the physical concept of a *time metric*, that is most useful for the functional specification of a system.

As was pointed out in Section 3.5, it is important that the levels of abstraction used in the design process be related to each other and to physical concepts. The *sequence domain* is a self-consistent abstraction that applies across several levels of system design—programming, organization, logic. However, sequence domain representations such as flowcharts and state diagrams do not say anything explicit about space, time, and other physical characteristics of a system. The value of abstraction to the design process is that it permits one to defer certain bindings to physical form. The hazard is that one can become so isolated from physics and economics as to produce elegant schemes that are unworkable in practice. Thus, an important goal of any study of timing is to devise and explain methods by which sequence and time can be systematically related.

Although the designer of an integrated system may think about a layout initially in topological terms, at some point it becomes necessary to think also about geometry. The cost of a design will eventually be measured in silicon area, and whether it will work depends on adhering to the geometrical design rules of the fabrication process. Likewise, the sequence domain conception of system behavior becomes coupled with time in two ways. First, most systems are designed to achieve performance objectives that can be traced back to human needs, expectations, or desires. Second, the electrical behavior of devices and wires is governed by physical laws that are expressed as partial differential equations in time.

Unfortunately, the world is full of examples of digital systems that, even when functionally correct, have disappointed their designers and users by being unreliable or too slow. Why is it that so many systems have "timing problems" or fail to achieve performance objectives?

As is the case with the spatial dimensions, the design problems in the third dimension result not from a lack of possible forms but rather from an overabundance. If one is to build a large scale integrated system with any hope of reliable operation, it is necessary to restrict oneself to a consistent style of design. The canonical forms in the time dimension are signaling conventions that are adhered to throughout the system and serve the function of establishing between all parts engaged in a communication an interval or sequence of intervals of time for this communication. If such a scheme is to be regarded as a discipline, it must be possible to state precisely the requirements that the signaling convention places on system interconnections and element timing.

Alternative disciplines of design in the time dimension can be characterized by the way in which they connect sequence and time. The products of two very different disciplines of design are described in this chapter: *synchronous systems* and *self-timed systems*. In the synchronous discipline of design, which has been used in a form with a two-phase clock in the designs presented previously in this book, sequence and time are connected by means of a system-wide clock signal. In the self-timed discipline, sequence and time are connected in the interior of parts called *elements*. The terminal behavior of a self-timed element must satisfy a sequence domain representation, which assures that correct sequential operation of a self-timed system is insensitive to element and wiring delays.

Since synchronous systems are by far the best known and most widely used, we take them as the starting point for the body of this chapter. However, synchronous systems possess some serious limitations, which are made even worse as λ is scaled down and as chips become larger.[1] Some of these limitations are physical in nature and relate to the difficulties of moving information from point to point within a single clock period. Another limitation is the difficulty of managing very large designs in a framework in which all system parts must operate together in "lockstep."

The same considerations of managing the design of very large integrated systems that provide a motivation for dividing a system into modular parts argue that the parts be independently timed. If the parts are each synchronous systems with *independent* clocks, information communicated from one part to another must be synchronized to the receiver's clock. Unfortunately, as we show in a later section, this synchronization cannot be accomplished with complete reliability. The reason is that synchronizing elements are bistable and have a metastable, or balanced, condition that occurs under the conditions in which synchronizers must operate. As was discussed in Section 1.14, there is no bound for the time the bistable element may remain in this metastable condition. There are many methods to reduce the probability that such a fault would crash a system, but they all cost time and so would reduce efficiency.

The limitations imposed by the synchronous discipline suggest that other disciplines be tried. The outline of the self-timed discipline presented in the final sections of this chapter describes one approach to system timing that scales well to VLSI, and that retains synchronous systems under exactly those conditions in which the synchronous discipline is workable. Self-timed elements can be designed as synchronous systems with an internal clock that can be stopped synchronously and restarted asynchronously. This type of clock allows synchronous elements to communicate reliably, because their clocks are partly dependent, i.e., not independent. Elements may also be designed as asynchronous or speed-independent circuits. There is no system-wide clock or time reference in a self-timed system. Instead, initiation of a given computational step depends on completion signals produced by its sequential predecessors. Thus self-timed systems operate at a rate determined *locally* by element and wiring delays, a rate that tends to reflect average, rather than worst case, delays.

The subject of self-timed logic has two principal facets: the design of elements and the design of systems of interconnected elements. Along the seam between those subjects are conventions for self-timed signaling. This bifurcation of the discipline is deliberate. The design of elements is difficult because it is here that logic, physics, and timing come together. However, the element designer can work within a domain in which physical and logical scale are both restricted to be small enough to make the design manageable. The design of systems is difficult because of the combinatorics of scale. However, the system designer can work within a domain similar to that of a programmer, in which many of the details of the underlying physical system have been suppressed and replaced by an abstraction that is free of hidden rules; namely, the sequence domain abstraction.

7.2 SYNCHRONOUS SYSTEMS

In the synchronous discipline of design, sequence and time are connected through the use of a system-wide *clock* signal. The clock signal serves two purposes – or one might say it serves two masters. The clock is a sequence reference and also a time reference. As a sequence reference, its *transitions* serve the *logical* purpose of defining successive instants at which system state changes may occur. As a time reference, the *period* or interval, either fixed or variable, between clock transitions serves the *physical* purpose of accounting for element and wiring delays in paths from the output to input of clocked elements.

The ability of the clock signal to serve two masters, logic and physics, has a certain compact elegance and conforms to an established tradition of parsimony in the use of active elements. However, the dual role of the clock binds the system sequencing and timing so closely that "timing" is the source of numerous difficulties in the design, maintenance, modification, and reliability of synchronous systems.

The logical model that synchronous systems resemble is the finite-state machine, a model that has been described in detail both in Section 3.11 and in Chapter 6. As illustrated in Fig. 7.1, any such system must satisfy a *topological requirement* that every closed signal path pass through a *clocked storage element*. Closed paths that do not pass through clocked storage elements are excluded as they may create nondeterministic behavior, either through oscillation or through asynchronous latching. There are several important consequences of this topological constraint on the logical design: (1) It assures deterministic behavior if the physical aspects of the design are also correct. (2) It relieves the designer of any requirement that the combinational logic be free of transients (static or dynamic hazards) on its outputs. The only dynamic characteristic of a combinational net that matters is its propagation delay time. (3) The storage or history dependence of the system resides entirely within the clocked storage elements, a fact that

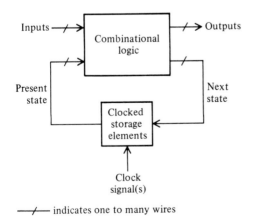

Fig. 7.1 Finite-state model.

 indicates one to many wires

simplifies the design process and often also the maintenance and testing of a system.

The clocked storage elements in a synchronous system may take any of a variety of forms, discussed below, depending on physical requirements such as speed, economy, or static operation. While these elements are distinguished as being the only recipients of clock signals, in practice there may be a number of timing signals derived as different phases or submultiple frequencies of the clock. In these cases it may be difficult to see the correspondence between the circuit and the finite-state model. Circuits such as shift registers are finite-state machines, but they already possess such a natural and regular structure that it would be pointless and awkward to describe their behavior with state diagrams. Control elements, such as the finite-state machine stoplight controller described in Section 3.11.1, are a case of imposed structure, in which the combinational logic and clocked storage elements can be patterned on the silicon in a form that mimics the usual block diagram of a finite-state machine.

Although clocked storage elements may be clocked by different schemes, they all are binary storage devices. The sort of physical device that has the property of storing information (also called memory, or history dependence) is one that stores energy, or it may be film or punched cards in which energy is required to change some detectable condition of the medium.

For semiconductor integrated circuits the energy represented by charge stored on circuit capacitances is the only practical mechanism for storing information. Inductance plays the same role in superconducting circuits. MOS circuits employ this mechanism very directly in the dynamic register introduced in Chapter 3 and used in designs throughout this book. In the dynamic register illustrated in Fig. 7.2, the output stored data follows the input as long as the enable input to the transfer gate is high. When the enable signal goes low, the charge stored on the node is very well isolated and so maintains the same voltage.

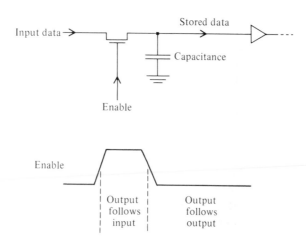

Fig. 7.2 Dynamic register (1-bit).

Unfortunately, this is not the whole story. While the charge is very well iso-lated, it is not perfectly isolated. Charge escapes by two different mechanisms, which scale differently. The principal leakage path for 1978 MOS technology is the reverse leakage current of the drain junction of the pass transistor. The time con-stant of this decay is in the order of a few seconds at room temperature but de-creases exponentially with temperature to a millisecond or so at 70°C. This leak-age path is a current per unit area and, because scaling down the circuit dimen-sions increases the capacitance per unit area due to decreased oxide thickness, the time constant of charge decay increases with reduced circuit dimensions. The fine details of this scaling are largely masked by the exponential temperature depen-dence, however, so the time constant of junction leakage is reasonably regarded as approximately independent of λ over the recent past and future. Subthreshold currents are expected to become the limiting factor in holding charge on a node as soon as threshold voltages are reduced to much below 1 volt. This effect of scaling down the dimensions of MOS circuits was discussed in Section 1.16. We refer to the time a node will reliably store a bit as the *refresh period*.

The decay of charge on a dynamic node is no problem so long as the charge is sensed and refreshed frequently enough; for example, in every clock period as is done in the OM design. In some cases this is not possible, and *static registers* are used instead. An assortment of static registers is shown in Fig. 7.3. The general idea evident in the circuits in parts (a) and (b) is to amplify and feed back the stored data so as to counteract the decay of charge. The part (a) circuit does this through a resistance that must be much larger than the effective R_{on} of the pass transistor, so that the storage node can be driven to the value desired. For vol-tages close to the switching threshold this circuit amounts to a negative resistance termination to V_{inv}, where the resistance is $-$(large R)/(voltage gain of the pair of inverters). This circuit can be used advantageously as a termination for buses to assure static operation. The diagram in part (c) is a storage circuit typical of bipo-lar families, in which low impedances make dynamic storage unattractive, so that static storage is the rule rather than the exception.

These static circuits are logically equivalent to the dynamic storage circuit, except that the stored information will remain indefinitely. The use of extra cir-cuitry for each node to accomplish this continuous refreshing is usually unneces-sary, and it will be omitted in the following discussion and figures, with the under-standing that it could in all cases be included where required.

As an aside, the reader may wonder about the statement above – that capaci-tance provides the mechanism for information storage. Is this true of the static and cross-coupled storage circuits as well? Many references on switching circuits leave the impression that the existence of two logically consistent stable states in these cross-coupled circuits is sufficient to ensure that the circuit will store a bit. Some references mention also that the circuit must have more than unity gain around the loop, which is indeed a necessary, but not sufficient, condition. Con-sider the consequence if the circuit capacitances were all taken to zero. The circuit

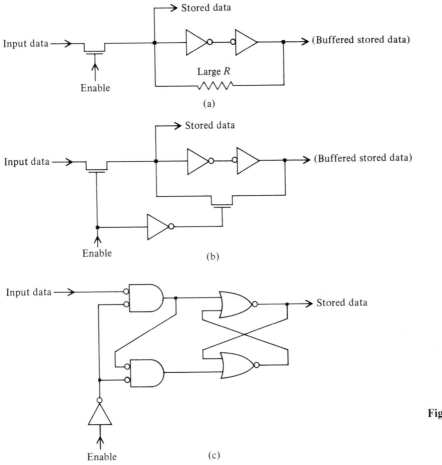

Fig. 7.3 Static registers.

would then be able to respond instantly to an excitation, which means that a current pulse of arbitrarily short duration could change the state of the circuit. If this arbitrarily small amount of energy could change the state of the circuit, one could not reasonably expect the circuit to remain in either state. Without capacitance, it cannot store information. This physical aspect of storage devices is discussed in detail in Chapter 9.

With this background on storage elements, let us proceed to clocking schemes. As we pointed out in the opening section of this chapter, it should be possible to state precisely the requirements the timing form places on system interconnections and on element timing. For synchronous systems, the requirement on interconnections is the topological requirement that all closed paths pass through one or more clocked storage elements. The requirements on element timing depend on the clocking scheme.

The storage devices described above, all logically equivalent to the dynamic register, may be used as clocked storage elements in a cheap, fast, and *risky* clocking scheme illustrated in Fig. 7.4. (We know of no example of this scheme being used in LSI circuits, but it was common in many of the early "transistorized" computers.) This scheme might best be termed "narrow pulse clocking," because it requires that the clock pulse be narrow compared to the delay of the combinational logic. The present state information changes a short time, about $R_{on}C_{in}$, after the leading edge of the clock. The delay through the combinational logic must be greater than the clock width, or else the change in the present state information will propagate through the combinational logic to change the next state information before the trailing edge of the clock.

The combinational logic must be designed so that its delay satisfies a *two-sided relation*—greater than the clock width and less than the clock period. As indicated above, the clock width must also satisfy a two-sided relation—greater than the time required to transfer charge to the present state inputs of the combination logic and less than the minimum delay of the combinational logic. The clock period is also involved in a two-sided relation, unless static registers are used. The relations that must be satisfied are summarized in Fig. 7.4. They are relations that apply in a worst-case sense, in spite of variations in temperature, power supply voltage, aging, and manufacturing.

This "narrow pulse clocking" scheme was abandoned because of the difficulty of satisfying so many two-sided relations simultaneously under so many conditions of variation. Also, the economies achieved due to the simplicity of the clocked elements were partly offset by the necessity to "pad out" the delay in many of the combinational nets. This clocking scheme is quite feasible for inte-

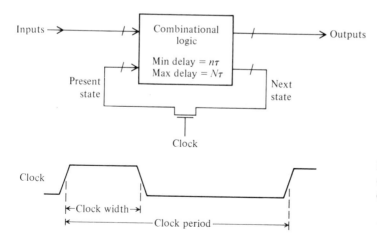

Fig. 7.4 Cheap, fast, and risky clocking scheme. R_{on} is the effective on resistance of the pass transistor, and C_{in} is the input capacitance of the combinational logic.

$$(R_{on}C_{in})_{max} < \text{clock width} < n\tau$$
$$N\tau < \text{clock period} < \text{refresh period}$$

grated systems, since it is inherent in their manufacture and operation that most of the variables will track if the clock signal is generated on-chip.

Two-sided relations on timing create enough difficulties in the design and maintenance of a system so that it is generally worthwhile to use more logic to make the timing relations one-sided. Elimination of the two-sided relation on clock width, or the complementary relation on combinational delay, requires the use of at least two clock phases. This minimum form occurs for much the same reason that a canal lock requires at least two watertight gates.

The two-phase clocking scheme illustrated in Fig. 7.5 includes four sequentially repeated epochs. During φ_1, previously stored information is applied to the present state inputs of the system's combinational logic. The φ_1 signal must remain high long enough to charge the present state input nodes, a process that incurs some delay on the order of $R_{on}C_{in}$, and that is called the *delay time* of the clocked storage element. Following this delay, if inputs are also available, the combinational logic starts setting up the outputs and next states, independent of when φ_1 may transit from high to low. This epoch is analogous to the operation of a canal lock releasing a ship. The gate must open before the ship leaves, and the ship must clear the gate before it is again closed. If the lock master chooses to leave the gate open for a while after the ship leaves, it does not slow down the ship.

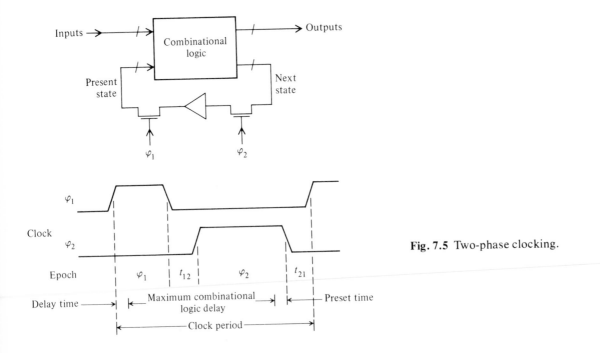

Fig. 7.5 Two-phase clocking.

What the lock master must never, never do is open both gates at once! The epoch, labeled in Fig. 7.5 as t_{12}, is an interval produced by the nonoverlapping phases of the clock. By analogy with the narrow pulse clocking scheme, it is clear that overlap less than the minimum delay of the combinational logic is harmless to correct operation. However, as the minimum delay of the combinational logic is ordinarily legislated to be zero (most people would agree it could not be less, and for circuits such as shift registers the delay does approach zero), the overlap period t_{12} must be greater than zero. In practical cases, because t_{12} does not represent ''dead time'' but time during which the combinational logic is working, this time is made as short as is convenient but not necessarily as short as is possible.

In the epoch during which φ_2 is high, the clocked element samples its input. The combinational outputs must be stable slightly before the trailing edge of φ_2, an interval called the *preset time* of the clocked storage element. Of course, φ_2 must be wider than the preset time. This φ_2 epoch is analogous to the entry of a ship into a lock. The ship cannot enter the lock until the gate is opened, and the gate should not be closed until the ship is completely inside.

Following φ_2 is another period of nonoverlap, t_{21}, during which the system is idle. The minimum clock period for correct operation is the maximum combinational delay, plus the maximum delay time and preset time, plus t_{21}. It is important therefore to make t_{21} as small as possible if one is designing for performance. As we show in the following section, t_{21} does serve a useful purpose of accommodating clock skew, a variation in the arrival time of the clock to different clocked storage elements, and so in some systems t_{21} can be reduced only as much as a clock skew allows.

The net result of this two-phase clocking scheme is that the clock period and its constituent epochs are, with static storage devices, involved in one-sided relations in which a region of reliable operation can always be found by making epochs longer. With dynamic storage devices, the relationship between clock epochs and refresh period does not practically limit choices of periods today but can be expected to rule out periods longer than a few hundred τ when the technology reaches its ultimate limits. The complementary relation on the propagation delay of the combinational logic is a simple maximum value.

Many variations on this basic scheme are found in different kinds of digital systems, more schemes than we could hope to describe individually. Many processors and storage systems are advantageously designed with more than two clock phases. In processors these phases (usually four or eight, and sometimes a variable number) delineate minor cycles that subdivide the major cycle. Multiple phases are also commonly used in systems that employ precharged pull-up, and other charge transport techniques such as CCD storage. Magnetic bubbles are made to move in response to a rotating magnetic field, a two-phase clocking scheme when viewed as two orthogonal fields with sinusoidal oscillation 90° out of phase with each other.

Some mention of variant forms with respect to inputs and outputs is also required. Inputs to synchronous systems must appear in synchrony with the clock, a requirement that is readily satisfied when the inputs are outputs of another system that shares the same clock. If an input does not assume its correct value until after the leading edge of φ_1, its worst-case delay relative to the leading edge of φ_1 is accounted for in the same way as the delay of the clocked storage element. The general procedure for checking compliance with timing bounds consists of marking nodes starting with clocked element outputs and system inputs with the latest time relative to the beginning of φ_1 that the signal will become stable. Programs to accomplish such checking are not trivial, as they may at first appear, because the program should account for different delays in different states and inputs. Otherwise, the large differences between delays for positive and negative transitions cannot be accounted for; nor can circumstances such as time of output determination in simple AND and OR circuits be dealt with simply, because these times are data dependent. Unfortunately, this checking problem is about as difficult as exhaustive simulation.

The form of finite-state machine model of synchronous systems used in the descriptions above of clocking schemes is the transition output (Mealy) machine. It is more general than the state output (Moore) machine since its outputs are functions both of its input and present state, while for the state output machine the output is a function only of the present state.[2] Transition output machines tend to be used in cascade arrangements in which economy is the principal design goal, while state output machines are characteristic of pipeline architectures in which high performance is sought.

Networks of transition output machines must be acyclic, the same requirement as for the combinational paths in a single machine. Combinational delays may accumulate on paths through many machines, so a more general checking procedure must be used. Networks of state output machines may be connected cyclically, and checking is confined to communicating pairs of machines. Checking can be made entirely local to each finite-state machine if communication between machines is performed in pipeline fashion through clocked elements. The finite-state machine described in Section 3.11 and shown in Figs. 3.21 and 3.22 is of this sort.

In a two-phase clocking scheme, φ_1 and φ_2 are symmetrical in the sequence sense, and perhaps also in time. This symmetry suggests that a reversal of roles between φ_1 and φ_2 is possible. For example, there is no reason in the finite-state machine structure shown in Fig. 7.5 not to replace the simple amplifier or double inverter with combinational logic. The general structure allowed in a two-phase clocking scheme is any composition of basic elements consisting of registers followed by combinational logic, in which outputs of elements clocked by φ_1 drive only inputs to elements clocked by φ_2, and vice versa. Please note that combinational delays must be checked across communicating pairs of machines. This scheme is similar to that used in both the OM2 and its controller and is well illustrated in Chapters 5 and 6.

7.3 CLOCK DISTRIBUTION

Readers who have been designing systems with catalog parts are now invited to stand up and object: "What's with all of these clock phases? My systems use a single-phase clock." That is a good question.

The clock supplied externally to catalog parts such as registers, counters, shift registers, and microprocessors is most often a single-phase clock because this approach is convenient, and the clock then uses only a single package pin. Internally, a two-phase clock, or its functional equivalent, is derived from the single-phase clock, either as part of each clocked element with the single-phase clock distributed through the chip or once for the chip with the derived two-phase clock signals distributed as required.

Figure 7.6(a) shows a relationship between a single-phase clock and a two-phase clock derived from it, and Fig. 7.6(b) shows a circuit that performs this function. In the following text this form will be taken as canonical, although other conventions of relating single-phase and two-phase clocks are possible. The single-phase clock is used in a trailing-edge-triggering discipline, so-called because system state changes occur following the trailing edge of the clock pulse. This mode of operation has the advantage over leading-edge triggering that the state variables are stable during the clock pulse, so one can perform logical operations between the clock and state variables in order to derive gated clock signals for selective register loading.

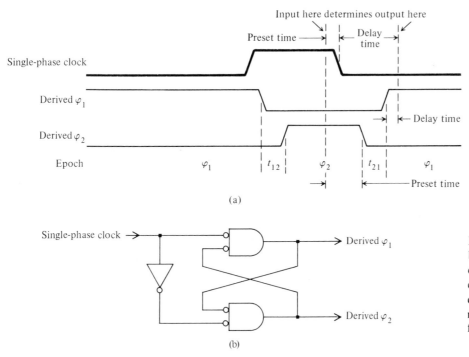

Fig. 7.6 (a) Relationship between a single-phase clock and a two-phase clock. (b) Logic circuit to derive a two-phase nonoverlapping clock from a single-phase clock.

The preset and delay time relations to the derived two-phase clock can be translated, as shown in Fig. 7.6(a), to be referred to the trailing edge of the single-phase clock. A study of this figure shows that, because of delays in deriving φ_1 and φ_2, preset times referred to the single-phase clock may be negative. Over a set of clocked storage elements the preset times will have maximum and minimum values, just as delay times will. The maximum preset time is called the *setup time*, and the minimum preset time, with its sign reversed, is called the *hold time*. In single-phase clocking arrangements, the input to a clocked storage element should become stable by the setup time before the clock edge and hold this value until at least the hold time after the clock edge. In the two-phase clocking arrangements most often used in MOS LSI, the preset time is always positive since it is referenced to the trailing edge of φ_2, so only its maximum value or setup time is critical to correct circuit operation.

It is an essential feature of clocking schemes with one-sided relations that there is a critical period during which the clocked storage element is actually storing two bits of information. For either single- or two-phase clocking this critical period begins at the preset time and ends at the delay time. One of the stored bits is the input value presented before the preset time; the other is the bit presented at the clocked storage element output. This critical period provides a built-in tolerance for clock skew. This term refers to a variation in the effective arrival time of the clock at different clocked elements. These variations may be due to a combination of several effects: different threshold voltages, signal propagation delays on wires, or variation in element delays such as is found when gated clock signals are used to control register loading.

Clock skew even within a chip can be a problem. The time required for signals to propagate on wires in MOS technologies is not limited by "speed-of-light" considerations, except for the very shortest wires, for which these times are negligible anyway. Rather, it is the resistivity of wires, together with their parasitic capacitance, that determines the rate at which the voltage driven onto a wire at one point will equalize across the length of the wire. This process is governed by a diffusion equation, and so it is referred to here as the *diffusion delay*. As was indicated in Section 1.11, the diffusion delay is approximately quadratic with length. The diffusion delay is independent of line width, since a wider line has lower resistance but proportionately higher capacitance per unit length. The term *wire delay* in what follows refers to the combined effects of the velocity of electromagnetic wave propagation and diffusion *within the wire itself* and should not be confused with the delay that the parasitic capacitance of a wire induces on the output of a logic circuit.

As was pointed out in Section 2.7, propagation of signals on poly lines, as they are of fairly high resistance, is a process for which the delays are not negligible. Given the distances and need for short delay in clock distribution, use of diffused wires for carrying the clock more than short distances is not recommended either. For example, the diffusion delay in wires 10 mm long is calculated, by the method presented in Section 1.11, and using the typical 1978 MOS electrical parameters given in Table 2.1, to be about 200 nsec in 50 $\Omega/\square$ poly wires and about 100 nsec in diffused wires. The diffusion delay over the same 10 mm distance on a metal line is calculated to be only 0.1 nsec.

As λ is scaled down, the resistance per unit length of a conductor scales up quadratically, since it is made not only narrower but also thinner in order to maintain the same relative surface flatness. Capacitance per unit length stays constant in scaling, since reduced capacitance due to decreased width just balances the increased capacitance per unit area due to thinner oxides. Thus in scaling λ down by a factor of ten from 3 microns to 0.3 microns, one can expect the diffusion delay to scale up by a factor of 100 for a wire of the same length in microns, or to remain constant for a wire of the same length in λ units. A metal line 10 mm long in this scaled technology would incur a delay of about 10 nsec, about 300τ for this 0.03 nsec transit time technology. In τ-relative terms, the metal conductors are much like today's poly or diffused wires.

Since the diffusion delay is quadratic with length, the delay can be reduced by placing repeaters at intervals along a long line. However, if clocks are to be distributed with a delay or skew of at most a few τ, clocks cannot be distributed over areas even approaching that of a chip. The area over which clocks could be distributed with tolerable skew does represent a lot of circuitry, more than an entire chip in 1978 technology. Process technology might be able to relieve this diffusion delay problem in clock distribution and the metal migration problems in power distribution by providing two or three relatively thicker additional layers of metal.

Clock skew is a serious and common problem in systems built from "families" of SSI and MSI circuits. The origin of the problem lies not entirely in clock distribution, but also in the manufacturers' inattention to preset and delay time bounds. It is easy to find examples in popular families of catalog parts of certain pairs of clocked elements in which the hold time (minimum negative preset time) specified for one part is greater than the minimum delay time of another part. Although the system may work if delays are "typical," some fraction of a production run can be expected not to work, or worse, may fail intermittently in service. Designers of families of circuits intended and advertised to work together should strive to make all preset and delay time characteristics consistent within a family, as variations from one part to another decrease the tolerance of the system to clock skew.

Let us return now to the typical MOS integrated system with a two-phase clock distributed as the signals φ_1 and φ_2. These are generally "popular" signals, second only to VDD and GND, and the capacitance accumulated in their distribution and gate connections is considerable. It is possible, of course, to drive φ_1 and φ_2 directly onto the chip from an external driver. Systems operating at the extremes of performance of the MOS technology benefit from this approach. The external clock driver can be made to switch fast, and to a higher voltage than VDD, so that transmission gate outputs can then transit all the way to VDD.

Because the saturation current varies as $(V_{gs} - V_{th})^2$, there is a large speed advantage on the chip from a higher voltage clock drive.

The performance benefits gained from driving φ_1 and φ_2 onto the chip from fast off-chip drivers are achieved at the expense of external components. Although on-chip clock drivers may limit the performance of a system slightly, they are generally used in the interests of economy. An integrated system that is large by today's standards may have a clock load on the order of $10^4 C_g$, where C_g is the gate capacitance of a minimum dimension transistor. The techniques developed in Section 1.5 for driving large capacitive loads are applicable to clock driving. The total delay in an exponential driving structure with the optimum fan-out of e, starting from a minimum energy signal to drive $10^4 C_g$, is only 25τ. However, the last stage of the driver would be impractically large; the gate width would be $2 \cdot 10^4 \lambda/e$. The trade-offs between performance and area may dictate a large fan-out for the final stage of the clock driver, say about 40, with smaller fan-outs for the drivers preceeding it. Most of the delay in this driver structure would be in the final stage.

A clock driver of this sort is illustrated schematically in Fig. 7.7(a), with size ratios indicated by the delay times. Waveforms expected for this circuit are shown in Fig. 7.7(b). The clock is assumed to originate from somewhere off the chip. There is no reason to use two pins for this function where one would serve, so the canonical scheme for production of a two-phase clock from a single-phase clock is used. The capacitance of the pin and package are so large compared to C_g that there is no point in starting from a minimum energy signal. The clock driver shown in Fig. 7.7(a) presents a load of about $30 C_g$. If the clock source were an on-chip clock generator, more stages would be used. The slight asymmetry between φ_1 and φ_2 is of course due to the inverter at the input.

The reader should take careful note of an important characteristic of this clock driver; namely, that nonoverlap of the clock phases is assured independent of clock loading. This desirable characteristic is the reason that it is the clock phases, rather than the NOR gate outputs, that are fed back in a cross-coupled fashion. The nonoverlap periods can be reduced somewhat at the expense of silicon area by cross-coupling at every stage. This same circuit trick can be extended as shown to assure nonoverlap independent of clock loading for gated clocks as well. The techniques used in these clock driver circuits to assure the existence of the nonoverlap period independent of the effects of loading on element timing are much in the spirit of those techniques used in speed-independent and self-timed systems, in that the circuit adapts its temporal behavior to conform to a sequencing constraint.

A question is sometimes raised as to whether clocks for different sections of a chip should be buffered separately, as is often done at a circuit board level for systems built from collections of circuit boards. On a physical basis, the answer is no. It is best for two reasons that the clock lines be common throughout the chip: (1) this approach minimizes clock skew, and (2), since one must pay the same

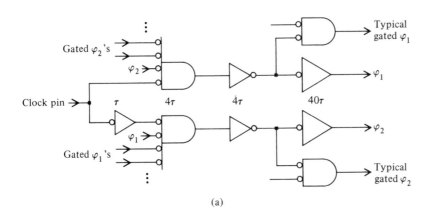

(a)

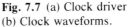

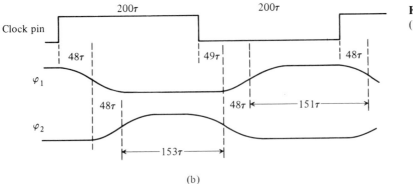

(b)

Fig. 7.7 (a) Clock driver (b) Clock waveforms.

transistor area whether the driver is lumped or distributed, optimum driver design locates the *stray* capacitance of the clock distribution wires where the largest signal energy is available, which is *after* the final driver. There may be organizational reasons for distributing the final driver stage to locations close to sections being driven when some logical operations are performed on the clock signal.

7.4   CLOCK GENERATION

Where great clock accuracy is required, or where a system clock is distributed to many circuits, the clock-generation task is external to the chip and is not the direct concern of its designers. The clock will ordinarily originate from an electronic oscillator circuit, the period of which is typically controlled by a crystal or some other resonant network. Process variation in integrated circuit fabrication does not allow accurate resonant networks to be fabricated by usual means, but it is perfectly feasible, indeed essential for self-contained VLSI systems, to generate clock signals on the chip. It is best in approaching this subject to forget about electronic oscillator circuits and instead to take a more basic approach originating with an understanding of what clocks are for.

As we have mentioned before – and this is a principle that bears repeating on every opportunity – the role of the clock in a synchronous system is to connect sequence and time. The interval between clock transitions, whether these transitions are on one wire or distributed over several wires, must be such as to permit enough time for the activities planned for that interval. When viewed in this way, a clock is more like a set of timers than like an oscillator. A *model* of the temporal behavior of the systems being clocked is built into the clock generator in the choice of times for the various timers.

The easiest way to build these timers is as chains of inverters. The propagation delay time of such a chain will of course vary with τ, according to the way in which the fabrication process, aging, temperature, and power voltage affect τ. However, these variations only make the inverter chain a better model of the system being clocked than a fixed timer would be, since on the same piece of silicon these variable factors are nearly the same for the clock and for the system.

It is helpful to distinguish between the two kinds of timers shown in Fig. 7.8. The first is called a *symmetrical delay*, because the propagation delay for positive

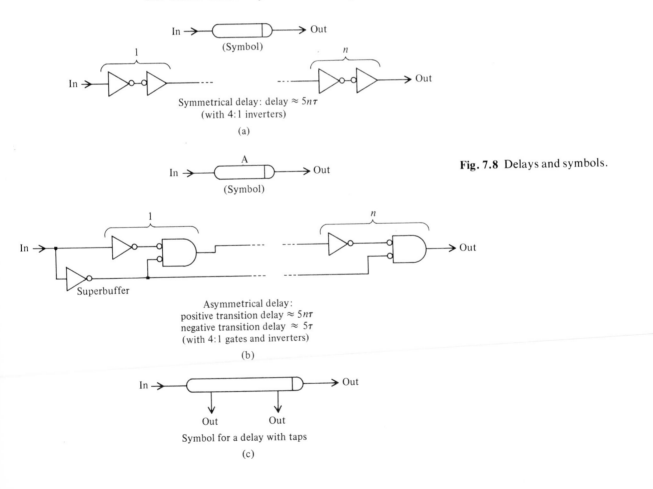

In → (Symbol) → Out

Symmetrical delay: delay $\approx 5n\tau$
(with 4:1 inverters)

(a)

In → A (Symbol) → Out

Fig. 7.8 Delays and symbols.

Superbuffer

Asymmetrical delay:
positive transition delay $\approx 5n\tau$
negative transition delay $\approx 5\tau$
(with 4:1 gates and inverters)

(b)

In → Out

Out   Out
Symbol for a delay with taps

(c)

and negative transitions at the input is about the same. The second is a logic network designed to produce as asymmetrical a delay as possible. A negative transition propagates through the delay in about 5τ independent of length. A complementary form of *asymmetrical delay* is also possible, but to simplify the figures and symbology in what follows, we shall use only the form in which a low input resets the delay and a positive transition propagates slowly. The symbols shown in the figure allow for *taps* at various points along the delay.

Clocks that employ these delays as timers are all elaborations of the *ring oscillator* circuit shown in Fig. 7.9(a). Rings of an odd number of inversions have no stable condition and will oscillate with a period that is some odd submultiple of the delay time twice around the ring. The oscillation of the largest period may eventually predominate following bringing power on, but the erratic clock signals produced during power-up could leave the system in a peculiar state. It is much better to produce an initialization signal that is held high during power-up and to use it to initialize the state *and the clock*. In the modification of the ring oscillator shown in Fig. 7.9(b), clock signals are suppressed during initialization and will start im-

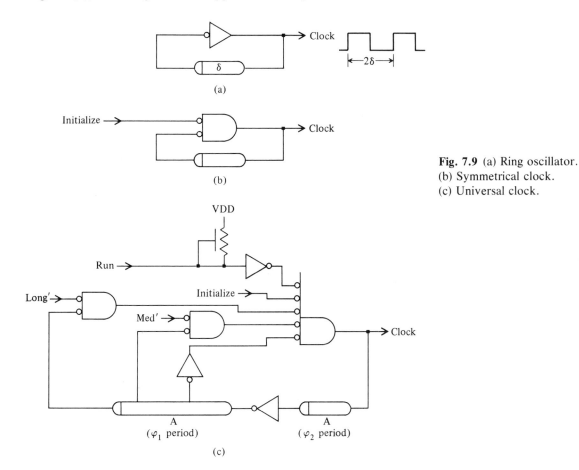

Fig. 7.9 (a) Ring oscillator. (b) Symmetrical clock. (c) Universal clock.

mediately following the negative transition of the Initialize signal. This circuit produces a symmetrical single-phase clock that can be converted to a two-phase clock by the circuit shown in Fig. 7.7(a).

Although the clock circuit shown in Fig. 7.9(b) would be adequate for some applications, many elaborations can be included that are shown together in Fig. 7.9(c). The width of the single-phase clock pulse, related to the φ_2 period, is determined by one *asymmetrical* delay, while the interval between clock pulses, related to the φ_1 period, is determined by an independent *asymmetrical* delay. Another feature of this universal clock is that it allows the system being clocked to select between a variety of periods, which can be changed on a cycle-by-cycle basis according to the combinational delay of the operation performed on that cycle. In order to visualize how this works, note that following the trailing edge (negative transition) of the clock signal, any high input to the 5-input NOR circuit has the effect of preventing the occurrence of the next clock pulse. The usual default case is with the Long', Med', and Run high, and Initialize low, resulting in a short period determined by the first tap on the period delay. If a decoding of the state indicates that a longer period is required for that cycle, the Med' or Long' lines must be driven low before the short default period is elapsed. If for example the Long' line were low, the period before the next clock pulse would be stretched to that determined by the full delay. Of course, this scheme may be generalized to any number of delay taps. Signals such as Med' or Long' can be derived either from function coding of the combinational sections whose modeled delay they match, or as microcode bits.

The Run line is a bus intended to generalize this cycle-stretching feature so that any part of the system being clocked may stop the clock synchronously and then permit it to restart asynchronously. If this cycle is ever to be stretched to more than the refresh time, static storage elements must be used (at least for the φ_1 part of the cycle for two-phase clocking). This technique of control over the clock is the basic mechanism exploited later in this chapter to allow asynchronous communication between synchronous systems.

7.5 SYNCHRONIZATION FAILURE

Jean Buridan (1295–? after 1366), a fourteenth century French philosopher often cited as a percursor of Isaac Newton for his priority in giving a technical definition of kinematic terms such as inertia and force, posed in his commentary on Aristotle's *De caelo* a paradox; that a dog could starve if placed midway between two equal amounts of food. The unfortunate creature placed in this position would be equally attracted to each source of food, that is, in a position of equilibrium. One twentieth century explanation of this paradox—if indeed it is a paradox at all—is that the structure consisting of the dog and the two sources of food is and behaves just as any other structure that can store a bit of information.

The analysis of the electrical behavior of cross-coupled circuits presented in Section 1.14 and developed in physical terms in Section 9.6 applies also to the situation described by Buridan. The equilibrium condition either for the dog or for the cross-coupled circuit is unstable, as any displacement from equilibrium brings about forces that tend to destroy rather than restore the equilibrium condition. An unstable equilibrium of this sort is called a *metastable condition*. Buridan was correct in believing that the dog could starve, since a characteristic of a metastable condition is that it *may* persist indefinitely. A functional definition of metastability applied to cross-coupled circuits is the occurrence under undriven conditions of an output voltage in a *range* around V_{inv} that cannot reliably be interpreted as either high or low.

A bistable element in a self-contained synchronous system never has the opportunity to reach a metastable condition, since satisfaction of the timing constraints assures that the output is driven to a voltage outside of the metastable range. But is any system really self-contained? A system such as a microprocessor may be entirely synchronous internally but cannot extend this synchrony indefinitely to encompass all of the external world with which it may interact. If asynchronous signals of external origin are allowed to enter a synchronous system as ordinary inputs, the timing constraints required to assure correct operation cannot be satisfied, since there is no known relationship between the timing of the asynchronous inputs and the clock.

Figure 7.10 illustrates in a small fragment of a larger synchronous system the consequence of ignoring synchronization altogether. Even if one employs a model of the clocked storage elements as having perfectly discrete outputs, the unequal delay in the paths from the asynchronous input X to the clocked storage elements allows the inputs to the clocked storage elements during state A to represent an illegal successor state for a period following a transition of the asynchronous in-

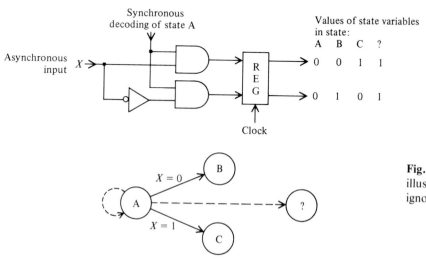

Fig. 7.10 Fragment of a logic circuit illustrating the consequence of ignoring synchronization.

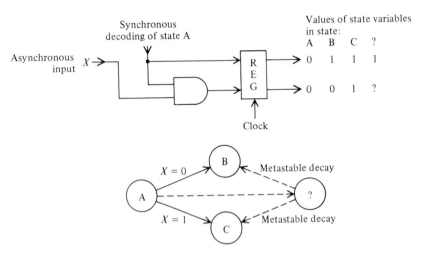

Fig. 7.11 Fragment of a logic circuit illustrating synchronizer that would work perfectly, if perfectly discrete bistable devices were possible.

put. If the clock happens to capture the inputs during this transitory period, one of the illegal state transitions shown in dashed lines on the state diagram will result.

So, a slightly smarter thing to do is to ensure that only one clocked storage element is affected by a given asynchronous input. A clocked storage element that is used in this way is called a *synchronizer*, since it is intended to produce an output signal that is in synchrony with the clock. Figure 7.11 shows a redesigned version of the previous fragment of a circuit modified in its state coding so that the asynchronous input X affects the input to only one clocked storage element. If it were only possible to build perfectly discrete bistable devices−indeed, if perfect discreteness exists in nature (see Chapter 9)−, this scheme would be perfectly reliable. Unfortunately, there is some probability of *synchronization failure*, since a transition of the asynchronous input at certain times relative to the clock will leave the synchronizer in a metastable condition, and the time required for the clocked storage element to get out of a metastable condition is unbounded. During the period in which the synchronizer output remains in a metastable condition, the logic cannot discriminate between states B and C, and if the condition persists for too long, an illegal or incorrect successor state can result.

It is part of the interesting history of the subject of synchronization that even long after the necessity to synchronize asynchronous inputs was recognized as a standard part of good engineering practice, the faith of logic designers in the discreteness of the outputs of clocked storage elements was so great that the very existence of synchronization failure was widely denied. Another curious aspect of the sociology of the problem is the many schemes proposed to "solve" the problem, which only move it to another location in a system or reduce its probability. The inevitability of the problem can be understood by seeing that perfect solutions require that a discontinuous function be produced, where physics is continuous. We suspect that synchronization of an input signal to a free-running clock cannot be accomplished with perfect reliability with finite circuits. At best, a synchronous system with a clock such as that shown in Fig. 7.9(c) permits a metastable synchronizer condition to be detected and used to postpone the next clock, thus synchronizing the system to the signal.[3,4] Synchronization failure was discovered independently by numerous researchers, designers, and engineers in the 1960s, some of whom published reports of their analyses or observations.[5,6,7,8,9,10] The work done at the Computer Systems Laboratory of Washington University by Thomas J. Chaney and Charles E. Molnar,[11] and by Marco Hurtado[12] has provided convincing demonstrations of the existence and fundamental nature of the problem.

It is fairly easy to estimate the probability of a synchronization failure with a simple mathematical model. If one observes that a bistable device is in a metastable condition at some time t, what is the probability that it will have left this condition by time $t + \delta$, in the limit as δ approaches zero? The only answer to this question that seems reasonable is that the probability is proportional to δ; let it be $\rho\delta$. This *assumption* produces a simple model in which the exit from a metastable condition is a Poisson process of rate ρ, and the probability that a clocked storage element will remain in the metastable condition, once in it, for a period D or longer is $e^{-\rho D}$. The prediction of this simple model has been verified experimentally and is consistent with analyses based on circuit models,[12] including the analysis presented in Section 1.14. The parameter ρ depends on circuit characteristics. A dynamic storage element is not an acceptable synchronizer, as the time evolution of its output in undriven conditions is possibly even *toward* the metastable region (see Section 9.6). One can identify ρ in the analysis in Section 1.14 with $1/\tau_0$. The time evolution of the voltage output of the cross-coupled circuit has the effect of transforming a uniform probability distribution of initial conditions to an exponential or Poisson distribution of exit events. For ratio logic with a ratio of r, τ_0 is about equal to the pair delay, $(r + 1)\tau$.

In order to estimate the probability of a fault in a synchronous system due to the nonvanishing probability that a synchronization will take longer than some bounded time, one must also calculate the probability that a synchronization event will put the synchronizer into a metastable condition. For most synchronizations, the asynchronous level to be synchronized will transit sufficiently far away from the time at which it is sampled that the clocked storage element will be overdriven in the usual way. Only over a rather narrow time aperture, denoted here as Δ, does the occurrence of a transition result in the synchronizer taking more than the usual delay time of the clocked storage element. The boundaries of this aperture are not sharp but may be treated as such, so that for a particular frequency of transitions of the asynchronous signal, f, the *probability* that a metastable condition will be produced in a single synchronization event is $f\Delta$. One may take this relation as the definition of Δ.

The overall probability of a system failure at each synchronization event, $f\Delta e^{-\rho D}$, depends on ρ and Δ, which are parameters of the clocked storage element used as a synchronizer; on D, which is a parameter of the synchronous system in which the synchronizer is used; and on f, which is a parameter of the asynchronous input signal. D is the time allowed in the synchronous system for the *decay* of the probability of metastability and is effectively like a delay. It corresponds to the *excess* delay allowed from clocked storage element outputs to inputs (see Fig. 7.5) and, even for zero combinational delay, cannot exceed the clock period less delay and preset times.

In order to get some feeling for the failure rates involved, consider a synchronous processor that is accepting data from, or sending data to, a disk storage unit at a 1 MHz rate. An asynchronous signal alerts the processor to the presence of, or need for, a new data item, but the processor is able to clear this signal synchronously. We can assume that ρ is about $1/(5\tau)$ for ratio logic with $r = 4$. If almost all of a fairly short 100τ clock period were available for the decay of the probability of metastability, D would be about 80τ. It is interesting that when D is expressed as a multiple of τ, the exponent in the formula is then independent of τ and so is independent of scaling circuit dimensions. The scaled down version of this system would allow less time for the decay of the probability of metastability, but the synchronizer would exhibit a proportionately higher metastable exit rate.

Experimental determinations of Δ for nMOS circuits[13] and experiments performed with several bipolar circuit families indicate that Δ is a small fraction of τ, say about $\tau/10$. This estimate agrees with the notion that Δ corresponds to the time required for a signal to transit through a small voltage range around the switching threshold, a time which is proportional to τ. For present values of τ, Δ is then approximately 30 picoseconds, and the probability of a system failure for a single synchronization event would be about $(10^6)(30 \ 10^{-12})e^{-16}$ or about $3 \cdot 10^{-12}$. So, about one in each $3 \cdot 10^{11}$ items transferred across this interface would be in some fashion mistreated.

Failure *rate* depends on the frequency at which the system samples asynchronous inputs. This frequency cannot be greater than the clock frequency and, as is clear from a careful study of Fig. 7.11, this frequency may be as low as the frequency of the states whose choice of successor depends on the asynchronous input. This figure is intended to suggest how synchronizers can be *sheltered* from needless synchronization events, as the synchronizer is here sheltered by ANDing the asynchronous input with a signal that indicates that the system is in state A. Synchronizers that are used directly on asynchronous input signals and whose outputs enter PLA or ROM structures may cause failures even when the system is in a state in which the successor does not depend on the asynchronous input.

For the disk example above, the synchronous processor must sample every transition of the asynchronous input in order to transfer every data item. If this processor were engaged in transferring data at a 1 MHz rate about one third of the time, synchronization failures would occur at a (Poisson) rate of once each 10^6 seconds, or about every 10 days. It is worth noting that the exponential relation in ρD makes the failure rate remarkably sensitive to ρD. This dependence may be particularly noticeable if the clock signal originates off-chip so that ρD depends on τ. A chip with a slightly larger than typical τ may exhibit a drastically higher than typical failure rate.

The probabilistic character of synchronization failures makes them exceedingly difficult to trace. Designers of synchronous systems who wish to avoid the curses and plagues that are the just reward for those that build secret flaws into human tools and enterprises should cultivate a rational conservatism toward this problem. The worst-case failure rate for a design should be calculated. If the failure rate is higher than some criterion, it can be reduced by techniques that increase D. Use of cascaded synchronizers is one technique for increasing D that does not require increasing the system clock period. Criteria for acceptable failure rates depend on many of the same economic and social factors that influence other aspects of system reliability.

One conservative failure rate criterion that can be supported by a physical argument is that the rate of synchronization failures should be on the order of the rate at which the bistable synchronizer will change state due to the random thermal motions of the electrons. This rate is shown in Section 9.7 to be $(1/\tau)e^{-(E_{sw}/kT)}$. If s is the frequency of synchronization events (often $s = f$), this physical criterion is

$$sf\Delta e^{-\rho D} = sf\Delta e^{-D/(r + 1)\tau} < (1/\tau)e^{-(E_{sw}/kT)}.$$

Solving for D yields

$$D > [E_{sw}/kT + \ln(sf\Delta\tau)](r + 1)\tau.$$

If one takes Δ as about $\tau/10$ and s and f in the order of $1/(100\tau)$, the second term in the brackets can be ignored, since the switching energy for today's circuits is about $10^8 kT$. Bistable devices are so reliable today that the time required to allow the probability of metastability to decay to achieve the same reliability is very large—about $5 \cdot 10^8\tau$, or 0.15 seconds! However, this criterion scales in a remarkable way. The ratio D/τ, which represents the number of transit times the criterion provides for the metastable exit, scales with the switching energy, which goes down as α^3. This scaling should not be interpreted as meaning that smaller devices have a higher probability of metastable exit per transit time. Rather, smaller transistors result in less reliable storage devices, which makes it possible to lower one's standards. Ultimately small transistors with channel lengths of about 0.25μ would allow circuits with a switching energy of about $10^4 kT$. Because of the significant subthreshold currents at the low threshold voltages implied by this scaling, CMOS circuits with $r = 1$ would have to be used. At these minimum dimensions, this criterion implies $D > 2 \cdot 10^4\tau$, and taking $\tau = 0.02$ nanoseconds, $D > 400$ nanoseconds. So, at this ultimate limit of MOS technology, one cannot disregard synchronization failure, but one would not expect it to limit designs for synchronization rates up to 1 MHz or so. Since this criterion represents the most conservative position that can still be rationally defended on physical grounds, it is known as the *Mead Criterion*.

7.6 SELF-TIMED SYSTEMS

The operation of a synchronous system is reminiscent of soldiers marching to the commands of a drill sergeant. The temporal control over a collective activity is centralized in a single authority, and the soldiers respond to known commands that are synchronous with the marching cadence. Lockstep control results in a particularly simple form of organized behavior that people seem to associate with the relentless efficiency of machines. However, lockstep control is certainly not the only way to coordinate the collective activity of many participants, nor is it particularly efficient unless the tasks of the participants are very well matched.

Self-timed systems are patterned on quite a different image of organized activity, one in which the temporal control is delegated to the participants. If one were to try to construct a mental image of self-timed behavior, it would be one in which an airplane could not depart until after all the passengers scheduled for the flight had gotten on board. One tries to assure that all system events occur in proper *sequence*, but nothing ever has to occur at a particular *time*.

Here, in essence, is how this trick is accomplished. Self-timed systems are interconnections of parts, which are called *elements*. Time and sequence are related *inside* elements, so that events such as signal transitions at the terminals of an element may occur only in certain orders. Elements can be thought of as performing computational steps whose *initiation* is caused by signal events at their inputs, and whose *completion* is indicated by signal events at their outputs. The sequencing of the computational steps is determined by the way in which elements are interconnected. The time required to perform a computation is determined by the delays imposed by the elements between initiation and completion, and by interconnection delays.

The term *sequence* is used here in its most general sense, to indicate an ordering that may be either total or partial. The ordering relation between occurrences of signal events is denoted as $x \leqslant y$, read as *x precedes y*, for which one might also say either that *x is before y* or that *y is after x*. These orderings represent sequencing rules, such as initiation *precedes* completion, or the arrival of the operands *precedes* the operation, and are brought about by circuit behaviors such as input change *precedes* output change, or in general, as cause *precedes* effect. Although the occurrences of all system events can be expected to be ordered with some others, highest performance and efficiency are usually achieved in designs in which ensembles of machines may perform relatively independent parts of a computation concurrently (see Chapter 8). Occurrences that are not ordered with each other by virtue of their relative independence are denoted as $x \nleqslant y$, read as *x is concurrent with y*. It is convenient to treat the relation $\leqslant$ formally as a partial ordering, so that it is reflexive: $x \leq x$; antisymmetric: $x \leq y \cap y \leq x \rightarrow x \equiv y$; and transitive: $x \leq y \cap y \leq z \rightarrow x \leq z$. The relation $x \equiv y$ in the definition above means that x and y are *identical* occurrences, not simultaneous. The notion of *simultaneity* is specifically regarded as meaningless and disallowed by the antisymmetric property of $\leq$.

7.6.1 Equipotential Regions

The physical meaning of these sequence domain relations must be interpreted with care, since an element necessarily has some physical extent. According to the principle of relativity, relations between occurrences of events at different points in space may be interpreted inconsistently by observers in different locations. In a medium such as a chip, on which related signals are carried on wires whose routing and relative delay may be uncertain, a relation that holds in a region close to where it is created may fail to hold elsewhere. It is necessary here to introduce an approximation to avoid complicating the discussion of self-timed signaling more than is justified by the actual situation. This approximation is analogous to applying the classical limit over volumes sufficiently small to approximate a point in a relativistic space.

Over small areas that are here called *equipotential regions*, one may treat a signal as identical at all points on a wire. This approximation is justified so long as the area is sufficiently small that the delay associated with equalizing the potential across any wire is small in comparison with switching delays or signal transition times. This approximation is roughly equivalent to an assertion that related occurrences are known to be sufficiently separated in time in comparison with wire delays that the relation will be observed to hold from any point of observation within the region.

How large should an equipotential region be? It must be admitted that there are no hard and fast answers to this question. The choice adhered to here is a determination of the maximum wire length within an equipotential region for which wire delays may be disregarded and do not introduce any performance limitations. It is generally not possible to design elements so that the related occurrences of signal events are separated by less than a few τ, so our choice of the maximum size for an equipotential region is that it be small enough that the potential on any wire within this area will equalize in less than τ. This condition places an upper limit on wire length within an equipotential region, generally different for the various layers, rather than a limit on area. For those wires that cross the boundary of an equipotential region, the limit applies only to the length within the region. Also, the limit is not absolute. Longer wires may be used if one is willing to account for their delay as if they were active elements, by treating different points on a long wire as the source of different signals.

Wire delay has already been discussed in connection with clock distribution. The factor that limits the way in which a wire approximates an equipotential now and in the future of MOS technology is the diffusion delay. A single chip is today a good approximation of an equipotential region, as defined here to make all wire delays less than one transit time, so long as there are no metal wires longer than about 17 mm, diffused wires longer than about 500μ, or poly wires longer than about 300μ.

As λ and τ are scaled down, the diffusion delay for a given wire length measured in λ units remains constant but in τ-relative terms scales up linearly. The distance in λ units over which the potential will equalize in one transit time accordingly scales down as the 1/2 power of the scaling factor. The maximum area of an equipotential region, which is some measure of the amount of circuitry that can fit in an equipotential region, scales down roughly linearly. Suppose that λ and τ were scaled down by a factor of 10 from today's technology. A signal on a metal wire will travel only a few hundred microns in one transit time. This maximum wire length is adequate for communication with negligible delay in an amount of circuitry that is comparable to only about one tenth of one of today's chips, and over about one thousandth of the area of the chip in the scaled technology.

In other technologies with smaller transit times and wires that behave more like transmission lines, such as Josephson junction device technology, the limiting factor on the area of an equipotential region is transmission line delay. The notion of an equipotential region can be defined also for systems of chips interconnected on circuit boards. For typical transition times on package pins of many nanoseconds, a circuit board up to a meter or so square is a good approximation of an equipotential region. However, communication delays in pin-driving structures are so large compared with internal transition times that circuitry in two different chips may not be in the same equipotential region.

The approximation of the equipotential region was introduced to assure consistent physical meaning for the relations that hold within it. It is necessary that an element be contained entirely within at least one equipotential region, but as is illustrated in Fig. 7.12, an element may reside in more than one equipotential region, or more than one element may be included in a single equipotential region. The set of equipotential regions of a system does not partition the area in which a system is constructed, as it would if we required these regions be disjoint. All that is required is that the set of equipotential regions *cover* all of the elements of a

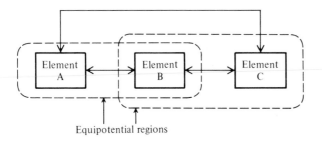

Fig. 7.12 Cover of equipotential regions.

system. Elements A and B, and also B and C, in Fig. 7.12 are related by a reflexive, symmetric, and intransitive compatibility relation.

The equipotential region is a formal expression in the self-timed discipline of the importance of mapping logical into physical *locality* and exerts a necessary and interesting influence on self-timed signaling and on the trade-off between the size and design difficulty of elements.

7.6.2 Self-timed Signaling Specifications

It is helpful for understanding self-timed signaling to be aware that the relations maintained at the terminals of an element or system are of two kinds: (1) relations that constrain the element to certain sequence domain behaviors, and which are called *functional* relations or constraints: and (2) relations that constrain the sequence domain inputs to those that the element is designed to accept, here called *domain* relations or constraints. There is an explicit duality here between the specifications of what the element is to produce at its outputs in response to inputs, its function, and the specifications of what the element requires at its inputs in response to its outputs, its domain. While explicit, this duality is hardly unique to the self-timed discipline. For example, the delay and preset times of a clocked storage element are functional and domain specifications, respectively. The terminology comes from that in mathematics for a function mapping from its domain. It is inevitable that physical devices used for particular purposes must restrict their domain by rules of usage.

It is characteristic of all self-timed signaling that elements or systems engaged in a communication are interconnected on at least one closed signal path. It is the closed-loop character of the signaling that allows domain constraints to be satisfied, for example, that an element or system will not be called upon to initiate an operation until after it has indicated that it has completed its previous operation. The functional constraints, because they are also sequence domain relations, allow an element to impose an *arbitrary delay* between the occurrence of input and output signal events. Relations between the occurrence of input and output signal events of an element are called *closed-loop relations*, and if in the form input ≤ output are functional, and if in the form output ≤ input are domain. *Open-loop relations* are either between the occurrence of two output signal events of an element, and are also functional, or between the occurrence of two input signal events of an element, and are also domain.

Self-timed signaling within or between equipotential regions differs precisely in the utility of open-loop relations. Self-timed communication between elements that reside in the same equipotential region is somewhat easier to accomplish than communication between elements not in the same equipotential region, because a sender may produce open-loop orderings at its outputs with assurance that these orderings are preserved at the inputs of the other elements in this equipotential region. For example, in one common self-timed signaling convention, the occurrence of a new data value on a set of wires at the sender precedes a transition on another wire indicating that the data are defined. Data-validity information in this case costs only a single wire. Communication between elements that are not in the same equipotential region, such as between elements A and C in Fig. 7.12, is more expensive. Any open-loop relation produced at the sender is not preserved once the signals leave its equipotential region. The once-ordered occurrences of signal events are concurrent at the receiver. Multiple bits cannot be sent in parallel, but only concurrently, and special encodings must be used to imbed data-validity information with data. One such encoding is illustrated in an example in Section 7.6.4.

In signaling between equipotential regions, an element sending data receives an indication that the data reached its destination only by signals that traverse a closed path back to the sender to initiate another operation. Since this closed path includes one or more elements, which may introduce arbitrary delay, it is clear that such signaling conventions accommodate arbitrary delay in wiring as well. The reader may wish to summarize his or her understanding by noting that the self-timed discipline tends to extremes in modeling wire delay. Within an equipotential region wires incur *negligible delay*, so that all relations produced anywhere in the region hold everywhere, and *equipotential self-timed signaling conventions* may be used. Wires between equipotential regions may incur *arbitrary delay*, no open-loop relations are preserved, and *delay-insensitive self-timed signaling conventions* must be used. These two forms of signaling conventions are functionally similar in their role of communicating data and differ only in details of encoding. Simple elements exist for translating between these forms.

7.6.3 Organizational Discipline

The physical aspects of the self-timed discipline discussed above are very closely meshed with an organizational discipline. As was described in the first section of this chapter, the bifurcation of the discipline is deliberate. The physical side is concerned with elements and the physical requirements on signaling and communication. The organizational side is concerned with systems of interconnected elements and the logical requirements on signaling.

The decision to confine time metrics to the interior of elements leads by physical argument to equipotential regions, and to the requirement that an element be contained entirely within an equipotential region. As a consequence, the scale of an element is restricted to be small enough physically that the timing behavior is simplified by negligible wire delays, and to be small enough logically to assure that the complexity of the design is manageable. These restrictions make it reasonable to expect that elements can be designed so that their physical and logical function can be certified to be correct.

Even given correctly functioning elements, the system designer is responsible for satisfying the two types of specifications whose duality was discussed above.

The system must perform its specified function as the collective behavior of its interconnected elements, but it must do so without violating the rules of usage of the elements. In the discussion of synchronous systems, the rules of usage were satisfied through timing and topological constraints. In a self-timed system, timing constraints are absent at the system level. Correct operation of a self-timed system composed of correctly functioning elements can then depend only on satisfying a topological constraint on the interconnection of elements.

The following definition serves as the basis for the interconnection of self-timed elements or systems and provides a rigorous framework for certifying correct sequence domain functioning of a system given only correct sequence domain functioning of its elements:

Definition. A self-timed system is either (1) a self-timed element, or (2) a legal interconnection of self-timed systems.

This recursive definition may be applied either as a rule of construction, by which a system satisfying this definition is built up from elements or systems, or as a rule of decomposition, by which a system can be analyzed for compliance with the constructive rules or for faults.

It is curious that the same organizational principles that here seem to follow from physical necessity—constraints on the size of elements to those that are managable, and constraints on their interconnection—are the same as others have arrived at about a decade ago through a different line of reasoning. The framework established by this definition is similar to that used by Dijkstra et al[14,15] for the structured programming discipline. The discipline described here in different terms is somewhat more ambitious, in that it seeks to include physical as well as algorithmic correctness, and is more open-ended, in that the set of legal interconnections is not limited to Dijkstra's pearls. The difficulties experienced in the combinatorics of scale for large programs suggests that logical complexity alone will make the design of VLSI systems sufficiently challenging, even without physical and timing complications. The organizational principles for self-timed systems are no more restrictive than those used successfully to organize large programs.

7.6.4 An Example

The problem of exposition that Dijkstra refers to in *Structured Programming* is evident here as well, that the objects of discussion, that is, very large systems, are beyond the scope of textbook examples. It is hoped that the following extended example can be extrapolated by the reader, will fill in some conceptual holes in a section in which a lot of ground was covered quickly, and will provide one demonstration that sets of elements, legal interconnections, and signaling conventions exist and are physically realizable.

This example is of the self-timed counterpart of a binary parallel adder. The general structure of the adder, illustrated in Fig. 7.13 for the addition of two 3-bit binary numbers yielding a 4-bit sum, is the usual unilateral iterative realization. Each cell is a "full adder," a circuit that forms the sum (exclusive-or) s_i and carry (majority) c_{i+1} functions of three inputs, the two bits a_i and b_i from the binary words to be added and the carry c_i from the previous cell. Although this particular network is illustrative of a performance characteristic of self-timed systems and is undoubtedly familiar to most readers of this book, in some of the following discussion a general result will be derived that is not dependent on the details of the cell functions and interconnection.

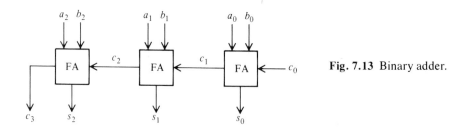

Fig. 7.13 Binary adder.

Let the cells be self-timed elements, each in its own equipotential region. The following convention of delay-insensitive signaling will be used. Each signal line has three conditions: zero (0), and one (1), and undefined (–). Transitions between 0 and –, or 1 and –, are allowed and transitions between 0 and 1 are disallowed. This ternary signaling scheme can be implemented physically either by three voltage ranges representing 0, –, 1, the undefined range being in the middle, or by a *double-rail code* in which two wires carry the codes 00 for undefined, 10 for zero, or 01 for one. In the double-rail code, which will be used here to illustrate the design of the element, the two wires used to represent the variable x are denoted as x^0, read as "x is zero," and x^1, read as "x is one." Since transitions between the 0 and 1 conditions are disallowed, the two wires are never in the process of switching at the same time.

The general signaling scheme for the elements, called with no particular significance to the reader the "weak conditions," is illustrated in Fig. 7.14. There is no reference to the switching functions produced by a full adder, since this is a signaling scheme that can be used for any combinational element or system. Any implementation of the full adder that satisfies these sequencing constraints will serve. This set of relations includes implicitly that inputs and outputs may become defined concurrently. The orderings labeled (1), (2), (4), and (5) in the figure are functional constraints, while the orderings labeled (3) and (6) are domain constraints. Notice that the second half of the full cycle, represented by the relations (4), (5), and (6), serve no purpose other than to return the entire length of every signal wire (pair) to its initial undefined condition. This example of delay-insensitive signaling is of a category variously referred to as return-to-zero, Mul-

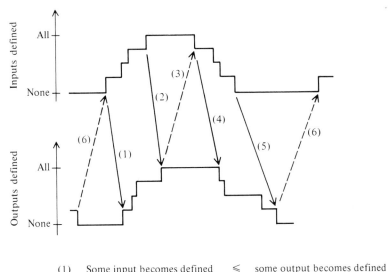

Fig. 7.14 Weak conditions.

(1)	Some input becomes defined	≤	some output becomes defined
(2)	All inputs become defined	≤	all outputs become defined
(3)	All outputs become defined	≤	some input becomes undefined
(4)	Some input becomes undefined	≤	some output becomes undefined
(5)	All inputs become undefined	≤	all outputs become undefined
(6)	All outputs become undefined	≤	some input becomes defined

ler, or 4-cycle signaling. The time and switching energy required to return the signals to their initial condition is circumvented in the category of signaling referred to as nonreturn-to-zero, transition, or 2-cycle signaling, but at the expense of additional circuitry and some difficulties with system initialization.

So, this example is concerned with a particular class of elements and systems, which will be abbreviated as CL for "combinational logic" that satisfies the weak conditions. Now, within the framework of the recursive definition of a self-timed system, it is necessary to say precisely how CL's may be interconnected, that is, what is a "legal interconnection." As tedious as this may seem at first, the discovery and certification of an interconnection as "legal" is just like the discovery and proof of a theorem. Fortunately, once stated and proved, such a theorem may be used over and over again in different designs. The following theorem is a particularly nice one for two reasons. First, it is broad enough to have wide application. Second, it is a demonstration of *closure*, or invariance, of a signaling convention under an interconnection rule. Some theorems are in the form "This system has property X if it is an element with property X, or if it is interconnection A of systems with property Y." In a closure theorem, X and Y are the same property.

The "weak conditions" theorem is stated as follows:

A CL is either a CL element, or is a finite set of U of CL's interconnected such that (1) each input of the CL's in U is either driven from an output of the

CL's in U or is an input to the interconnection (no dangling inputs); (2) each output of the CL's in U either drives an input of the CL's in U or is an output of the interconnection (no dangling outputs); and (3) there are no closed signal paths.

This theorem is of course applicable to the adder shown in Fig. 7.13 and means that this interconnection of elements that satisfy the weak conditions itself satisfies the weak conditions. Implicit in the theorem statement are also the domain or rule-of-use constraints, that if the domain constraints of a CL are satisfied, then they are satified also for the CL's of which it is composed in the interconnection specified.

This is not a difficult theorem to prove, but it is fairly lengthy. The critical section of the proof centers on relations (2) and (5). To prove that relation (2) is satisfied for the interconnection, take the set U_1 of CL's to which undefined inputs to the interconnection are connected. At least one output of each U_1 is undefined. Either these outputs are outputs of the interconnection, thus satisfying relation (2), or else they are inputs to a set U_2 of CL's. U_1 and U_2 are disjoint by the acyclic requirement, and U_2 is not empty. Define $\hat{U}_n = \hat{U}_{n-1} \cup U_n$ and proceed by induction. If $\hat{U}_n = U$, then the undefined outputs of U_n are outputs of the composition.

Why go to such lengths to prove that the systems one designs have certain properties? Suppose you had a 32-bit adder composed of self-timed full adder elements. It is not practically possible to assure oneself by combinatorial methods that this interconnection satisfies any sequence domain relations. Interconnection rules based on theorems of legal interconnections are not a matter of going to lengths, but of taking a shortcut, and serve also to structure the design process in a hierarchic way. Of course, this approach rests initially on being able to demonstrate by any means at one's disposal that the elements of which the system is composed satisfy certain sequence domain constraints. The combinatorics of dealing with systems as compositions of FET's and wires is already fairly difficult at the level of elements, let alone VLSI chips, both because of the number of parts and their relatively complex physical behavior. Elements define an enclosure in which the physical complexities inside are abstracted to the simplest timing concept on which system design can be based: sequence.

Figure 7.15 is a PLA-like design for the full adder element of our example. The gate labeled "C" is a Muller C-element, a very handy device for the design of self-timed elements. Its output becomes 0 when all of its inputs are 0 and becomes 1 when all of its inputs are 1, and otherwise the output stays in whatever condition it was. The pair of C-elements is required to satisfy the requirement of the weak conditions that at least one output of the element remains defined until all the inputs have become undefined. This is a remarkable circuit for the reason that correct sequence domain behavior at its terminals is absolutely independent of the speed of its gates, so long as wire delay is negligible. Circuits of this sort are called *speed-independent*, a discipline of digital system design developed in the 1950s by David Muller,[16] and from which much of the self-timed discipline has evolved.

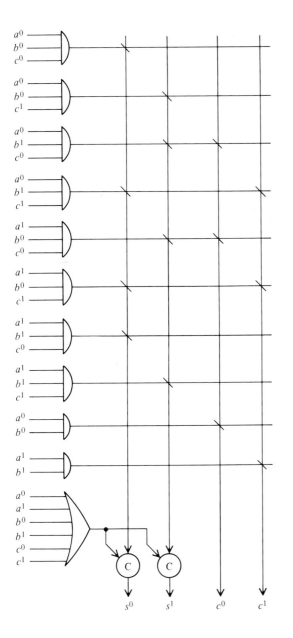

Fig. 7.15 Full-adder element.

Another thing to notice about this element is that the carry-out function c_{i+1} is generated as soon as the operands a_i and b_i become defined if they are in either the 00 (carry-kill) or 11 (carry-generate) conditions. The time required to perform an addition is generally limited by the worst-case carry propagation through the entire length of the adder. However, this case occurs only rarely. For operands chosen at random, it can be shown that the average maximum number of consecutive stages in the 01 or 10 (carry-propagate) condition is bound by $\log_2 n$, where n is the word length. So, the adder based on this element will have the sum completely defined in an average time that varies as $\log_2 n$ after the operands become defined. On average, this is a very fast adder, faster even than can be achieved in MOS technologies by carry-skip and carry-lookahead schemes. Some of the interest in "circuits that generate completion signals"[17] is motivated by the greed for speed.

7.7 SELF-TIMED SIGNALING

The most elementary signal event that can be used to compose self-timed signaling conventions is a transition. A "pulse," an occurrence of a level for some fixed *time* interval, is not a satisfactory elementary signal event, since regardless of what width is chosen for the pulse, the possibility exists that some element could be made that is so slow that it could not detect a pulse of the chosen width. Communication conventions with the same closed-loop character as self-timed signaling, referred to in the colloquial of hardware designers as "handshaking" conventions, often fail to provide a clean interface definition precisely because they mix time and sequence concepts. One may expect self-timed signaling conventions to be described exclusively in terms of sequence domain relations between the occurrences of signal transitions.

A distinction has already been drawn between *equipotential* and *delay-insensitive* signaling conventions, which is only one axis of variability. The goal of this section is to describe briefly a distinction based on the number of transitions associated with each use of an element.

Since there are time and energy costs with driving a transition onto a wire, it pays to use as few transitions as possible in self-timed signaling conventions. Ignoring data wire transitions for the moment, it is clear that there must be at least two transitions for each operation performed by an element, one to initiate the operation, and carried on a wire generally labeled Request, and one to indicate completion of the operation, and carried on a wire generally labeled Acknowledge. These occurrences must alternate. Suppose that the system were initialized so that all request and acknowledge wires are in the zero state. An operation in progress is then indicated by the request/acknowledge pair for an element or system being in opposite states, and an element or system can indicate completion by driving its acknowledge wire to the same state as the request wire.

Figure 7.16 illustrates the sequence domain relations for this signaling scheme, which is variously called transition, 2-cycle, or nonreturn-to-zero (NRZ) signaling. This diagram includes the necessary open-loop relations between input data and Request, and between output data and Acknowledge. Relations indicated by solid arrows are functional constraints, and those indicated by dashed arrows are domain constraints. Crosshatched areas of input and output data denote the intervals in which data values may be changing; otherwise, data are stable and defined.

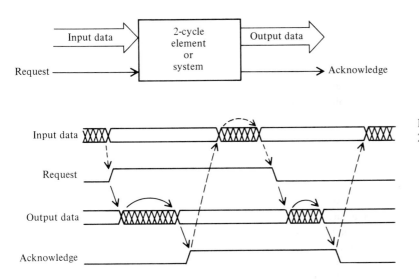

Fig. 7.16 (Equipotential) 2-cycle signaling.

The 2-cycle signaling scheme illustrated is directly related to a delay-insensitive functional equivalent form in which the request is imbedded in the data. The idea here is similar to the double-rail code, in that two wires are required for each data bit; however, a transition is driven into one wire, or the other, to indicate a zero or a one. All four states of the two wires are used, and there is no equivalent ternary form on a single wire. Although the request signal is distributed across many wires, the signaling convention is still effectively 2-cycle, since to the environment of the element the set of transitions travels only a single trip to and from the element. The 2-cycle scheme that satisfies the weak conditions follows only the first three of the six relations depicted in Fig. 7.14.

The good news about 2-cycle signaling is that it is as fast and as energy-efficient as possible. There is some bad news, however. Since logic devices tend to be sensitive to levels or to transitions in a particular direction, the detection of transitions that may occur in either direction requires some extra logic and state information in each element. Most transition logic designs use a lot of exclusive-or tests. For example, in the delay-insensitive form, the parity of each double-rail pair is 1 for the first piece of data, 0 for the second, 1 for the third, and so on. It is not possible to know what wire changed without some storage of the pair's previous condition, but it is possible to tell when the transitions have arrived by detecting a reversal in the parity of all pairs.

The only real alternative to 2-cycle signaling is a form first discovered by Muller and used in many of his examples of speed-independent circuits. It is sometimes referred to as Muller signaling, or by the terms 4-cycle or return-to-zero (RZ) signaling. This form has already been well illustrated in its delay-insensitive form in the example in the previous section. The return-to-zero character of 4-cycle signaling tends to result in very simple and natural circuit implementations but requires twice as many transitions as 2-cycle signaling, and requires signals to make two trips to and from an element for each operation. Whenever wire delay is a substantial fraction of the operation time, this extra trip is a serious performance penalty. However, when wiring delay is relatively small, it can be argued that speed-independent elements require the return-to-zero part of the 4-cycle to rid themselves of the information stored in gate and wiring delays in preparation for the next cycle. There are many organizational approaches for concealing this return-to-zero interval by making it concurrent with other operations. For the sake of completeness, one form of equipotential 4-cycle signaling is illustrated in Fig. 7.17.

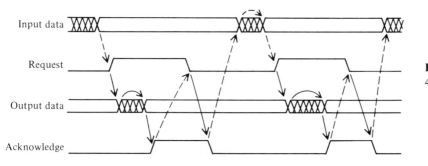

Fig. 7.17 (Equipotential) 4-cycle signaling.

Given the advantages and disadvantages of each form of signaling, one suspects that 2-cycle signaling will be used exclusively for long distance communication, particularly from chip to chip. A general one-to-many bus transmission scheme called the TRIMOSBUS[18] is a good example. The circuit economy of 4-cycle signaling favors its use in local communication, particularly with speed-independent elements.

7.8 SELF-TIMED ELEMENTS

The study of the design of self-timed elements is an extensive subject and is depicted here only by a series of illustrations of elements in nMOS technology.

7.8.1 Muller C-element

The Muller C-element, introduced in the circuit shown in Fig. 7.15, is also sometimes called a "rendezvous," "join," or "last-of" circuit. It is a bistable device that provides an action similar to hysteresis, in that its output becomes 1 only after *all* of its inputs are 1, and becomes zero only after *all* of its inputs are zero. A 2-input C-element can be made by connecting the output of a 3-input majority circuit back to an input, as shown in Fig. 7.18, both diagrammatically and as an nMOS circuit. A many-input C-element can be made either by an extension of the series and parallel structures in the circuit shown, or by the connection shown in

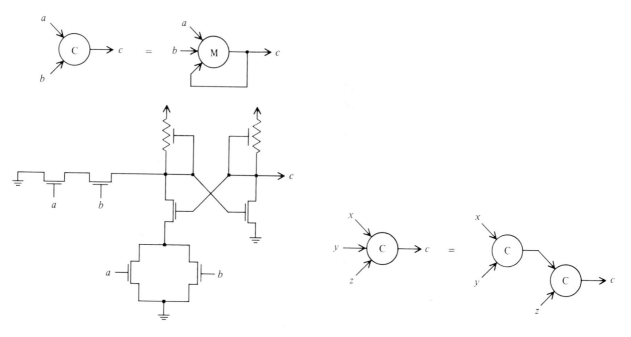

Fig. 7.18 Muller C-element.

Fig. 7.19 Associativity of C-element function.

Fig. 7.19. In the domain in which C-element inputs transit only to the condition complementary to the output, the sequential function of the C-element is commutative and associative like the combinational functions AND and OR.

When used by itself as a self-timed element, the C-element serves as a "rendezvous" or "join" for producing a request signal after all of a number of acknowledge signals are received, and does this job for either 2-cycle or 4-cycle signaling, and for either positive or negative logic conventions. When used as a component of a self-timed element, such as in the circuit of Fig. 7.15, the C-element serves as a latch that responds to the "last of" a set of signals changing in the same direction.

7.8.2 Combinational Elements

The style of combinational element shown in Fig. 7.15 has the advantages of speed-independence and operates directly on inputs in the double-rail code. Variants on this circuit with precharged pull-ups are an excellent fit to nMOS technology.

Where economy in area is more important than the advantages of speed-independence, one can use the very simple scheme illustrated in Fig. 7.20. The addition of a delay to any ordinary combinational net converts it to a self-timed combinational element operating on single-rail data. If 2-cycle request/acknowledge signaling is used, the delay should be symmetrical, and for 4-cycle signaling, asymmetrical. Depending on the data-validity coding scheme employed, the outputs may need to be loaded in latches by a signal produced as Request and not Acknowledge. The reliability and certification of this device as a self-timed element depends upon how accurately and conservatively the delay models the timing behavior of the combinational net.

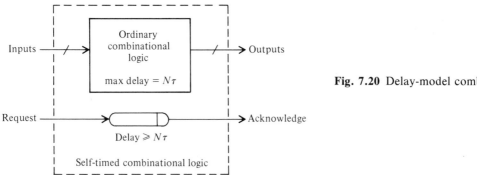

Fig. 7.20 Delay-model combinational element.

The most straightforward method for implementing speed-independent nets with single-rail data is to include the Request signal in every AND in the AND-plane and to use the conversion element described below at the output. Similarly, the delay-model style of implementation with double-rail data can use conversion elements at input and output.

7.8.3 Conversion elements

Figure 7.21 shows elements for converting between 4-cycle forms of single-rail plus request equipotential signaling and double-rail delay-insensitive signaling. Part of the element on the right can be recognized as an extended C-element, which for rendezvousing very large numbers of bits will require intermediate buffering.

The circuit of Fig. 7.22 provides a method for another form of conversion, to transfer data from an on-chip equipotential region to a larger τ circuit board equipotential region. The circuit functions by comparing the desired and actual states of the signals at the pads and produces an acknowledge signal only after they match. The delay in this circuit is required to allow time after data arrival for the comparison circuit to indicate inequality. The interval imposed between the Request and Acknowledge varies both according to the capabilities of the drivers and the package-pin loading. This technique can be used, for example, to drive data onto circuit board buses that are terminated by negative resistance termination to the switching threshold, such as the circuit of Fig. 7.3(a). After the transfer is acknowledged, the drivers can be disabled and a transition driven on another pin to indicate availability of the data.

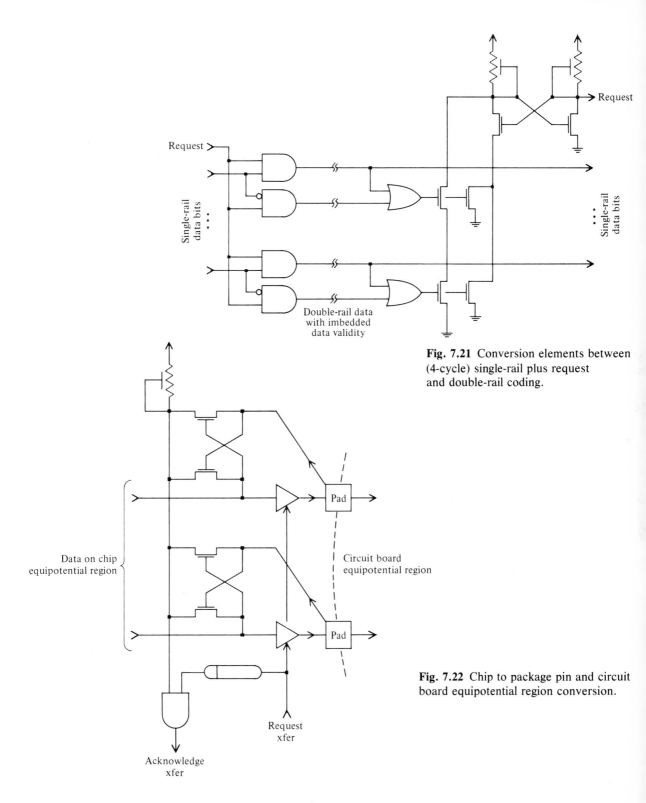

Fig. 7.21 Conversion elements between (4-cycle) single-rail plus request and double-rail coding.

Request

Single-rail data bits

Single-rail data bits

Double-rail data with imbedded data validity

Data on chip equipotential region

Circuit board equipotential region

Pad

Pad

Request xfer

Acknowledge xfer

Fig. 7.22 Chip to package pin and circuit board equipotential region conversion.

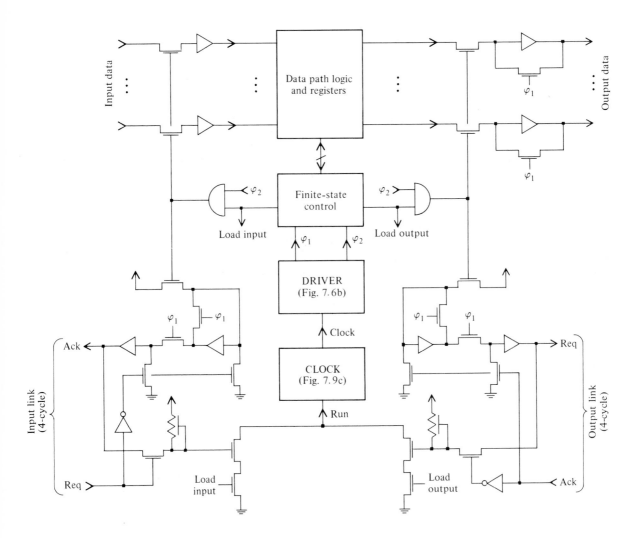

Fig. 7.23 Pipeline processor element implemented as an internal clock synchronous system.

7.8.4 Synchronous elements

Self-timed elements can be designed as synchronous systems with an internal clock. If this clock is of the variety shown in Fig. 7.9(c), in which the clock can be stopped synchronously and restarted asynchronously, the design can also be made free of synchronization failure. This scheme has been used in proprietary designs by the author since 1968 and is the subject of an excellent 1976 paper by Pĕchouček.[3] It is not unnatural to imbed synchronous systems as elements in self-timed systems. If reasonable constraints on the size of an equipotential region, hence of an element, are adhered to, the equipotential environment assures that performance and correct operation will not be compromised by wire delay in signals or in clock distribution.

An example of this style of design applied to a pipeline processing element is shown in Fig. 7.23. This element includes multiple request/acknowledge pairs, or *links,* one or more associated with its input data and one or more with its output data. One can think of the finite-state control as first obtaining an input word, then performing some sequence of operations on it, and finally placing the result in its output register. The way in which this synchronous system interacts with its asynchronous links is the main point of this example and involves a curious flow of control through the clock. In the clock step, say, in which the input register is to be loaded, the input Req that indicates the presence of input data is not tested directly. If it is 0, or if the Ack signal is still 1, the system stops in its φ_1 epoch. Notice that the Run signal becomes 0 in known time relation or in synchrony with the clock, so there is no problem of synchronization failure in the clock circuit. When the conditions Req and not Ack are achieved on the input link, the clock restarts asynchronously, loading the input data on the φ_2 cycle, then acknowledges receipt of the data at the beginning of the next φ_1 cycle. The remainder of the 4-cycle is accomplished asynchronously by the circuit shown. The output circuitry operates in an analogous fashion. Dynamic input registers can generally be used, since the synchronous system can complete its operation on the input data in a known time interval. However, as is shown in Fig. 7.23, static output registers are required since there is no bound on the time the element may spend in either its input or output loading steps.

7.8.5 Queue (FIFO) Elements

In pipeline processes in which operation times are variable, increased throughput can be achieved by interconnecting the processing elements through queues. A queue could be designed as a synchronous element on the same general plan as illustrated in Fig. 7.23. However, much faster and more area-efficient queues such as shown in Fig. 7.24 can be made with simple asynchronous controls. The inner cell is intended to be replicated as many times as the number of words the queue is to be able to store, and the same control will operate a queue of any word length. This design is by no means speed-independent, but it is illustrative of the opposite extreme of asynchronous design. This queue uses 3/2 rules, which means that one may expect misoperation if particular sets of 3 gates have a smaller cumulative

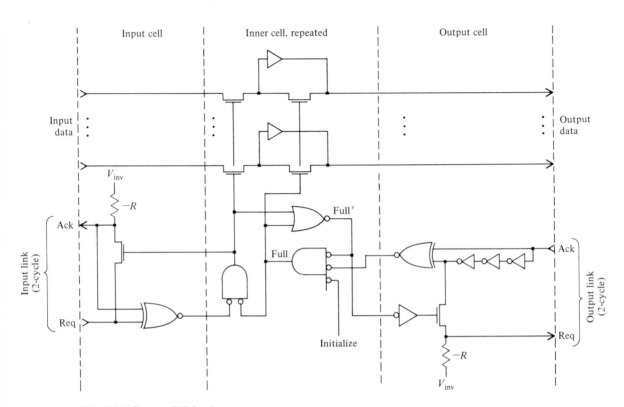

Fig. 7.24 Queue (FIFO) element.

propagation delay time than other sets of 2 gates. Circuits of this sort must be certified by careful analysis and simulation based on parameters calculated from circuit layouts, much as is the case with dynamic storage cells, ratioless shift registers, and other designs in which relative sensitivity of operation to layout and circuit values is accepted in order to achieve density and speed.

7.8.6 Interlock element

The interlock element, shown in Fig. 7.25, operates on two control links, represented by Req1/Ack1 and Req2/Ack2, which follow the 4-cycle signaling convention. While the requests may occur concurrently, the acknowledges are restricted to be mutually exclusive; hence, the alternative name of this device is a *mutual exclusion circuit*. Interlocks are used either alone or as components of more elaborate elements called *arbiters* to allow multiple processes to access a single shared resource such as a large random-access store.

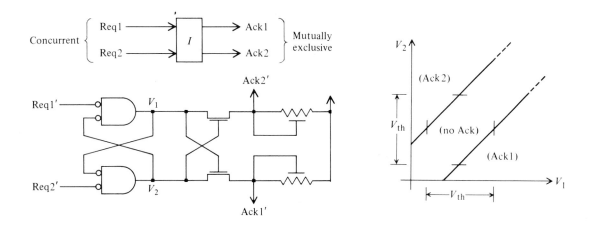

Fig. 7.25 Interlock element (4-cycle).

Circuit operation is most easily visualized starting with the neither-requesting input condition, both inputs high, V_1 and V_2 both near 0 volts, both outputs high, and then tracing circuit operation from one input changing to low. Since the requests may occur concurrently, the cross-coupled NOR structure can "hang" for an indeterminate period in a metastable condition. However, both outputs will remain high until one NOR output differs from the other by greater than V_{th}.

ACKNOWLEDGEMENTS

In connection with the exposition of the self-timed discipline, the technical contributions and encouragement of Anatol W. Holt, Wesley A. Clark, Charles E. Molnar, Ivan E. Sutherland, Dan Cohen, Martin Rem, and my students in the courses at the University of Utah in 1972 and at Caltech starting in 1977, is gratefully acknowledged. Drafts of this chapter have benefited from the scrutiny of Carver Mead, Lynn Conway, Bob Sproull, Ivan Sutherland, Dan Cohen, Charlie Molnar, Carlo Séquin, Dick Lyon, Alan Bell, Chuck Thacker, and Wayne Wilner. Support for research at Caltech in self-timed systems under an ARPA contract, N00123-78-C-0806, is gratefully acknowledged.

REFERENCES

1. Charles L. Seitz, "Self-timed VLSI Systems," *Proceedings of the Caltech Conference on VLSI*, January 1979.
2. Zvi Kohavi, *Switching and Finite Automata Theory*, New York: McGraw-Hill, 1970, Chapter 9.
3. Miroslav Pěchouček, "Anomalous Response Times of Input Synchronizers," *IEEE Transactions on Computers*, vol. C-25, no. 2, February 1976.

4. M. J. Stucki and J. R. Cox, Jr., "Synchronization Strategies," *Proceedings of the Caltech Conference on VLSI*, January 1979.

5. H. J. Gray, *Digital Computer Engineering*, Englewood Cliffs, New Jersey: Prentice-Hall, 1963, pp. 198–201.

6. Ivor Catt, "Time Loss Through Gating of Asynchronous Logic Signal Pulses," *IEEE Transactions on Electronic Computers*, vol. EC-15, no. 1, February 1966, pp. 108–111.

7. W. M. Littlefield and T. J. Chaney, "The Glitch Phenomenon," Technical Memorandum #10. Washington University Computer Systems Laboratory, St. Louis, Missouri, December 1966.

8. G. R. Couranz, "An Analysis of Binary Circuits Under Marginal Triggering Conditions," Technical Report #15, Washington University Computer Systems Laboratory, St. Louis, Missouri, November 1969.

9. C. L. Seitz, "Graph Representations of Logical Machines," MIT Ph.D. thesis, January 1971, preface.

10. T. J. Chaney; S. M. Ornstein; and W. M. Littlefield, "Beware the Synchronizer," COMPCON-72 IEEE Computer Society Conference, San Francisco, California, September 1972.

11. T. J. Chaney and C E. Molnar, "Anomalous Behavior of Synchronizer and Arbiter Circuits," *IEEE Transactions on Computers*, vol. C-22, no. 4, April 1973, pp. 421–422.

12. M. Hurtado, "Dynamic Structure and Performance of Asymptotically Bistable Systems," Washington University D.Sc. dissertation, 1975.

13. Thomas J. Chaney and Fred U. Rosenberger, "Characterization and Scaling of MOS Flip-Flop Performance in Synchronizer Applications," *Proceedings of the Caltech Conference on VLSI*, January 1979.

14. O.-J. Dahl; E. W. Dijkstra; and C. A. R. Hoare, *Structured Programming*, New York: Academic Press, 1972.

15. E. W. Dijkstra, *A Discipline of Programming*, Englewood Cliffs, New Jersey: Prentice-Hall, 1976.

16. Raymond E. Miller, *Switching Theory*, vol. 2, New York: Wiley, 1965. Chapter 10 is a review of David Muller's work on speed-independent circuits and includes a bibliography.

17. Stephen H. Unger, *Asynchronous Sequential Switching Circuits*. New York: Wiley, 1969.

18. Ivan E. Sutherland; Charles E. Molnar; Robert F. Sproull; and J. Craig Mudge, "The TRIMOSBUS," *Proceedings of the Caltech Conference on VLSI*, January 1979.

8 HIGHLY CONCURRENT SYSTEMS

8.1 INTRODUCTION

How can the properties of VLSI be exploited to build computational structures? Our discussion to this point has concentrated primarily on principles for structuring circuits and wires on the chip rather than on the application of VLSI to solve interesting computational problems. Although the OM example described in Chapters 5 and 6 shows an elegant use of the structuring principles in the design of a conventional processor, we are left with an intriguing question: Does VLSI offer more than inexpensive implementations of conventional computers?

This chapter answers the question with a resounding YES! Because both processing elements and memory elements can be easily implemented in VLSI, we are encouraged to find structures that use a great deal of *concurrency*—a large number of calculations occurring at the same time. Although we can clearly design VLSI structures that have many sites at which processing is performed, how are these structures to be applied? Some applications may require different sorts of concurrent processing than others. Are there any principles or theories that will guide us in the design of highly concurrent systems? (For an introduction to the promises and problems of VLSI and concurrency, see Sutherland and Mead.[1]) Unfortunately, we lack experience in designing systems of this sort. As a consequence, this chapter can offer no complete designs that have been applied in real system applications. Instead, we offer several glimpses of the possibilities available with VLSI, and of its limitations.

The chapter is organized into four quite separate sections; although they are designed to be read sequentially, they may also be read concurrently! Section 2 reviews the problems that conventional computer designs present when implemented in VLSI and summarizes efforts to achieve concurrency in general-purpose computers. Section 3 takes up a particular sort of concurrent organization, the array of identical processors, and shows its application to matrix arithmetic. Section 4 examines hierarchically organized machines—in this case, machines structured as a binary tree—and demonstrates how they can be programmed to perform several tasks. Finally, Section 5 presents a nascent theory of planar computational structures. It links the topological and electrical properties of VLSI elements to the structure of computations.

8.2 COMMUNICATION AND CONCURRENCY IN CONVENTIONAL COMPUTERS

The architectures of conventional computers suffer from two difficulties that we try to avoid when designing VLSI computational structures. (1) A processor is separated from its memory by long communication paths such as buses. These buses are long enough to slow substantially the transmission of information between a processor and memory. (2) The "von Neumann machine" provides only a single processor that sequentially fetches and executes instructions; it offers few opportunities for concurrent processing activity. In this section, we survey some of the attempts to reduce communication costs and to use several processors concurrently. Although designs using a great deal of concurrency have been cumbersome to implement in the past, VLSI makes these designs considerably more attractive because of the ease with which memory and processing elements can be placed in close proximity.

Human organizations, like computer organizations, suffer if communication costs are high or if concurrent processing cannot be exploited. In fact, a human brings to an organization what VLSI brings to a circuit: both combine processing and memory effortlessly! Analogies with human structures may help to suggest the kinds of behavior we might achieve in computational structures.

Humans struggle to reduce communication costs, because the cost is often measured in large quantities of time. Consider a student assigned to write a research paper that requires the use of a large library. Each time the student needs to consult a book, he or she could make a trip to the library, climb into the stacks to retrieve the book, read a few relevant paragraphs, and replace the book. Then the student would head home to write the sentence that depends on the information just acquired. Both libraries and people recognize the inefficiency of this approach and allow students to borrow books. The student will take several dozen books home and store them on a handy short shelf. Now the communication cost required to find information is reduced, provided the item lies within the group of books selected. A student who finds it difficult to select a small number of sufficient books may move to a carrel in the library to work, again in order to reduce communication costs with the large library "memory." The human strives to keep the information supply close to the processing task.

Concurrency is widely exhibited in human organizations. Henry Ford introduced the production line as a way to exploit concurrency in a well-understood manufacturing process. This is a particularly simple structure, in which information and goods flow rigidly along the production line. A more prevalent, general-purpose approach to concurrency in organizations is the *hierarchy*: the president of a company supervises several subordinates, each of whom in turn supervises a like number of sub-subordinates, and so forth until we reach the lowest level workers.

Two goals of the hierarchy are to keep everyone about equally busy and to allow adequate information flow in the organization. A supervisor must generate enough commands to keep several subordinates busy; otherwise it would not be possible to build large organizations at all. In addition, each subordinate requires a certain amount of attention from the supervisor. These requirements limit the number of subordinates who can be assigned to a single supervisor—ten underlings can run the most diligent supervisor ragged. Supervisors gather information to make decisions by querying their subordinates. In a badly organized hierarchy, supervisors may confer frantically with their superiors to find answers needed for crucial decisions. Meanwhile workers stand idle, waiting for directions from above. While it is not possible in general to have all needed information available from one's own subordinates, concurrent systems require this *locality* property to reduce interference from too much communication.

The design of computers and of algorithms has yet to show the ingenuity reflected in human organizations. This failure is not for want of cleverness in designers, but rather because the technologies used to implement computers are much less flexible than the human beings used to implement corporations. VLSI offers more flexibility than earlier technologies because memory and processing structures can be implemented with the same technology, and in close proximity.

8.2.1 Communication Costs in Computers

The archetypal computer consists of a single "processor" (the CPU or "central processing unit"), connected to a large, homogeneous memory (Fig. 8.1). The processor fetches an instruction from memory, decodes it, executes it, and repeats the cycle. Many instructions will cause additional references to memory in order to fetch operands or to store results. The performance of such a computer depends critically on the speed with which memory can be accessed.

Figure 8.1

A very simple argument can be developed to determine the speed of the memory. If a memory of M bits is implemented on a single chip in a two-dimensional array, wires approximately $M^{1/2}$ long are required to transmit data between a memory cell and the processor. (We are concerned with relative units of length and time because we intend only to compare different designs, not to determine absolute execution speeds.) The time required for data transmission is proportional to this length: the longer the wire, the greater the distance the signal must propagate and the greater the wire's capacitance, slowing propagation. In addition to slowing the memory, long wires also consume a great deal of chip space and require substantial power to drive. In present implementations of large computers, performance is further decreased by the several levels of packaging required to provide a memory of significant size: chip, printed-circuit board, backplane. The wiring on chips and printed-circuit boards grows as $M^{1/2}$, but backplane wiring grows linearly with memory size.

The organization shown in Fig. 8.1 is also rather wasteful of resources since most of the memory and memory wiring is idle most of the time. For a typical large memory, M might be $32*10^6$, but only a 16- or 32-bit word will be delivered to the processor with each memory reference. If the memory is organized as an array of 10^6 bits for each bit in the word, only 2 of the 2000 wires needed to address the array are used in a given reference (1000 select wires running horizontally, and 1000 data wires running vertically). Vast areas of memory thus lie idle because the amount of information extracted on a single reference is small compared to the size of the entire memory. The costs of communication are thus exorbitant in today's computers. Most of the expense, time, and energy required to compute are consumed by the communication of data over large distances.

8.2.1.1 Memory Locality

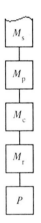

Computer designers have recognized the difficulty of communicating with a very large memory and have taken steps to utilize the memory more effectively. The result is a *memory hierarchy*, outlined in Fig. 8.2. The processor communicates with a series of memories, whose size increases and speed decreases as they become farther from the processor. The closest memory (M_r) provides high-speed "registers" or "accumulators" that are used very frequently, usually to contain intermediate results of arithmetic calculations. Next comes "cache" memory (M_c), designed to hold data and instructions that are referenced frequently. The "primary" memory (M_p) is similar to the large memory of several million bits (Fig. 8.1). Finally, a "secondary" memory (M_s) of some sort is provided, usually implemented with disks.

Figure 8.2

The average time required to reference a memory element will depend on which piece of the memory hierarchy holds the desired element. The intent is that fast, small memories be referenced more frequently than the slow, large ones. This desire is reflected in the design of the instruction set of the computer: referencing "registers" is usually encouraged by the structure of the instruction set; referencing primary memory (or cache) is supported by the instruction set, but perhaps in less flexible ways than for register access; finally, accessing a disk is not directly supported by instruction sets at all, but requires complicated "I/O control."

It is instructive to formulate a crude model to estimate the performance of the memory hierarchy. We need to assign representative values to the frequency with which each memory is accessed and to the size of each memory:

$$
\begin{aligned}
M_r &\approx 16 &&; & f_r &\approx .6 && \text{Frequency of access to registers } (M_r)\\
M_c &\approx 10^3 &&; & f_c &\approx .38 && \text{Frequency of access to cache } (M_c)\\
M_p &\approx 10^5 &&; & f_p &\approx .02 && \text{Frequency of access to primary memory } (M_p)\\
M_s &\approx 10^{10} &&; & f_s &\approx .000005 && \text{Frequency of access to secondary memory } (M_s)
\end{aligned}
$$

Using our model of memory access time, the time required to access memory on the average is $f_r\sqrt{M_r} + f_c\sqrt{M_c} + f_p\sqrt{M_p} + 100 f_s\sqrt{M_s}$, measured in arbitrary units. (The factor of 100 arises because disk access times are substantially worse than our memory wiring model indicates.) It is instructive to note the relative contributions of the separate memories: 2.4, 12, 6, 50, for a total of ≈ 70. The cost of access to the slowest memory, the disk, is the most important contribution to the average.

The memory hierarchy is an improvement over the homogeneous memory of Fig. 8.1. The time to reference a single memory of size 10^5 is 320 units. The time to reference a three-level hierarchy of about the same size (M_r, M_c, M_p, with frequencies shown above) is a mere 20 units.

The effectiveness of the memory hierarchy depends on *locality* of the memory references. Cache algorithms copy large chunks (8–32 words) of primary memory into the cache, hoping that additional memory references will occur in the neighborhood of the first reference. A similar hope is attached to transfers from secondary memory. If an application arises in which most of the memory references do *not* go to the fast register memory, the memory hierarchy will perform poorly.

Locality can also be viewed as a function of size. If a program and its data can reside in primary memory for the duration of execution and do not require secondary memory, the average memory access time will drop from 70 to 20. If the program is small enough to fit in the small cache memory, access time will drop further to 14.

8.2.1.2   Concurrency in Computers

Not content with the increases in speed due to a memory hierarchy, computer designers have also sought to increase the concurrency in computer designs. A number of different approaches have been tried[2]; we shall illustrate *pipeline* structures and *multiprocessor* structures.

Pipelined processors. Pipelined processors are patterned after the production line found in manufacturing: a portion of the processing is performed by each of several processors and then handed to the next processor in the line. Starting from Fig. 8.1, the designer reasons that two processors could function concurrently, each assigned to half the original memory (Fig. 8.3); a communication path is provided so that the first processor can transmit results to the second.

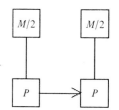

Figure 8.3

The two-processor pipeline more than doubles the processing power available. If we neglect the cost of interprocessor communication, the time required to execute an instruction is $(\frac{1}{2}) (M/2)^{1/2}$, or about one third the time required by the uniprocessor (Fig. 8.1). The improvement comes from two effects: doubling the number of processors doubles the speed, but reducing the memory size also increases speed.

A special case of pipelining is illustrated by *instruction-fetch overlap* in computers. One processor is responsible for fetching an instruction from memory; it then passes on to the second processor information required to execute the instruction; the second processor actually performs the execution. In Chapter 6, we saw this technique applied in OM: while one microinstruction is being executed, the controller is fetching the next microinstruction. *Execution overlap* allows the execution itself to be pipelined. (See Ramamoorthy and Li[3] for more pipelining structures.)

Pipelined structures are perhaps most effective in special-purpose applications that can utilize a large number of processors. Signal processing is a particularly good example: a signal is sampled digitally to generate a *stream* of signal data. This data is pipelined through processors to perform corrections, correlations, frequency analysis, etc. (Section 2 of this chapter illustrates the application of pipelines to matrix arithmetic of various sorts.)

Unfortunately, it is not always possible to cast problems in a framework suited to execution on pipelined computers. If the workload is not divided evenly among the processors, some will stand idle, reducing the effective speed increase. But it is the rigid communication discipline that most severely restricts the application of pipelines.

Multiprocessors. Another important class of concurrent computers are *multiprocessors*. Unlike the pipeline, these structures provide switching structures that allow each processor to communicate with each other processor. The hope is that those algorithms not suited to pipelines because of their communication requirements can be executed on multiprocessors.

Figure 8.4 shows a dual-processor configuration, again adapted from Fig. 8.1. Each processor communicates primarily with a memory half the size of the original. In addition, a common "bus" is provided to allow each processor to reference the other's memory.

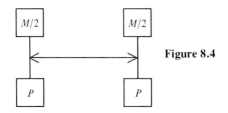

Figure 8.4

Two problems with the dual-processor arrangement are immediately apparent. (1) If each processor references memories at random, the two will interfere often and vitiate some of the speed gain. (2) Can we assure that the sequential program suited to the uniprocessor architecture of Fig. 8.1 can be adapted to the dual-processor configuration? Putting aside for the moment the problems of programming a multiprocessor, we shall examine its performance.

Let us construct a crude model of the time required to execute an instruction on the dual processor. Assume that each processor references its own memory with probability $(1-f)$, and the other's with probability f. Further, assume that the useful duty cycle of each processor is d. If both processors can be productively employed at all times, d will be 1. However, if the two processors must occasionally wait for each other, i.e., must "synchronize," d may fall below 1. We can identify three cases:

1. P_a references M_a and P_b references M_b; probability is $(1-f)^2$.
2. P_a and P_b both reference M_a (or equivalently M_b);
   probabilities sum to $2f(1-f)$.
3. P_a references M_b and P_b references M_a; probability is f^2.

We also need to model the time required to complete each of the three cases. A processor references its own memory, of size $M/2$, in time $(M/2)^{1/2}$. When a reference is made to a neighbor's memory, we assume the time for communication on the bus and referencing the memory sum to $(M)^{1/2}$, as if it were addressing the entire memory as one array. The times required for the three cases then become

1. $(M/2)^{1/2}$,
2. $(M/2)^{1/2} + M^{1/2}$,
3. $M^{1/2} + M^{1/2}$.

From these estimates we calculate the expected instruction execution time, remembering that $2d$ processors are available:

$$\text{time} = M^{1/2}(1/d)(2^{1/2}/4 + f - f^2/2).$$

This expression is plotted in Fig. 8.5, assuming $d = 1$.

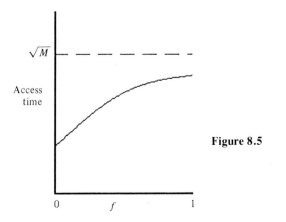

Figure 8.5

The simple model of a dual-processor configuration is suggestive of behavior we can expect from multiprocessor systems that require global communication. We observe that if $f = 0$, execution speed is more than twice that of the uniprocessor illustrated in Fig. 8.1. Just as in the pipeline, doubling the number of processors contributes a factor of two, but additional speed is achieved because each processor addresses a smaller memory.

The model also illustrates the importance of locality in the use each processor makes of its memory. If f is allowed to grow too large, the factor of two contributed by two processors is erased by interference between the processors when accessing the common memory.

Perhaps the most important parameter is d, which is determined by our ability to adapt algorithms to multiprocessor configurations. Some applications seem to decompose nicely for execution on concurrent hardware, and some offer difficulties. In human organizations we have become resigned to *always* attacking large problems in a concurrent way. We will, no doubt, have to do the same with computer programs.

8.2.2 Summary

The schemes we have illustrated that reduce communication costs and try to exploit concurrency can be combined in various ways in computer structures. The table below summarizes the speedup effect that these techniques offer, as derived from our crude models (n denotes the number of processors used):

Technique	Typical speedup factor
Memory hierarchy	10
Pipelining	
instruction overlap	2
special-purpose	n
Multiprocessors	$< n$

The processor-memory structures and algorithms presented in the remainder of this chapter all attempt *to use as many processors as can be kept simultaneously productive and to locate them as close as possible to the data they require*. These are the considerations exhibited by our simple models of memory hierarchies, pipelines and multiprocessors. The examples presented here by no means exhaust the topic of concurrent computation; the interested reader will find literatures on computer architecture,[2,4] parallel processors and processing,[3,5,6,7] performance evaluation,[2] and algorithm design.[8,9,10,11,12]

8.3 ALGORITHMS FOR VLSI PROCESSOR ARRAYS*

8.3.1 Introduction

"And the smooth stream in smoother numbers flows."

Alexander Pope

The developments in microelectronics have revolutionized computer design. Integrated circuit technology has increased the number and complexity of components that can fit on a chip or a printed-circuit board. Component density has been doubling every one-to-two years, and already a multiplier can fit on a very large scale integrated (VLSI) circuit chip. As a result, the new technology makes it feasible to build low-cost, special-purpose, peripheral devices to rapidly solve sophisticated problems. Reflecting the changing technology, this section proposes new multiprocessor structures and parallel algorithms for processing some basic matrix computations.

We are interested in high-performance parallel algorithms that can be implemented directly on low-cost hardware devices. By performance, we are not referring to the traditional operation counts that characterize classical analyses of algorithms, but rather, the throughput obtainable when a special-purpose peripheral device is attached to a general-purpose host computer. This implies that time spent in I/O, control, and data movement as well as arithmetics must all be considered. VLSI offers excellent opportunities for inexpensive implementation of high-performance devices. Thus, in this section the cost of a device will be determined by the expense of a VLSI implementation. "Fit the job to the bargain components" (Blakeslee, p. 4).[13]

VLSI technology has made one thing clear. Simple and regular interconnections lead to cheap implementations and high densities, and high density implies both high performance and low overhead for support components. (Sutherland and Mead[1] contains a discussion of the importance of having simple and regular geometries for data paths.) For these reasons, we are interested in designing parallel algorithms that have simple and regular data flows. We are also interested in using pipelining as a general method for implementing these algorithms in hardware. By pipelining, processing may proceed concurrently with input and output, and consequently overall execution time is minimized. Pipelining plus multiprocessing at each stage of a pipeline should lead to the best-possible performance. In the following, we demonstrate some simple and regular VLSI processor arrays that are capable of pipelining matrix computations with optimal speed-up.

*Contributed by H. T. Kung and Charles E. Leiserson, Department of Computer Science, Carnegie-Mellon University. The first version of Section 8.3, including results reported in Sections 8.3.3 thru 8.3.6 of the present version, was written in April 1978 for submission as a paper to the Symposium on Sparse Matrix Computations and Their Applications, which was held in Knoxville, Tennessee, November 2–3, 1978. The paper was presented at the Symposium.

In Section 8.3.2, we describe the basic hardware requirements and interconnection schemes for the proposed VLSI processor arrays and discuss the feasibility of building these networks. Section 8.3.3 deals with the matrix-vector multiplication problem. Multiplication of two matrices is considered in Section 8.3.4. In Section 8.3.5, we show that essentially the same interconnection scheme and algorithm as those used for matrix multiplication in Section 8.3.4 can be applied to find the LU-decomposition of a matrix. Section 8.3.6 is concerned with solving triangular linear systems. We show that this problem can be solved by almost the same network and algorithm for matrix–vector multiplication described in Section 8.3.3. Section 8.3.7 discusses applications and extensions of the results presented in the previous sections. The applications include the computations of finite impulse response filters, convolutions, and discrete Fourier transforms. Some concluding remarks are given in the last section.

The size of each of our processor array networks is dependent only on the band width of the band matrix to be processed and is independent of the length of the band. Thus, a fixed-size processor array can pipeline band matrices with arbitrarily long bands. The pipelining aspect of our algorithms is, of course, most effective for band matrices with long bands. For this reason most of the results in this paper will be presented in terms of their applications to band matrices. It is important to note, however, that all the results apply equally well to dense matrices since a dense matrix can be viewed as a band matrix with the maximum possible band width.

8.3.2   The Basic Components and Array Structures

8.3.2.1   *The Inner Product Step Processor*

The single operation common to all the algorithms considered in this section is the so-called inner product step, $C \leftarrow C + A \times B$. We postulate a processor that has three registers R_A, R_B, and R_C. Each register has two connections, one for input and one for output. Figure 8.6 shows two types of geometries for this processor. Type (a) geometry will be used for matrix–vector multiplication and solution of triangular linear systems (Sections 8.3.3 and 8.3.6), whereas type (b) geometry will be used for matrix multiplication and LU-decomposition (Sections 8.3.4 and 8.3.5). The processor is capable of performing the inner product step and is called the *inner product step processor*. We shall define a basic time unit in terms of the operation of this processor. In each unit time interval, the processor shifts the data on its input lines denoted by A, B, and C into R_A, R_B, and R_C, respectively; computes $R_C \leftarrow R_C + R_A \times R_B$; and makes the input values for R_A and R_B together with the new value of R_C available as outputs on the output lines denoted by A, B, and C, respectively. All outputs are latched and the logic is clocked so that when one processor is connected to another, the changing output of one during a unit time interval will not interfere with the input to another during this time interval. This is not the only processing element we shall make use of, but it will be the work-

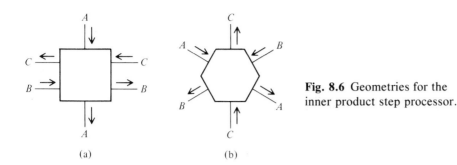

Fig. 8.6 Geometries for the inner product step processor.

horse. A special processor for performing division will be specified later when it is used.

8.3.2.2   Processor Arrays

A device is typically composed of many interconnected inner product step processors. The basic network organization we shall adopt for processors is mesh-connected and all connections from a processor are to neighboring processors. (See Fig. 8.7).

The most widely known system based on this organization is the ILLIAC IV.[14] If diagonal connections are added in one direction only, we shall call the resulting scheme *hexagonally mesh-connected*, or *hex-connected* for short. We shall demonstrate that linearly connected and hex-connected processors are natural for matrix problems.

Processors lying on the boundary of the processor array may have external connections to the host memory. Thus, an input/output data path of a boundary processor may sometimes be designated as an external input/output connection for the device. A boundary processor may receive input from the host memory through such an external connection, or it may receive a fixed value such as zero. On the other hand, a boundary processor may send data to the host memory through an external output connection. An output of a boundary processor may sometimes be ignored; this will be designated by omitting the corresponding output line.

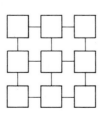

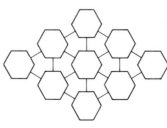

Fig. 8.7 Mesh-connected processor arrays.

(a) Linearly connected   (b) Orthogonally connected   (c) Hexagonally connected

Throughout Section 8.3 we assume that the processors in an array are synchronous as described in Section 8.3.2.1. However, it is possible to view the processors as being asynchronous, each computing its output values when all its inputs are available as in a data flow model. For the purposes of this section we believe the synchronous approach to be more direct and intuitive.

The hardware demands of the VLSI processor arrays described here are readily seen to be modest. The processing elements are uniform, interprocessor connections are simple and regular, and external connections are minimized. It is our belief that construction of these processor arrays will prove to be cost-effective.

8.3.3   Matrix–Vector Multiplication

We consider the problem of multiplying a matrix $A = (a_{ij})$ with a vector $x = (x_1, \ldots, x_n)^T$. The elements in the product $y = (y_1, \ldots, y_n)^T$ can be computed by the following recurrences:

$$
\begin{aligned}
y_i^{(1)} &= 0, \\
y_i^{(k+1)} &= y_i^{(k)} + a_{ik}x_k, \\
y_i &= y_i^{(n+1)}.
\end{aligned}
$$

Suppose A is an $n \times n$ band matrix with band width $w = p+q-1$. (See Fig. 8.8 for the case when $p = 2$ and $q = 3$.) Then the above recurrences can be evaluated by pipelining the x_i and y_i through w linearly connected processors. We illustrate the algorithm for the band matrix–vector multiplication problem in Fig. 8.8. For this case the linearly connected network has four processors. See Fig. 8.9.

The general scheme of our pipelining algorithm can be viewed as follows: The y_i, which are initially zero, move to the left while the x_i are moving to the right and the a_{ij} are moving down. All the moves are synchronized. It turns out that each y_i

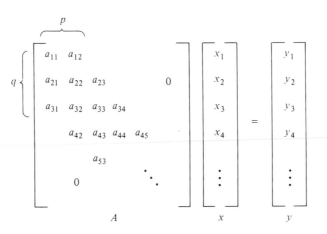

Fig. 8.8 Multiplication of a vector by a band matrix with $p = 2$ and $q = 3$.

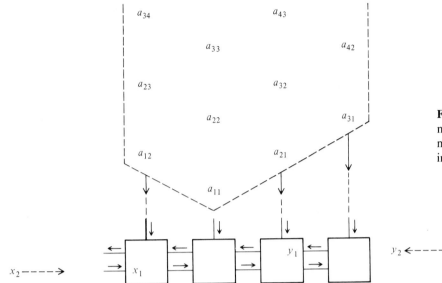

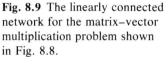

Fig. 8.9 The linearly connected network for the matrix–vector multiplication problem shown in Fig. 8.8.

is able to accumulate all its terms, namely, $a_{i,i-2}x_{i-2}$, $a_{i,i-1}x_{i-1}$, $a_{i,i}x_i$, and $a_{i,i+1}x_{i+1}$, before it leaves the network. Figure 8.10 illustrates the first seven steps of the algorithm.

Note that when y_1 and y_2 are output they have the correct values. Observe also that at any given time alternating processors are idle. Indeed, by coalescing pairs of adjacent processors, it is possible to use $w/2$ processors in the network for a general band matrix with band width w.

We now specify the algorithm more precisely. Assume that the processors are numbered by integers $1, 2, \ldots, w$ from the left-end processor to the right-end processor. Each processor has three registers, R_A, R_x and R_y, which will hold entries in A, x and y, respectively. Initially, all registers contain zeros. Each step of the algorithm consists of the following operations (but for odd-numbered time steps, only odd-numbered processors are activated, and for even-numbered time steps, only even-numbered processors are activated):

1. Shift.

   R_A gets a new element in the band of matrix A.

   R_x gets the contents of register R_x from the left neighboring node.
   (The R_x in processor 1 gets a new component of x.)

   R_y gets the contents of register R_y from the right neighboring node.
   (Processor 1 outputs its R_y contents and the R_y in processor w gets zero.)

2. Multiply and Add.
   $R_y \leftarrow R_y + R_A \times R_x$.

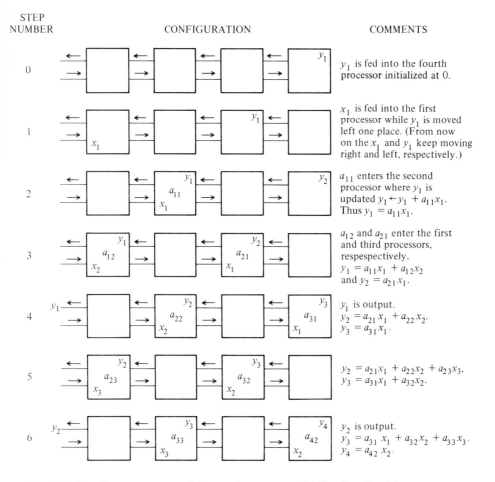

STEP
NUMBER   CONFIGURATION   COMMENTS

0   y_1 is fed into the fourth
     processor initialized at 0.

1   x_1 is fed into the first
     processor while y_1 is moved
     left one place. (From now
     on the x_1 and y_1 keep moving
     right and left, respectively.)

2   a_{11} enters the second
     processor where y_1 is
     updated $y_1 \leftarrow y_1 + a_{11}x_1$.
     Thus $y_1 = a_{11}x_1$.

3   a_{12} and a_{21} enter the first
     and third processors,
     respespectively.
     $y_1 = a_{11}x_1 + a_{12}x_2$
     and $y_2 = a_{21}x_1$.

4   y_1 is output.
     $y_2 = a_{21}x_1 + a_{22}x_2$.
     $y_3 = a_{31}x_1$.

5    $y_2 = a_{21}x_1 + a_{22}x_2 + a_{23}x_3$.
     $y_3 = a_{31}x_1 + a_{32}x_2$.

6   y_2 is output.
     $y_3 = a_{31}x_1 + a_{32}x_2 + a_{33}x_3$.
     $y_4 = a_{42}x_2$.

Fig. 8.10 The first seven steps of the matrix–vector multiplication algorithm.

Using the type (a) inner product step processor postulated in Section 8.3.2.1, we
note that the three shift operations in step 1 can be done simultaneously, and that
each step of the algorithm takes a unit of time. Suppose the bandwidth of A is w. It
is readily seen that after w units of time the components of the product $y = Ax$
start shifting out from the left-end processor at the rate of one output every two
units of time. Therefore, using our network all the n components of y can be
computed in $2n + w$ time units, as compared to the $O(wn)$ time needed for the
sequential algorithm on a uniprocessor.

8.3.4   Matrix Multiplication on a Hexagonal Array

This subsection considers the problem of multiplying two $n \times n$ matrices. It is
easy to see that the matrix product $C = (c_{ij})$ of $A = (a_{ij})$ and $B = (b_{ij})$ can be

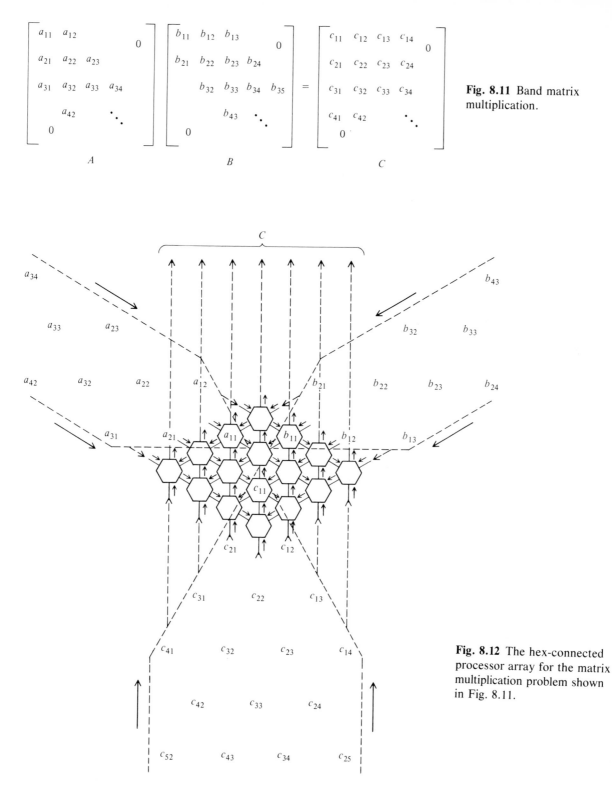

Fig. 8.11 Band matrix multiplication.

Fig. 8.12 The hex-connected processor array for the matrix multiplication problem shown in Fig. 8.11.

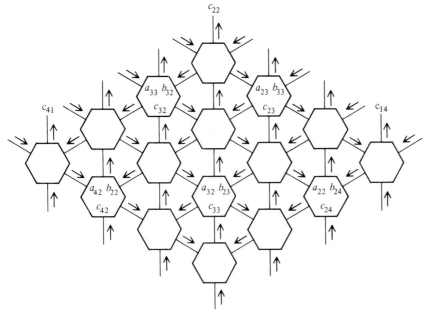

(a)

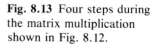

Fig. 8.13 Four steps during the matrix multiplication shown in Fig. 8.12.

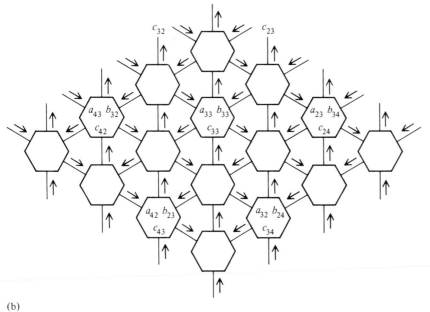

(b)

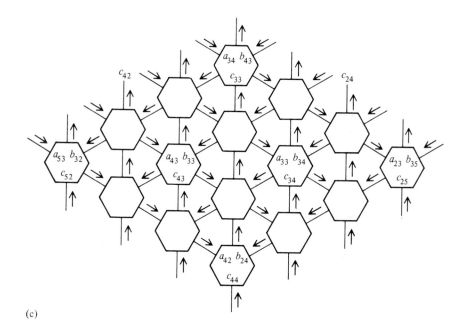

(c)

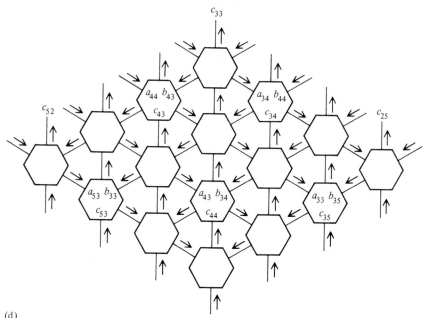

(d)

computed by the following recurrences:

$$c_{ij}^{(1)} = 0,$$
$$c_{ij}^{(k+1)} = c_{ij}^{(k)} + a_{ik}b_{kj},$$
$$c_{ij} = c_{ij}^{(n+1)}.$$

Let A and B be $n \times n$ band matrices of band width w_1 and w_2, respectively. We show how the recurrences above can be evaluated by pipelining the a_{ij}, b_{ij}, and c_{ij} through an array of $w_1 w_2$ hex-connected processors. The algorithm uses the same principle as the one in Section 8.3.3. We illustrate the general scheme by considering the matrix multiplication problem depicted in Fig. 8.11. The diamond-shaped interconnection network for this case is shown in Fig. 8.12, where processors are hex-connected and data flows are indicated by arrows.

The elements in the bands of A, B, and C move through the network in three directions synchronously. Each c_{ij} is initialized to zero as it enters the network through the bottom boundaries. One can easily see that with the type (b) inner product processors described in Section 8.3.2.1, each c_{ij} is able to accumulate all its terms before it leaves the network through the upper boundaries. Figure 8.13 shows four consecutive steps in the execution of the algorithm. The reader is invited to study the data flow of this problem more closely by making transparencies of the band matrices (shown in the figures), and moving them over the network picture as described in the algorithm.

Let A and B be $n \times n$ band matrices of band width w_1 and w_2, respectively. Then a network of $w_1 w_2$ hex-connected processors can pipeline the matrix multiplication $A \times B$ in $3n + min(w_1, w_2)$ units of time.

Note that in any row or column of the network, out of every three consecutive processors, only one is active at any given time. It is possible to use about one third of the $w_1 w_2$ processors in the network for multiplying two band matrices with band widths w_1 and w_2.

8.3.5. The LU-Decomposition of a Matrix on a Hexagonal Array

The problem of factoring a matrix A into lower and upper triangular matrices L and U is called LU-decomposition. Figure 8.14 illustrates the LU-decomposition of a band matrix with $p = 4$ and $q = 4$.

Once the L and U factors are known, it is relatively easy to invert A or solve the linear system $Ax = b$. (We deal with the latter problem in Section 8.3.6.) This section describes a parallel LU-decomposition algorithm that has hex-connected data paths.

We assume that matrix A has the property that its LU-decomposition can be done by Gaussian elimination without pivoting. (This is true, for example, when A is a symmetric positive-definite, or an irreducible, diagonally dominant matrix.) The

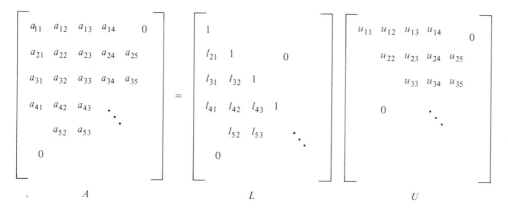

Fig. 8.14 The LU-decomposition of a band matrix.

triangular matrices $L = (l_{ij})$ and $U = (u_{ij})$ are evaluated according to the following recurrences:

$$a_{ij}^{(1)} = a_{ij},$$

$$a_{ij}^{(k+1)} = a_{ij}^{(k)} + l_{ik}(-u_{kj}),$$

$$l_{ik} = \begin{cases} 0 & \text{if } i < k, \\ 1 & \text{if } i = k, \\ a_{ik}^{(k)} u_{kk}^{-1} & \text{if } i > k, \end{cases}$$

$$u_{kj} = \begin{cases} 0 & \text{if } k > j, \\ a_{kj}^{(k)} & \text{if } k \leqslant j. \end{cases}$$

We show that the evaluation of these recurrences can be pipelined on a hex-connected processor array. A global view of this pipelined computation is shown in Fig. 8.15 for the LU-decomposition problem depicted in Fig. 8.14. The processor array in Fig. 8.15 is constructed as follows: The processors below the upper boundaries are the standard type (b) inner product step processors and are hex-connected in exactly the same way as the matrix multiplication network presented in Section 8.3.4. The processor at the top, denoted by a circle, is a special processor. It computes the reciprocal of its input and passes the result southwest and also passes the same input northward unchanged. The other processors on the upper boundaries are again type (b) inner product step processors, but their orientation is changed: the ones on the upper left boundary are rotated 120 degrees clockwise; the ones on the upper right boundary are rotated 120 degrees counterclockwise.

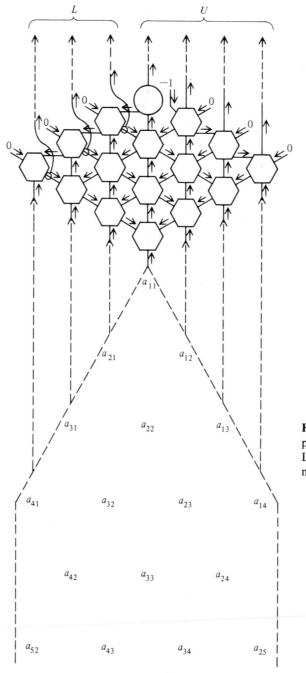

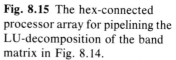

Fig. 8.15 The hex-connected processor array for pipelining the LU-decomposition of the band matrix in Fig. 8.14.

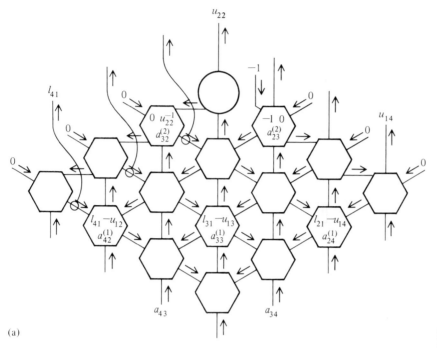

(a)

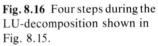

Fig. 8.16 Four steps during the LU-decomposition shown in Fig. 8.15.

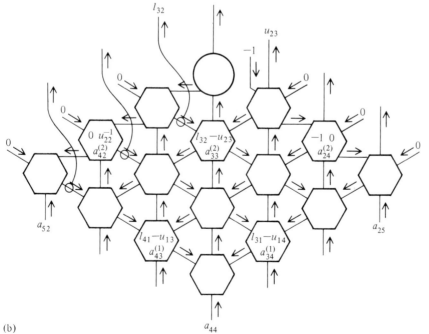

(b)

(*Continued*)

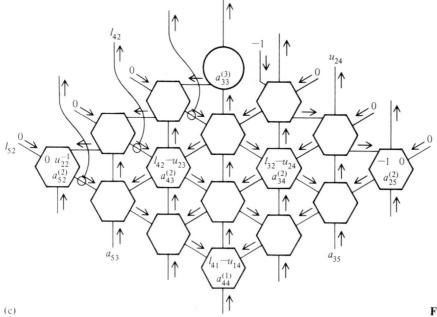

(c)

Figure 8.16 (*cont.*)

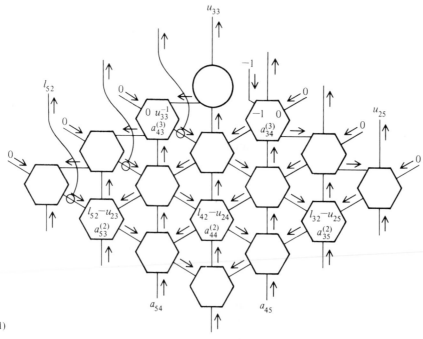

(d)

The flow of data on the network is indicated by arrows in the figure. As in the matrix multiplication algorithm, each processor only operates every third time step. Figure 8.16 illustrates four consecutive steps during the execution of the algorithm. Note that in the figure, because A is a band matrix with $p = 4$ and $q = 4$, we have $a_{i+3,i}^{(k)} = a_{i+3,i}$ and $a_{i,i+3}^{(k)} = a_{i,i+3}$ for $1 \leq k \leq i$ and $i \geq 2$. Thus a_{52}, for example, can be viewed as $a_{52}^{(2)}$ when it enters the network.

There are several equivalent networks that reflect only minor changes to the network presented in this section. For example, the elements of L and U can be retrieved as output in a number of different ways. Also, the ``$-1$'' input to the network can be changed to a ``$+1$'' if the special processor at the top of the network computes minus the reciprocal of its input.

If A is an $n \times n$ band matrix with band width $w = p + q - 1$, a processor array having no more than pq hex-connected processors can compute the LU-decomposition of A in $3n + min(p,q)$ units of time. If A is an $n \times n$ dense matrix, this means that n^2 hex-connected processors can compute the L and U matrices in $4n$ units of time, which includes I/O time.

The remarkable fact that the matrix multiplication network forms a part of the LU-decomposition network is due to the similarity of their defining recurrences. In any row or column of the LU-decomposition network, only one out of every three consecutive processors is active at a given time. As we observed for matrix multiplication, the number of processors can be reduced to about $pq/3$.

8.3.6   Triangular Linear Systems

Suppose that we want to solve a linear system $Ax = b$. Then after having done with the LU-decomposition of A (e.g., by methods described in Section 8.3.5), we still have to solve two triangular linear systems, $Ly = b$ and $Ux = y$. This section concerns itself with the solution of triangular linear systems. An upper triangular linear system can always be rewritten as a lower triangular linear system. Without loss of generality, this section deals exclusively with lower triangular linear systems.

Let $A = (a_{ij})$ be a nonsingular $n \times n$ band lower triangular matrix. Suppose that A and an n-vector $b = (b_1, \ldots, b_n)^T$ are given. The problem is to compute $x = (x_1, \ldots, x_n)^T$ such that $Ax = b$. The vector x can be computed by the following recurrences:

$$y_i^{(1)} = 0,$$

$$y_i^{(k+1)} = y_i^{(k)} + a_{ik}x_k,$$

$$x_i = (b_i - y_i^{(i)})/a_{ii}.$$

Suppose that A is a band matrix with band width $w = q$. (See Fig. 8.17 for the case when $q = 4$.) Then the above recurrences can be evaluated by the algorithm and network almost identical to those used for band matrix–vector multiplication in

Section 8.3.3. (Observe the similarity of the defining recurrences for these two problems.) We illustrate our result by considering the linear system problem in Fig. 8.17. For this case, the network and the general scheme of the algorithm are described in Fig. 8.18.

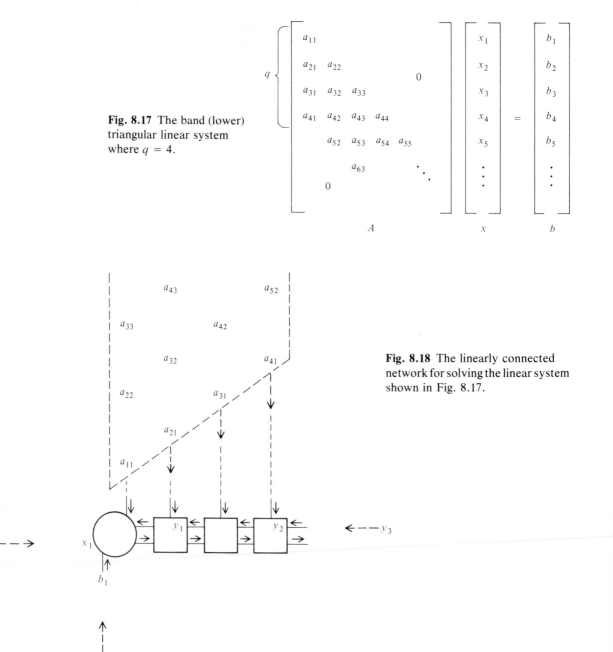

$$q \left\{ \begin{bmatrix} a_{11} & & & & & & \\ a_{21} & a_{22} & & & & & \\ a_{31} & a_{32} & a_{33} & & & 0 & \\ a_{41} & a_{42} & a_{43} & a_{44} & & & \\ & a_{52} & a_{53} & a_{54} & a_{55} & & \\ & & a_{63} & & & \ddots & \\ & 0 & & & & & \end{bmatrix} \right. \begin{bmatrix} x_1 \\ x_2 \\ x_3 \\ x_4 \\ x_5 \\ \vdots \\ \vdots \end{bmatrix} = \begin{bmatrix} b_1 \\ b_2 \\ b_3 \\ b_4 \\ b_5 \\ \vdots \\ \vdots \end{bmatrix}$$

$$\quad\quad\quad A \quad\quad\quad\quad\quad\quad x \quad\quad\quad b$$

Fig. 8.17 The band (lower) triangular linear system where $q = 4$.

Fig. 8.18 The linearly connected network for solving the linear system shown in Fig. 8.17.

The y_i, which are initially zero, move leftward through the network while the x_i, a_{ij}, and b_i are moving as indicated in Fig. 8.18. The left-end processor is special in that it performs $x_i \leftarrow (b_i - y_i)/a_{ii}$. (In fact, the special processor introduced in Section 8.3.5 to solve the LU-decomposition problem is a special case of this more general processor.) Each y_i accumulates inner product terms in the rest of the processors as it moves to the left. At the time y_i reaches the left-end processor, it has the value $a_{i1}x_1 + a_{i2}x_2 + \cdots + a_{i,i-1}x_{i-1}$, and consequently the x_i computed by $x_i \leftarrow (b_i - y_i)/a_{ii}$ at the processor will have the correct value. Figure 8.19 demon-

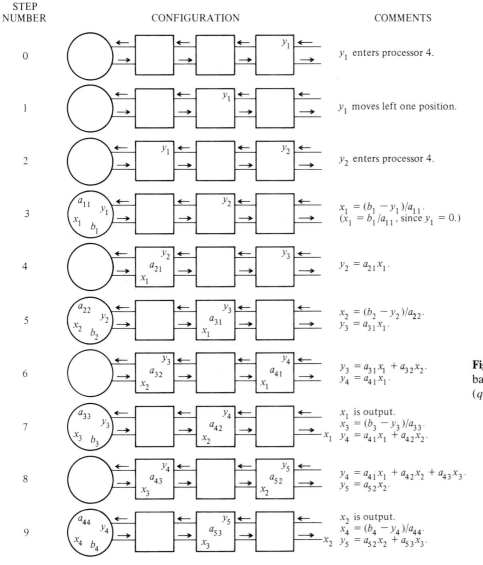

Fig. 8.19 Solving a lower band triangular system ($q = 4$).

strates the first seven steps of the algorithm. From the figure one can check that the final values of x_1, x_2, x_3, and x_4 are all correct. With this network we can solve an $n \times n$ band triangular linear system with band width $w = q$ in $2n + q$ units of time. As we observed for the matrix–vector multiplication problem, the number of processors required by the network can be reduced to $w/2$.

8.3.7   Applications and Comments

8.3.7.1   *Variants of the Algorithms and Networks*

Variants of the basic algorithms and networks presented above will often be used in actual practice. No attempt is given here for listing all the possible variants; it is important that the reader understand the basic principles used so that he or she can construct appropriate variants for specific problems.

As pointed out in Section 8.3.1, although most of our illustrations are done for band matrices, all the algorithms work for the regular $n \times n$ dense matrix. In this case the band width of the matrix is $w = 2n - 1$. If the band width of a matrix is so large that a corresponding algorithm requires more processors than a given network provides, then one should decompose the matrix and solve each subproblem on the network. For instance, the matrix multiplication of two $n \times n$ matrices or the LU-decomposition of an $n \times n$ matrix can be done in $O(n^3/k^2)$ time on a $k \times k$ array, for $k < n$.

One can often reduce the number of processors required by an algorithm if the matrix is known to be sparse or symmetric. For example, the matrices arising from a set of finite differences or finite elements approximations to differential equations are usually "sparse band matrices." These are band matrices whose nonzero entries appear only in a few of those lines in the band that are parallel to the diagonal. In this case by introducing proper delays to each processor for shifting its data to its neighbors, the number of processors required by the algorithm in Section 8.3.3 can be reduced to the number of those diagonal lines that contain nonzero entries. This variant is useful for performing iterative methods involving sparse band matrices. Another example is concerned with the LU-decomposition problem considered in Section 8.3.5. If matrix A is symmetric positive-definite, then it is possible to use only the left portion of the hex-connected network, since in this case U is simply DL^T, where D is the diagonal matrix $(a_{kk}^{(k)})$.

The optimal choice of the size of the network to solve a particular problem depends upon not only the problem but also the memory bandwidth to the host computer. For achieving high performance, it is desirable to have as many processors as possible in the network, provided they can all be kept busy doing useful computations.

It is possible to use our algorithms and networks to solve some nonnumerical problems when appropriate interpretations are given to the addition ($+$) and multiplication ($\times$) operations. For example, some pattern-matching problems can be viewed as matrix problems with comparison and Boolean operations. It can be instructive to view the + and × operations as operations in an abstract algebraic structure, such as a semi-ring, and then to examine how our results hold in these abstract settings.

8.3.7.2 Convolution, Filter, and Discrete Fourier Transform

There are a number of important problems that can be formulated as matrix–vector multiplication problems and thus can be solved rapidly by the algorithm and network in Section 8.3.3. The problems of computing convolutions, finite impulse response (FIR) filters, and discrete Fourier transforms are such examples. If a matrix has the property that the entries on any line parallel to the diagonal are all the same, then the matrix is a Toeplitz matrix. The convolution problem is simply the matrix–vector multiplication where the matrix is a triangular Toeplitz matrix (see Fig. 8.20).

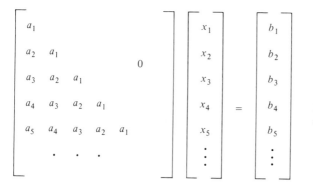

Fig. 8.20 The convolution of vectors a and x.

A p-tap FIR filter can be viewed as a matrix–vector multiplication where the matrix is a band upper triangular Toeplitz matrix with band width $w = p$. Figure 8.21 represents the computation of a 4-tap filter.

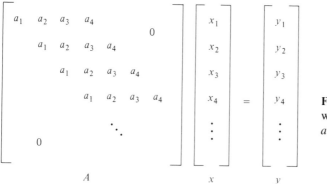

Fig. 8.21 A 4-tap FIR filter with coefficients a_1, a_2, a_3, and a_4.

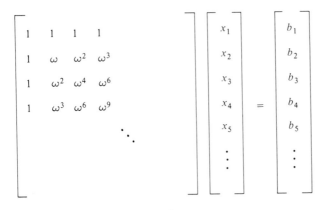

$$\begin{bmatrix} 1 & 1 & 1 & 1 \\ 1 & \omega & \omega^2 & \omega^3 \\ 1 & \omega^2 & \omega^4 & \omega^6 \\ 1 & \omega^3 & \omega^6 & \omega^9 \\ & & & \ddots \end{bmatrix} \begin{bmatrix} x_1 \\ x_2 \\ x_3 \\ x_4 \\ x_5 \\ \vdots \end{bmatrix} = \begin{bmatrix} b_1 \\ b_2 \\ b_3 \\ b_4 \\ b_5 \\ \vdots \end{bmatrix}$$

Fig. 8.22 The discrete Fourier transform of vector x.

On the other hand, an n-point discrete Fourier transform is the matrix–vector multiplication, where the (i,j) entry of the matrix is $\omega^{(i-1)(j-1)}$ and ω is a primitive nth root of unity (see Fig. 8.22).

Therefore, using a linearly connected network of size $O(n)$ both the convolution of two n-vectors and the n-point discrete Fourier transform can be computed in $O(n)$ units of time, rather than $O(n \log n)$ as required by the sequential FFT algorithm. Moreover, note that for the convolution and filter problems each processor has to receive an entry of the matrix only once, and this entry can be shipped to the processor through horizontal connections and can stay in the processor during the rest of the computation. For the discrete Fourier transform problem each processor can in fact generate on-the-fly the powers of ω it requires. As a result, for these three problems it is not necessary for each processor in the network to have the external input connection on the top of the processor, as depicted in Fig. 8.9.

In the following we describe how the powers of ω can be generated on-the-fly during the process of computing an n-point discrete Fourier transform. The requirement is that if a processor is i units apart from the middle processor, then at time $i + 2j$ the processor must have the value of ω^{j^2+ij}, for all i, j. This requirement can be fulfilled by using the algorithm below. We assume that each processor has one additional register R_t. All processors except the middle one perform the following operations in each step, but for odd- (respectively, even-) numbered time steps, only processors that are odd (even) units apart from the middle processor are activated. For all processors except the middle one the contents of both R_A and R_t are initially "0".

1. Shift. If the processor is in the left- (respectively, right-) hand side of the middle processor, then

    R_A gets the contents of register R_A from the right- (respectively, left-) neighboring processor.

    R_t gets the contents of register R_t from the right- (respectively, left-) neighboring processor.

2. Multiply.

$$R_A \leftarrow R_A \times R_t.$$

The middle processor is special; it performs the following operations at every even-numbered time step. For this processor the contents of both R_A and R_t are initially "1".

1. $R_A \leftarrow R_A \times R_t^2 \times \omega.$
2. $R_t \leftarrow R_t \times \omega.$

8.3.7.3   The Common Memory Access Pattern

Note that all the algorithms given in this section store and retrieve elements of the matrix in the same order. (See Figs. 8.9, 8.12, 8.15, and 8.18.) Therefore, we recommend that matrices always be arranged in memory according to this particular ordering so that they can be accessed efficiently by any of the algorithms.

8.3.7.4   The Pivoting Problem, and Orthogonal Factorization

In Section 8.3.5 we assume that the matrix A has the property that there is no need of using pivoting when Gaussian elimination is applied to A. What should one do if A does not have this nice property? (Note that Gaussian elimination becomes very inefficient on mesh-connected processors if pivoting is necessary.) This question motivated us to consider Givens's transformation (see, for example, Hammering[15]) for triangularizing a matrix, which is known to be a numerically stable method. It turns out that, like Gaussian elimination without pivoting, the orthogonal factorization based on Givens's transformation can be implemented naturally on mesh-connected processors, although a pipelined implementation appears to be more complex. (Results on Givens's transformation will be reported elsewhere.) (Sameh and Kuck[16] considered parallel linear system solvers based on Givens's transformation, but they did not give solutions to the processor communication problem considered here.)

8.3.8   Concluding Remarks

Research in interconnection networks and algorithms has been traditionally motivated by large scale parallel array computers such as ILLIAC IV.[6,17,18] The results presented here were, however, motivated by the advance in VLSI, though they are certainly applicable to parallel array processors. We have shown that many basic computations can be done very efficiently by special-purpose multiprocessors, which may be built cheaply using VLSI technology. The important feature common to all of our algorithms is that their data flows are very *simple* and *regular*, and they are *pipeline algorithms*. We have discovered that some data flow patterns are fundamental in matrix computations. For example, the two-way flow on the linearly connected network is common to both matrix–vector multiplication and solution of triangular linear systems (Sections 8.3.3 and 8.3.6), and the three-way flow on the hexagonally mesh-connected network is common to both matrix multiplication and LU-decomposition (Sections 8.3.4 and 8.3.5). A practical implication of this fact is that one device may be used for solving many different problems. Moreover, we note that almost all the processors needed in any of these devices are the inner product step processor postulated in Section 8.3.2. A careful design for this processor is desirable since it is the workhorse for all the devices presented.

For the important problem of solving a dense system of n linear equations in $O(n)$ time on $n \times n$ mesh-connected processors, we have improved upon the recent results of Kant and Kimura[19]. The basis of their results is a theorem on determinants that was known to J. Sylvester in 1851. Their algorithm requires that the matrix be "strongly nonsingular" in the sense that *every* square submatrix is nonsingular. It is sufficent for our algorithms that the matrix be symmetric positive-definite or irreducible diagonally dominant.

Hoare[20] and Thurber and Wald[7] describe some matrix multiplication algorithms on an orthogonally connected processor array. Unlike our results, their algorithms require that one or more of the three matrices involved in matrix multiplication have to stay in the array statically during the computation. This means extra I/O time and extra logic in each processing element in the network. Because of the use of hexagonal connection for the array, we are able to pipeline all three matrices through the network.

Inter-processor communications will likely continue to dominate the cost of parallel algorithms and systems. Communication paths inherently take more space and energy than processing elements in many problems of practical interest. We regard the problem of minimizing communication costs as fundamental. We hope the results of this section have demonstrated that the communication problem in parallel algorithms is not only tractable but also interesting. We expect that a large number of algorithms having small communication costs will be discovered in the future.

8.4 HIERARCHICALLY ORGANIZED MACHINES

We know that human organizations use hierarchical structure to extract the greatest possible benefit from the daily activities of tens of thousands of individuals. We know that complex systems can be constructed by subdividing them into less complex systems, which are again subdivided, as many times as necessary, until the resulting systems are simple enough to construct easily. In Section 8.5 we show that the organization of real estate on the silicon surface dictates a hierarchical communication system for any devices that must support global communication. Such hierarchical communication exists in conventional computers only in a limited way. Are there new machine structures that communicate hierarchically, that support systems consisting of an arbitrary hierarchy of subsystems, and that can coordinate the activities of any number of submachines?

8.4.1 Binary Trees

Consider any number of processors physically arranged as a binary tree. Each processor has two subprocessors that it can control. These subprocessors, in turn, have two sub-subprocessors, and so on. A possible layout of such a binary processor tree is shown in Fig. 8.23. At the lowest level a small array of ordinary memory cells, labeled M_0, is accessed by the lowest level processors, labeled P_0. The combination of one lowest level processor with its associated memory is the element of computing power. These units are grouped together in pairs and accessed by the next level processor, labeled P_1. Two P_1's with their associated lower level units are grouped together and accessed by the next level higher processor, labeled P_2. This arrangement is repeated recursively until an entire silicon chip is covered by the processor-memory hierarchy. The rate at which information can be transferred within a processor is independent of the level of the processor. As the wires within a processor get longer, the drivers must become proportionately larger to drive them. The highest level processor that communicates off the silicon chip to the outside world has large drivers and hence is able to drive off-

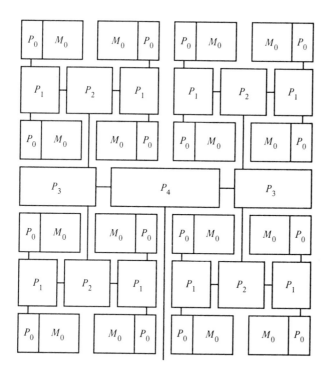

Fig. 8.23 Layout for a binary processor tree.

chip without suffering a severe performance penalty. Such a machine can thus be extended to a large number of individual chips and still maintain the full speed of the individual processors within it.

A conventional computer is a special case of this organization, consisting of a memory array and a bottom-level processor. Also, there is another way to map a conventional computer onto a binary tree of processors. View the highest level processor as a cpu and load all subprocessors with programs that merely decode requests for the memory below them. Loaded with these programs, the structure between the two extreme levels becomes a memory decoder tree between a conventional cpu and its memory.

More importantly, this binary tree structure is a completely general, concurrent processing engine and can be used for problems decomposed in an arbitrary hierarchical way. If a problem requires more than two subprocessors at any level, a subtree of physical processors can be operated as one logical processor, matching the problem's structure. Algorithms for constructing logical processors of any size are given in the next section. The tree has the inherent ability for all processors to compute concurrently and hence has a vastly larger potential computing power than a conventional machine using a similar amount of silicon real estate.

Since the number of processors decreases exponentially with the level, the total bandwidth available, whether processing or communication, decreases exponentially with the level. Half of the total bandwidth of the system is concentrated at level 0, one quarter at level 1, one eighth at level 2, etc. A particular computation is well matched to such a processor if its bandwidth requirements are concentrated at the lowest levels. If an algorithm requires more communication at any level than the structure provides, it will not be able to take advantage of all the processing power of the structure. An extreme example of this sort is the von Neumann machine where all computation occurs at the highest level processor, and where the lower level processors are used only one at a time as an ordinary memory. Such a machine requires equal bandwidth at each level of the hierarchy and is an exponential waste of the resources of the machine.

It is also clear that a tree structure is testable if a single processor is testable. Each supervisor merely loads a test program into its two subordinates and exercises them. Once it has established that both work correctly, it loads each with the program it just used to test them. A tree of N levels can thus be tested in N times the time necessary to test one processor.

It is difficult to predict how any radically different machine structure will perform in a real computing environment. Ideally, we should implement a number of complete systems, spanning a large range of user requirements, in order to gain experience with the strengths and weaknesses of any given scheme. Failing that, we can at least map certain algorithms onto our machine in the hope that they will shed light on its capabilities and its problems. Several such mappings are presented in the next section. (We plan to develop others and we hope our readers will contribute still more for subsequent editions of this text.)

8.4.2 Algorithms for the Tree Machine*

Suppose we apply the notion of hierarchical design to interprocessor communication paths. That is, we use a binary tree as a model for the interconnections between a collection of processors. This section presents some problems that map onto this architecture nicely, and take advantage of the concurrency provided in the tree.

A Word About Notation

The processors in the tree have some characteristics that must be emphasized by the notation we use to describe them.

Each processor is a general computing machine with some amount of local store. We want to describe a template for both the program and the data that will characterize the processor. This template will be instantiated as the many nodes of the tree.

We want to limit communication paths between each processor, its parent, and its children, to explicitly defined entry points. That is, there is no omnipotent processor that is able to oversee and influence the actions of other processors except as explicitly described. Each processor can expect to have local sovereignty and can only be affected by communication it expects.

And perhaps most importantly, we want to encourage locality in the problem solutions. Communication between processors requires synchronizing their actions, limiting the amount of concurrency that can be utilized.

The notation that embodies these criteria is the class construct described in Dahl, Dijkstra, and Hoare.[21] The class allows us to define as a single entity both a data structure and the procedures that operate on it. Thus the implementation details are known only to the class itself. Each instance of the class, an object, can be thought of as a machine capable of local computation but responding to well-defined orders from the outside world.

The most widely known programming language that incorporates the class construct is SIMULA 67.[22] Classes are also available in various flavors in languages such as Alphard,[23] CLU,[24] and Smalltalk.[25]

We will use a modified version of the SIMULA syntax to describe the nodes in the processor tree. The syntax for a class declaration can be described in BNF as follows:

<class declaration> :: = class <class identifier>;
                                <formal parameter part>;
                                <attribute part>;
                                <class body>

<class body> :: = <statement>

(Statements in SIMULA use the syntax of Algol 60. SIMULA is in fact a superset of Algol 60.)

*Contributed by Sally A. Browning, California Institute of Technology.

Because we are describing highly concurrent algorithms, we need to get around the sequential nature of SIMULA statements. For this reason we expand the meaning of the semicolon symbol. In vanilla SIMULA, semicolon is used to terminate a statement. We use semicolon to make a statement about the execution as well. Read semicolon as "At this point, all statements in progress must be terminated before advancing to the next statement." Linefeed will be used to indicate syntactic end of the statement. In other words, linefeeds are used to separate statements; semicolons are used to separate groups of statements that can execute concurrently. E. W. Dijkstra introduced this semicolon convention.[26]

A Word About Branching Ratios

Although the physical structure of our tree restricts each processor to two descendents, we can impose a logical structure that allows an arbitrary branching ratio. Each logical processor consists of several physical processors, enough to provide the desired number of offspring. A logical node with N children is built from $N-1$ physical nodes and is log N levels deep. Figure 8.24 shows some sample logical processors.

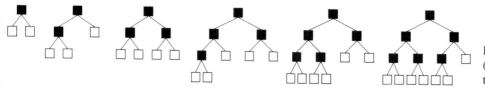

Fig. 8.24 Logical nodes (solid boxes) with two to seven descendents.

We will now describe the process of mapping our logical structure onto the physical tree in SIMULA. We define two CLASSes: a node and a processor. A node represents the physical entity. It has exactly two descendents. A processor will refer to the logical entity, with an arbitrary number of children.

In the following SIMULA definitions, N represents the number of descendents desired. As we build the logical node, we attempt to keep it balanced. That is, all available physical nodes on a given level of the tree will be used before a new level is added. Nodes on a given level are added to the logical processor from left to right, as in Fig. 8.24. Note that CLASS Processor is a refinement of CLASS Node that knows how to choose one of N descendents.

```
CLASS Node(n); INTEGER n;
BEGIN
    REF(Node)left,right;
    !init code to build logical node;
    IF n>2 THEN left:-NEW Node((n+1)//2);
    IF n>3 THEN right:-NEW Node(n//2);
END of CLASS Node;
```

```
Node CLASS Processor;
BEGIN
    REF(Processor) PROCEDURE Son(s); INTEGER s;
    BEGIN REF(node)p;
        p:-IF s<=(n + 1)//2 THEN left ELSE right;
        WHILE p IN Processor DO
            p:-IF s<=(p.n+1)//2 THEN p.left ELSE p.right;
        Son:-p;
    END of PROCEDURE Son;
END of CLASS Processor;
```

8.4.2.1 Algorithms with Polynomial Complexity

One of the traditional approaches to solving a problem that is too large or too complex when considered as a whole is to recursively break the problem into pieces that are manageable. In a structure with many interconnected processors, we attempt to concurrently apply as many processors to the problem as possible in order to reduce execution time. We will look at two algorithms that use this approach — sorting and matrix multiplication. While both of these problems are solved nicely on cellular arrays, it is instructive to map them onto a machine with different communication properties.

Sorting in Linear Time

We use a binary tree with depth log N to sort N numbers. The sort is accomplished as a by-product of loading the numbers into memory and then reading them out again. The numbers themselves are never in sorted order internally but come out of the tree in the desired order.

Sorting is a particularly interesting example because it illustrates a fundamental issue in concurrency. It is well known that sorting on a sequential machine can be done with $O(N \log N)$ comparisons. However, it has been shown on very fundamental grounds that if communication is restricted to two nearest neighbors, at least N^2 comparisons are required.[27] The apparent advantage of the $O(N \log N)$ algorithms comes as a direct result of longer communication paths. It is also clear that no scheme will be able to produce an ordered set of numbers until all numbers are loaded into the machine. For this reason, the best we can expect is to use N processors for $O(N)$ cycles. The following algorithm is an implementation of heap sorting, one of the well-known techniques used in sequential machines.[12]

The algorithm that runs in each processor has a procedure for loading the tree called *Fillup* and a procedure invoked during the output cycle called *Passup*.

Fillup keeps the largest number seen to date and passes the smaller one to the right or left child, keeping the tree balanced by alternating sides.

Passup returns this processor's current number and refills itself with the larger of the numbers stored in its descendents. This action is pipelined so that the largest number is always available in the root.

Here is a SIMULA description of the heap sort algorithm running in each processor. The variable *number* holds the number stored in this processor. The Boolean symbol *empty* reflects the validity of that number. *Balanced* is a Boolean identifier that is used to keep the tree balanced as it is loaded.

```
CLASS processor;
BEGIN

    INTEGER number;
    BOOLEAN balanced,empty;
    REF(processor)left,right;


    PROCEDURE fillup(candidate); INTEGER candidate;
    BEGIN
        IF empty THEN
        BEGIN
            number: =candidate
            empty: =FALSE;
        END
        ELSE
        BEGIN
            IF candidate>number THEN !swap;
            BEGIN INTEGER t;
                t: =candidate;
                candidate: =number;
                number: =t;
            END;
            IF balanced
            THEN left.fillup(candidate)
            ELSE right.fillup(candidate);
            balanced: =NOT balanced;
        END;
    END of procedure fillup;


    INTEGER PROCEDURE passup;
    BEGIN
        passup: =number;
        IF left==NONE AND right==NONE THEN empty: =TRUE !its a leaf;
        ELSE
        IF left.empty THEN
        BEGIN
            IF right.empty THEN empty:=TRUE !both subtrees empty;
            ELSE number: =right.passup; !fill from right son;
        END
```

```
ELSE
IF right.empty THEN number: =left.passup !fill from left son;
ELSE number: =IF left.number>right.number
    THEN left.passup ELSE right.passup;
    !take the larger of the two;
END of procedure passupnumber;


!init code;
empty: =TRUE;
balanced: =TRUE;
!left and right set;


END of class processor;
```

This sort algorithm is bounded by the time it takes to load and remove the numbers. Thus it has time complexity O(N). It requires N processors as well.

P. N. Armstrong[28] has proposed a special-purpose machine that uses N comparitors (processors) to sort N numbers in linear time. His design is a hardware implementation of the bubble sort algorithm. While we have chosen a different algorithm, namely heap sort, we can sort as quickly on our general-purpose machine. The two structures are functionally equivalent when applied to the sorting problem.

Matrix Multiplication

We now examine the problem of multiplying two $N \times N$ matrices together. The brute-force attack on the problem yields a solution in O(N^3) time. We present two algorithms to do the multiplication. Each is a brute-force approach. They differ in the way they subdivide the matrix and in communication requirements. The first approach we will take is to divide the matrix into progressively smaller matrices until we are multiplying single elements together. The second method divides the multiplicand into rows and the multiplier into columns and then pipelines the loading and multiplication of pairs of single elements.

After presenting the two matrix multiplication algorithms, we discuss the difference in complexity of the two. We also look at the possibility of doing chain multiplication and matrix exponentiation. Finally, we compare the performance of the tree machine with the hexagonal array of Kung and Leiserson (Section 8.3).

Subdivision into submatrices.   Suppose we have two $N \times N$ matrices to multiply together. Using the divide and conquer approach, we can break the problem down into smaller and smaller submatrices until they can be easily multiplied together. There is an algorithmic way of combining the product submatrices to produce the correct solution to the complete problem.

The simplest matrices to multiply are of size 1×1; we stop the decomposition at this point. Note that each of the N^2 elements in the product is generated by summing N pairwise products, with one operand contributed by the multiplicand matrix and the other from the multiplier matrix. N 1×1 matrices are multiplied for each of the N^2 elements of the product, for a total of N^3 multiplications.

The tree machine that uses this recursive decomposition has N^3 leaves, one for each pairwise product of single elements from the operands. The other nodes in the tree are used to decompose and reassemble the matrices.

We use the following rule to subdivide the problem.

Let A, B, and C be $N \times N$ matrices such that $AB = C$. We subdivide all three into four $(N/2) \times (N/2)$ submatrices. If $A = (A_{ij})$ and $B = (B_{ij})$, then

$$C = (A_{i1}B_{1j} + A_{i2}B_{2j}), \qquad i,j = 1,2.$$

We will consider matrices whose size N equals 2^M without loss of generality. A tree to multiply two matrices of size 2^M will have M levels of processors that add two matrices together, M levels that split and assemble the matrix, and one level (the leaf nodes) that multiply two numbers together. Thus the tree is $2M + 1$ logical levels deep.

Each adder node has two descendents, and each split/assemble node has four descendents. Thus the physical structure will use two levels to simulate the 4-way branching, and the tree will in fact be $3M$ levels deep. That is, the depth is $3 \log N$. Accordingly, it requires N^3 leaf nodes and $2N^3 - 1$ processors to do the computation.

Let us look at the communication requirements between nodes of the tree. The root node must be prepared to store the entire matrix. The adder nodes in level one (the root is level 0) will each deal with a quarter of the original matrix, as will the split/assemble nodes in level three. The further down the tree you go, the smaller the matrix the node must store and transfer.

However, note that each of the N^2 elements must travel the entire length of the tree and back again during the execution of the algorithm. While communication requirements are low at the leaves, they are extremely high (like N^2 numbers to receive and $(N/2)^2$ to pass down to each descendent) at the root.

In the algorithm given below, the add operation takes $O(N^2)$ time. By doing the adding in parallel either by row, using N processors, or by pairs of elements, using N^2 processors, we can make this operation linear or constant in time. However, the problem is still limited by the split/assemble process that requires each element to travel the height of the tree. That is, the best time performance we can achieve with this algorithm is $O(N^2)$.

We now present a SIMULA representation of a matrix and use it in the algorithm that follows. The algorithm uses two kinds of processors, the adders and the split/assemble nodes. Each matrix is divided into submatrices by procedure

Quarter as follows:

$$\begin{pmatrix} 1 & 2 \\ 3 & 4 \end{pmatrix}$$

```
CLASS Matrix(n); INTEGER n;
BEGIN
   INTEGER ARRAY val[1:n,1:n];

   REF(matrix) PROCEDURE quarter(select); INTEGER select;
   BEGIN REF(matrix)aq; INTEGER i,j,k,l;
      aq:-NEW matrix(n//2);
      i:=j:=1;
      IF select=2 THEN j:=n//2+1
      ELSE IF select=3 THEN i:=n//2+1
      ELSE IF select=4 THEN i:=j:=n//2+1;
      FOR k:=1 STEP 1 UNTIL aq.n DO
         FOR l:=1 STEP 1 UNTIL aq.n DO
            aq.val[k,l]:=val[i+k-1,j+l-1];
      quarter:-aq;
   END of procedure quarter;

   REF(matrix) PROCEDURE compose(a,b,c,d); REF(matrix)a,b,c,d;
   BEGIN INTEGER i,j;
      FOR i:=1 STEP 1 UNTIL a.n DO
         FOR j:=1 STEP 1 UNTIL a.n DO
            val[i,j]:=a.val[i,j];
      FOR i:=1 STEP 1 UNTIL b.n DO
         FOR j:=1 STEP 1 UNTIL b.n DO
            val[i,j+n//2]:=b.val[i,j];
      FOR i:=1 STEP 1 UNTIL c.n DO
         FOR j:=1 STEP 1 UNTIL c.n DO
            val[i+n//2,j]:=c.val[i,j];
      FOR i:=1 STEP 1 UNTIL d.n DO
         FOR j:=1 STEP 1 UNTIL d.n DO
            val[i+n//2,j+n//2]:=d.val[i,j];
      compose:-THIS matrix;
   END of procedure compose;
END of class matrix;

CLASS processor(size);
BEGIN
   REF(matrix)mat;
   REF(processor)one,two,three,four;
```

```
REF(matrix) PROCEDURE multiply(a,b); REF(matrix)a,b;
BEGIN REF(matrix)c;
    c:-NEW matrix(a.n);
    IF c.n=1 THEN c.val[1,1]:=a.val[1,1]*b.val[1,1];
    ELSE
    c.compose(one.mult&add(a.quarter(1),b.quarter(1),
    a.quarter(2),b.quarter(3)), two.mult&add(a.quarter(1),
    b.quarter(2),a.quarter(2),b.quarter(4)),
    three.mult&add(a.quarter(3),b.quarter(1),a.quarter(4),
    b.quarter(3)), four.mult&add(a.quarter(3),b.quarter(2),
    a.quarter(4),b.quarter(4)));
    multiply:-c;
END of procedure multiply;

REF(matrix) PROCEDURE mult&add(a,b,c,d); REF(matrix)a,b,c,d;
BEGIN REF(matrix)c1,c2; INTEGER i,j;
    c1:-one.multiply(a,b); c2:-two.multiply(c,d);
    FOR i:=1 STEP 1 UNTIL c1.n DO
        FOR j:=1 STEP 1 UNTIL c2.n DO
            c1.val[i,j]:=c1.val[i,j]+c2.val[i,j];
    mult&add:-c1;
END of procedure mult&add;
END of class processor;
```

Subdivision by row and column.   By subdividing the matrices differently, by row and column rather than as submatrices, we can perform the multiplication using fewer processors and taking greater advantage of the concurrency provided by the tree. This algorithm uses $2N^2 - 1$ processors and runs in $O(N^2)$ time. This is an order of N improvement in time over the sequential brute-force method.

Suppose we have a tree that has a branching ratio of N at each node and is two levels deep. That is, the root node has N descendents, each controlling N leaves of the tree. Then there are N^2 leaves and a total of $2N^2 - 1$ processors.

Each child node of the root, hereafter called a row supervisor, will represent a row of the multiplicand matrix and produce a row of the product matrix. Each of its N descendents will hold one element of the row it represents.

The multiplier matrix is loaded into the tree one element at a time, by column. The root hands each element to all row supervisors, which send it to their appropriate leaf: the first element in any column goes to the first child of each row supervisor, the Nth element to the Nth child. That child multiplies the multiplier element by the multiplicand element it holds and returns the product to the row supervisor. When an entire column of the multiplier has been loaded into the tree, each row supervisor takes the N products generated in its children, adds them, and returns one element in the corresponding column of the product matrix. That is, when the first column of the multiplier has been loaded into the tree, the first column of the product matrix is available, and so on.

This process can be pipelined to take $O(N^2)$ time. Thus the time it takes to load the N^2 elements of the matrices dominates the time complexity of the problem.

In the following SIMULA presentation of the algorithm described above, two different processor CLASSes are described. Each is a refinement of the processor that supported arbitrary branching ratios given in an earlier section. CLASS *Rowsupervisor* provides a template for the supervisory nodes that distribute the multiplier and sum the individual element products to produce an entry in the product matrix. CLASS *Leaf* defines the lowest level of the tree, capable of representing an element of the multiplicand and multiplying it by a number provided by the row supervisor. Both of these classes have two routines defined in them. *Load* is used to load the multiplicand matrix. *Multiply* is used to load the multiplier, perform the multiplication, and produce the product matrix.

```
Processor CLASS Rowsupervisor;
BEGIN

    !the matrix size, N, is an attribute of CLASS Processor, and is available
    to us;
    REAL product;
    INTEGER count;

    PROCEDURE Load(element); REAL element;
    BEGIN
        count: = count + 1;
        son[count].load(element);
        IF count = N THEN count: = 0;
    END of procedure Load;

    REAL PROCEDURE Multiply(element); REAL element;
    BEGIN
        count: = count + 1;
        product: = product + son[count].multiply(element);
        IF count = N THEN
        BEGIN
            multiply: = product;
            count: = 0;
            product: = 0.0;
        END;
    END of procedure Multiply;

    !initialization;
    count: = 0;
    product: = 0.0;
END of class Rowsupervisor;
```

```
Processor CLASS Leaf;
BEGIN

    REAL rowelement;

    PROCEDURE Load(element); REAL element;
    BEGIN
       rowelement:=element;
    END of procedure Load;

    PROCEDURE Multiply(element); REAL element;
    BEGIN
       multiply:=rowelement*element;
    END of procedure Multiply;

END of class Leaf;
```

A discussion of the two algorithms.   Let us reflect a moment on these two so-lutions to the matrix multiplication problem. The tree machine is basically a recur-sive structure. Yet the most obvious recursive solution to our problem, the sub-division into smaller matrices, does not yield the least complex algorithm. The second approach does not use the recursive decomposition of the matrix, but capitalizes instead on the independence and overlap of the sums and products of the solution matrix. We make use of the fact, for example, that the first column of the multiplier is used to generate the first column of the product, and never again. We do not need to preload the tree machine with both of the operand matrices, but we can overlap the loading of the multiplier with the generation of the product. This allows us to remove a factor of log N from the time complexity of the prob-lem.

What we learn from a comparison of these two algorithms is that mapping a recursive algorithm onto a recursive structure does not always yield an optimal solution. That is, the recursive algorithm is not necessarily maximally concurrent. By examining the individual steps of the process we want to accomplish, and finding the minimum set of states that require synchronization, we can do better.

Because the data elements travel around the tree less in the second algorithm, we can examine the possibility of chain multiplication, that is, multiplying more than two matrices together to produce the desired product matrix. Remember that one column of the product is generated for each column of the multiplier that is loaded into the machine. Each row supervisor contributes one element to that column. After all N columns have been generated, the jth row supervisor has produced all of the elements in the jth row of the product. If we were to load the product into the tree, we would want the jth row supervisor to receive the jth row.

If we introduce a mechanism for the row supervisor to keep the product elements as they are generated and dispense them to the appropriate leaf, we can avoid loading each intermediate matrix into the tree.

Suppose we use another memory cell in each leaf to store the value that will be used as the multiplicand element in the next multiplication. We also need a piece of state information that tells us when the multiplication in progress finishes and that it is the time to use the new multiplicand. Then when the jth column of the multiplier is being processed, the jth child of each row supervisor will be given the product element generated as its next value.

The time complexity for chain multiplication as described above is of the same order as if the intermediate products were generated explicitly: generated, read out, and reloaded as the multiplicand. However, there is a considerable saving in input/output operations if the chain multiply is done in place.

Suppose we want to multiply a chain of five $N \times N$ matrices together. Generating explicit intermediate products requires loading the tree with the N^2 elements eight times. If the multiplication is done in place, as our algorithm suggests, only the five operand matrices need be loaded. In general, a chain of M matrices can be multiplied with M loads in place, but it requires $2(M - 1)$ loads if intermediate values are generated explicitly.

The hexagonal array of processors described in Section 8.3 does matrix multiplication in $O(N)$ time using $3N^2 - 3N + 1$ processors. The linear time complexity is achieved by loading an entire row or column of the matrix in parallel.

If we widen the communication paths in our tree to handle an entire column in one burst, as indeed we must in order to make a fair comparison of the two architectures, we see that our tree can accomplish the matrix multiplication in linear time as well. Only one third of the processors in the hexagonal array are busy in a given cycle. Thus both the tree machine and the processor array use about the same number of active processors to solve the problem.

As in the sorting example, we see that a proper decomposition of the problem yields a solution on our general computing structure that is comparable to the performance of a special-purpose machine.

8.4.2.2 Solutions to NP-complete Problems

Complexity theory[9,10] has established a context within which it is possible to make certain statements concerning the inherent complexity of computations. These statements are universally couched in the terminology of sequential machines. There is, however, a class of problems for which the possibility of large-scale concurrency has been addressed.

Consider a computation in which there are N conceptual steps. At each step, Q alternative branches may be taken. Such a computation may be viewed as a tree with Q^N possible outcomes. If at each step there is enough information available to decide which branch to take, a sequential machine will be able to complete the computation in KN cycles, where K is the average number of cycles spent on each step. The dependence of the number of machine cycles upon the number of conceptual steps is thus *linear in N or of order N,* written O(N).

In many computations, not enough information has been generated by previous steps to determine which branch to take. Later steps will generate this information, but we cannot execute the later steps until after the earlier steps! In such cases, the sequential machine must simply try one branch at random. If it concludes after executing subsequent steps that the particular branch taken was wrong, it must backtrack to the original point and try another route.

In a wild flight of fancy, we might become frustrated with this behavior and wish we had a machine that was so smart that it could tell if it was on the right path, even if there was no possibility of choosing such a path with the information on hand. It would make an arbitrary choice at each branch—and always be right! Of course, such a machine cannot be built with real logic operating with real programs. However, we can imagine such a machine in much the same way we imagine a spaceship traveling faster than the speed of light. Machines of this sort are called *nondeterministic*, since there is no way a machine exhibiting this behavior can be specified on rational grounds.

Returning to our problem, it is clear that a sequential nondeterministic machine could solve the problem in O(N) cycles. Problems that can be solved by such an imaginary nondeterministic machine in a number of cycles that is bounded by some fixed power of N are said to be *Nondeterministic–Polynomial*, abbreviated *NP*.[9,10]

It is quite clear that the behavior of a nondeterministic machine can be simulated by a set of concurrent deterministic machines. Each machine can simply follow a separate path through the tree. At the end, there will be Q^N processors, representing each possible outcome of the computation. Although different problems will have different branching ratios (Q) and different depths (N), all can be mapped onto the tree machine using techniques described earlier.

It has been shown that there is a class of problems of this sort where there are no shortcuts. Working one path through to the end gives no clue concerning the outcome of another path. Such problems, called *NP-complete,* are, in some sense, maximally difficult among the problems in NP.

A great deal of lore has developed concerning NP-complete problems. It has been shown that, in a certain sense, they are all "equivalently hard."[29] Suppose machine Y can solve a single kind of NP-complete problem. The equivalence property states that there is an algorithm that can run on an ordinary sequential machine in a polynomial number of cycles that transforms a description of a problem into the one solvable by Y. If Y can solve its NP-complete problem in polynomial time, then it can be used to solve *any* NP-complete problem in polynomial time. If Y requires exponential time, *any* NP-complete problem will also require exponential time on machine Y.

The methods we use to describe trees of different branching ratios mapped onto a binary tree are similar to the methods used to map one NP-complete problem onto a machine that solves another. When a tree with branching ratio greater than 2 is mapped onto a binary tree, the depth of the tree increases. In a similar fashion, the algorithm that transforms NP-complete problems may increase the number of alternative branches (Q) and decrease the number of conceptual steps (N), or vice versa. Thus the mappings that establish the equivalence class of NP-complete problems are exactly like the mappings from trees of one branching ratio to another.

The theory that establishes the NP-complete equivalence class offers direct guidance in mapping such problems into a highly concurrent structure. Because we can solve any one problem in our concurrent tree machine, and because we know a mapping from an arbitrary NP-complete problem into this one, we can solve the arbitrary problem.

The traditional approach to solving the class of problems that grow non-polynomially has been to recognize space or processing power as a limited resource. The problems have exponential time complexity because the solutions proceed sequentially.

As VLSI becomes a reality, however, it is interesting to treat processors as an unlimited resource and look at the time complexity of these problems when they take advantage of ultraconcurrency. We emphasize, however, that while the time complexity is significantly reduced, problems require an exponential number of processors. A problem of reasonable size will use an enormous number of processors.

In a later section, an example is worked for an NP-complete problem that grows as N^N. The problem uses a graph of 4 nodes, and our concurrent solution requires 95 processors. A graph of 10 nodes could use as many as $2*10^{10}$ processors!

We will examine two NP-complete problems. The clique problem has time complexity of $O(2^N)$ when the possible cliques are considered sequentially. The color-cost problem is $O(N^N)$. By taking advantage of the parallel consideration of possible solutions, we present solutions to these two problems that take polynomial, in fact $O(N^2)$, time.

The Clique Problem

A clique is a complete subgraph. That is, given an undirected graph G, a clique C contained in G is a graph such that for all nodes n,m in C, there is an edge (n,m). Finding the largest clique in an arbitrary graph is an NP-complete problem.

Given a graph G with N nodes, numbered from 1 to N, we will consider each node sequentially and generate potential cliques. Ignoring the edges for a moment, a collection of M nodes leads to $2^M - 1$ potential cliques. This, interestingly enough, is the number of nodes in a binary tree of depth M. We will use this fact to generate the cliques in our graph incrementally.

Each level in the tree represents the addition of another node to be considered. Each processor at a given level will spawn two descendents. The left child will consider the subgraph consisting of the new node and all but the last node of the parent subgraph. The right-child's subgraph will add the new node to the complete parent subgraph. In this manner, we generate all possible subgraphs for a graph of N nodes. Figure 8.25 is an example for $N = 4$.

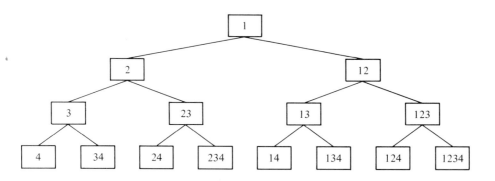

Fig. 8.25 Systematic generation of subgraphs in a graph of four nodes.

If each node stores an edge list, the tree can be pruned of subgraphs that are not cliques. The number of processors required is reduced, but the worst-case behavior is identical. A maximum of $2^N - 1$ processors is required to solve the problem for a graph of size N. Our solution, which uses pruning, requires $O(N^2)$ time.

Each processor stores the edge list as a Boolean matrix called *edge,* and an integer *size* that holds the size (number of nodes) of the clique this processor represents. An array called *clique* contains the numbers of the nodes that form the clique.

When a processor is activated by a call to procedure *Findclique,* it will already have a clique assigned to it. *Findclique*'s purpose is to generate cliques for its descendent nodes. It does this according to the method described above. That is, if the subgraph that contains the new node and all of the nodes in *clique,* except the last one, is a clique, it will be assigned to the left child. Likewise, if the addition of *node* to *clique* yields a complete subgraph, the right child will represent it. If either of the subgraphs is not complete, the descendent will not be generated.

The tree of all cliques is generated iteratively by considering each node of the graph in turn. In the main program given below, p is a reference to the root processor. Each processor in the tree will pass up the largest clique among its children. Thus the root returns the size of the largest clique known to date.

We now present the clique algorithm followed by a simple example.

```
CLASS processor;
BEGIN
    REF(processor)left, right;
    BOOLEAN ARRAY edge[1:n,1:n];
    INTEGER ARRAY clique[1:n];
    INTEGER size;

    BOOLEAN PROCEDURE IsClique;
    BEGIN INTEGER i,j;
        IsClique:=TRUE;
        FOR i:=1 TO size DO
            FOR j:=1 TO size DO
                IF NOT edge[i,j] THEN IsClique:=FALSE;
    END of procedure IsClique;

    REF(processor) PROCEDURE FindClique(node);
    BEGIN INTEGER i; REF(processor)l,r;
        l:=r:=THIS processor;
        IF size=0 THEN !then this is the root node;
        BEGIN
            clique[1]:=node
            size:=1
            FindClique:-THIS processor;
        END
        ELSE
        BEGIN
            IF left.size=0 THEN
            BEGIN
                FOR i:=1 TO size-1 DO left.clique[i]:=clique[i]
                left.clique[size]:=node
                left.size:=size;
                IF NOT left.IsClique THEN left.size:=0;
            END
            ELSE l:-left.FindClique(node);
            IF right.size=0 THEN
            BEGIN
                FOR i:=1 TO size DO right.clique[i]:=clique[i]
                right.clique[size+1]:=node
                right.size:=size+1;
                IF NOT right.IsClique THEN right.size:=0;
            END
            ELSE r:-right.FindClique(node);
            IF l.size>size THEN
            BEGIN
                IF l.size>r.size
                THEN FindClique:-l
                ELSE FindClique:-r;
            END
```

```
        ELSE IF r.size>size
    THEN FindClique:-r
        ELSE FindClique:-THIS processor;
    END;
  END of procedure FindClique;
  size:=0;
  !left and right set up correctly;
  !read in edge list;
END of class processor;

!main program to start it all up;
BEGIN REF(processor)largest; INTEGER i
  FOR i:=1 TO n DO largest:-p.findclique(node);
END of main;
```

Figure 8.26 gives a sample graph of six nodes. Figure 8.27 shows the processor tree that is built and used to find the cliques in the graph. The tree has height 6, and the largest cliques have 4 nodes. Each processor in the tree represents a clique in the graph.

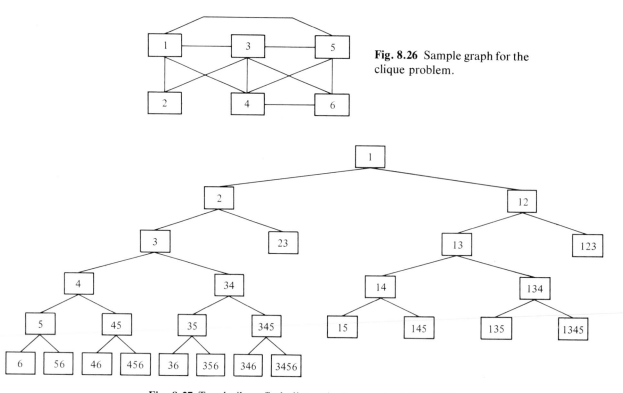

Fig. 8.26 Sample graph for the clique problem.

Fig. 8.27 Tree built to find cliques in the graph of Fig. 8.26.

The Color-Cost Problem

This NP-complete problem is an adaptation of the K-colorability problem. Given an undirected graph G of N nodes and a set of N colors, each with an associated cost, we want to find a minimum cost coloring of the graph such that no nodes sharing an edge are the same color.

There are N^N possible colorings of the graph. Evaluating them sequentially produces a solution in time $O(N^N)$. We present a parallel algorithm of order N^2.

In this problem we make use of the ability to simulate arbitrary branching ratios on our binary tree. We discuss the problem in terms of logical nodes with up to N descendents. An earlier section describes the method of mapping logical structures onto the physical one.

As in the clique problem, each level in the processor tree represents the consideration of another node. That is, level one shows possible colors for the first node, level two colors the second node based on the choices made for level one, and so on. We will describe the generation of the potential colorings.

Each node has an edge list called "edge" and a list of costs indexed by color number called "colorcosts." There is an array called "coloring" that reflects the color choices for preceding nodes, and there is a Boolean array called "colors" that is used to generate the possible colorings for this node.

The algorithm given in procedure "color" begins by assuming that all colors yield valid colorings. The array "coloring" is used to eliminate those colors that have been used to color nodes that share an edge with this node. This reduced set of colors, all of which are legal colorings, is used to spawn descendents, one for each coloring of this node.

When the tree is N levels deep, all the legal colorings have been generated. The leaf nodes calculate a cost for the coloring they represent, and each parent node takes as its cost the least cost among its children. Thus the minimum cost coloring is stored at the root.

Here is the color cost algorithm that will run in each processor.

```
CLASS processor;
BEGIN

    BOOLEAN ARRAY edge[1:n,1:n],colors[1:n];
    INTEGER ARRAY coloring[1:n],colorcosts[1:n];
    INTEGER cost;

    PROCEDURE color(node); INTEGER node;
    BEGIN INTEGER i;
      IF node>n THEN
      BEGIN
        cost:=0;
        FOR i:=1 TO node-1 DO cost:=cost+colorcost[coloring[1]];
      END
```

```
        ELSE
        BEGIN
          FOR i: =1 TO node-1 DO IF edge[1,node] THEN
          colors[coloring[i]]:=FALSE;
          FOR i: =1 TO n DO
             IF colors[i] THEN
             BEGIN
                son(i).coloring[node]:=i
                son(i).color(node+1);
             END
             ELSE son(i):-NONE;
          cost: =maxcost;
          FOR i: =1 TO n DO
             IF (IF son(i) = NONE THEN FALSE ELSE cost>son(i).cost)
             THEN cost: =son(i).cost;
        END;
    END of procedure color;

    REF(Processor) PROCEDURE Son(s); INTEGER s;
    BEGIN REF(node)p;
      p:-IF s<=(n+1) // 2 THEN left ELSE right;
      WHILE p IN Processor DO
          p:-IF s<=(p.n+1) // 2 THEN p.left ELSE p.right;
      Son:-p;
    END of PROCEDURE Son;

END of class processor;
```

Let us work a small example. We will use the graph and color set given in Fig. 8.28. Color Plate 16 shows the colorings and costs arrived at by the algorithm. Each level of the tree represents a node of the tree. That is, if the root is level 0, the first node is colored in level 1, and level 4 represents potential colorings for the fourth node. Besides representing a part of a coloring, each node also contains the minimum cost coloring found among its descendent colorings.

We see that there are two equivalent colorings that yield the minimum cost of 3. Coloring nodes (1,2,3,4) either (green,blue,red,blue) or (red,blue,green,blue) gives us a coloring with minimum cost.

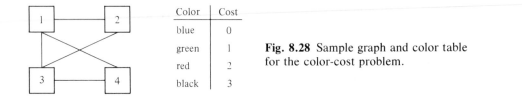

Color	Cost
blue	0
green	1
red	2
black	3

Fig. 8.28 Sample graph and color table for the color-cost problem.

8.4.3 Conclusions

The tree of processors we have described is a general computing structure. Each node in the tree is a processor with general computing capability. It is not designed with a specific problem or class of problems in mind.

The most dramatic results are achieved when the machine is applied to a problem that can take advantage of the ultraconcurrency provided by the tree of processors. We have presented solutions to four problems that, in varying degrees, have this characteristic.

The four examples presented can be summarized by citing the execution time and number of processors required. Note that the total chip area of a tree machine is related to the number of processors.

Problem	Time	Processors
Sorting	N	N
Matrix Multiplication	N^2	$2N^2 - 1$
Clique	N^2	$2^N - 1$
Color Cost	N^2	$2N^N - 1$

If an algorithm exhibits exponential growth, as do the clique and color-cost problems, the lower bound on time complexity is N. A tree with an exponential number of leaves will be $O(N)$ deep. Again, our solutions do not realize this lower bound. The loading of the edge matrix is an $O(N^2)$ operation. Additionally, each node of the graph is considered in turn and causes the traversal of a tree of depth up to N. This too is of $O(N^2)$ in time. Are there better algorithms that can achieve the lower bound complexity? And can a statement be made about the concurrency of the NP-complete problems in general?

Because we are used to designing machines for a sequential environment, we do not yet understand the effect that ultraconcurrency will have on the conceptualization of problem solutions. An open question is to characterize those problems that can benefit from the concurrency provided by our tree of processors. Are the communication paths of the tree adequate for this class of problems? Can we design algorithms with the traditional programming notations, or does their sequential nature hide the concurrency? What are the problem-independent procedures that should reside in each processor to facilitate communication among the nodes? And, given a fixed number of processors, can we algorithmically divide the problem to maximally use the nodes available to it? These are just a few of the interesting questions that arise from the study of a concurrent environment.

8.5 HIGHLY CONCURRENT STRUCTURES WITH GLOBAL COMMUNICATION*

This section presents an analysis of the constraints placed by physical laws on a VLSI system in which information must be communicated from any location to

---

*Adapted from a paper by Carver Mead and Martin Rem.[30]

any other. We will analyze in detail the requirements that global communication places on the design of such a computing structure.

There has previously been no adequate theoretical basis for optimizing the overall organization of systems implemented in the VLSI technology. Conventional complexity theory is inadequate because its measure of cost is the number of steps taken by a *sequential* machine to complete the computation. No account is taken of the size of the machine (and hence the time required for each step). Possible concurrency is ignored, thereby ruling out the most important potential contribution of the silicon technology. Traditional switching theory is also inadequate: it provides a beautiful formalism for describing elementary logic functions but its optimization methods concern themselves with logical operations rather than communication requirements. Even in today's integrated circuits, the wires required for communicating information across the chip account for most of the area. Driving these wires accounts for most of the time delay and energy dissipation. In very large scale integrated systems, the situation becomes even more extreme, a point considered in detail in Chapter 7. In this section, we describe a method by which the conceptual organization of a large chip can be analyzed and a lower bound placed on its size, cycle time, and energy dissipation, before a detailed design is undertaken. The results of this analysis suggest rather general guidelines for the organization of all large integrated systems.

8.5.1 Metrics of Space, Time, and Energy

8.5.1.1 Physical Properties

Devices used to construct monolithic silicon integrated circuits are universally of the charge-controlled type. A charge Q placed on the control electrode (gate, base, etc.) results in a current $I = Q/\tau$ flowing through the device. The transit time τ is the time required for charge carriers to move through the active region of the device.

All times in an integrated system can be formulated as simple multiples of τ. For one transistor to drive another identical to it, a charge Q must flow through its active region, requiring time τ. If the capacitance C_L of the load driven is K times the gate capacitance C_g of the driving transistor, a time $K\tau = (C_L/C_g)\,\tau$ is required. Likewise, the elementary energy associated with the signal charge Q on the gate capacitance C_g is $E_0 = C_g V^2/2$. A load capacitance KC_g requires an energy KE_0. Since wires have a minimum width, their capacitance is directly proportional to their length. Thus the energy required to transmit a signal from one point on the chip to another is proportional to the distance separating the two points. As the unit of length we employ the minimum spacing of two conducting paths. For the unit of time we choose the time it takes a minimum-sized transistor to charge a wire of unit length plus another transistor like itself. One unit of time is thus slightly larger than the transit time of a transistor.

8.5.1.2 Advantages of hierarchical structures

We are considering large integrated systems in which it is necessary to communicate information throughout the entire system. As an example, consider a bit of information stored on the gate of a minimum-sized transistor in a random-access memory. The bit must be communicated to the memory bus of a CPU. Since there are many words of data in the memory, there are many possible sources for each wire in the memory bus. Figure 8.29 illustrates two possible approaches to organizing such a bus. In part (a), a transistor associated with each bit drives the bus wire directly. If the bus wire has a capacitance C_W, the time required to drive the bus wire is $t = \tau (C_w/C_g)$. In a typical computer memory C_w is many orders of magnitude larger than C_g, and the delay introduced by such a scheme is very long. Since C_w is proportional to the length of the wire, it is also proportional to S, the number of driver transistors connnected to the wire, and to b_0, the spacing between transistors. The time is thus,

$$t = b_0\tau S. \tag{8-1}$$

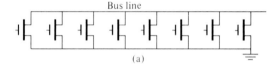

(a)

Fig. 8.29(a) A bus driven directly by memory cells.

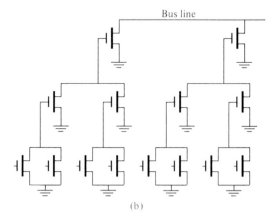

(b)

Fig. 8.29(b) A ``conceptual'' bus driver tree.

A second scheme is shown in part (b). Here each transistor drives a wire only long enough to reach its neighbor. Each such wire is connected to the gate of a transistor twice as large as the transistor driving it. The arrangement is repeated upward until the top level where all sources have a path to the bus. In this scheme the delay in driving the lowest level wire is approximately $2\tau b_0$. The delay introduced by the wires at each level is the same, since each driver transistor and each wire is twice as large as those driving it. Hence the delay in driving the bus line is $2\tau N b_0$ where N is the number of levels in the structure. Since there are $S = 2^N$ transistors at the lowest level, the delay may be written:

$$t = 2\tau b_0 \log_2 S. \tag{8-2}$$

Comparing Eqs. (8–1) and (8–2), we see that for large S the delay has been made much shorter by using a hierarchical structure.

8.5.1.3   A Cost Criterion

A hierarchy such as that shown in Fig. 8.29(b) may use any integral number, α, of transistors driving each wire. We refer to α as the *branching ratio* of the driver hierarchy. The driver transistors will in general be α times the size of those driving them. The delay for such a structure is $t = \alpha \tau b_0 \log_\alpha S = b_0 \tau (\alpha/\log \alpha) \log S$, dependent upon the branching ratio of the hierarchy. This delay is plotted in Fig. 8.30, normalized to its minimum value which is attained at $\alpha = e$.

While dramatic improvements in the performance of integrated structures can be achieved by a hierarchical organization, a penalty is always paid in the area required for wires. In the simple case shown, a bus requiring one wire when driven directly requires $\log_\alpha S$ wires when organized as a hierarchy. For this reason it is not possible to optimize a design without a cost function involving both area and

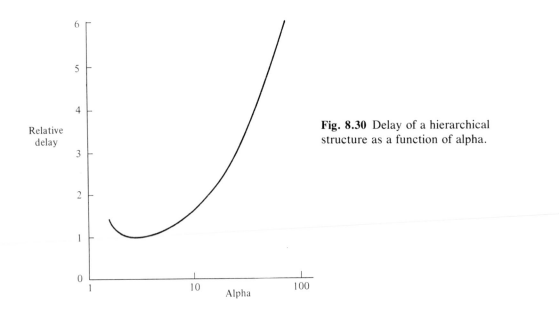

Fig. 8.30 Delay of a hierarchical structure as a function of alpha.

time. We will use the area-time product as an example of such a cost function. Other cost functions may be more appropriate under some circumstances. For the above simple example, the cost function is area*time $= b_0^3 \tau S (\log S)^2 \alpha/(\log \alpha)^2$. This cost is minimized for $\alpha = e^2 \approx 7.4$.

8.5.1.4 Hierarchical Computing Systems

The analysis given above suggests a very general structure for computing systems. Lowest level cells are grouped together into modules in such a way that α cells drive their outputs onto an output wire. Each output wire is connected to a driver transistor that is α times as large as those driving the wire. Modules are grouped in such a way that α of those module's drivers are connected to an inter-module communication wire. This wire in turn is connected to a driver transistor α^2 times as large as the lowest level transistors. This process is continued until the appropriate size system has been realized. Notice that the area of the driver transistor for each wire in such a structure is proportional to the area of the wire. For this reason, we compute only the area of the wires. The drivers somewhat enlarge the unit of wire area, but do not change the functional form of the solutions.

8.5.2 Random-Access Memory—An Example

In this section we discuss the design of a large random-access memory (RAM) of S bits. We will apply a rigid structural discipline to our design, and compute the cost and performance of the resulting memory.

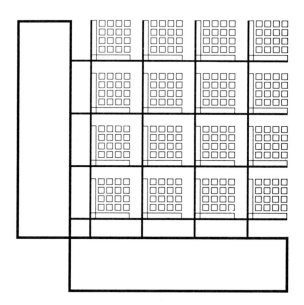

Fig. 8.31 Three levels of a memory hierarchy with alpha = 4.

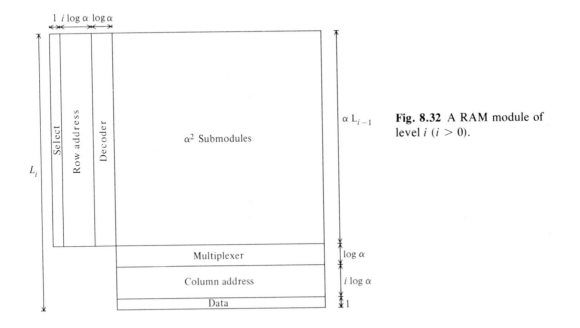

Fig. 8.32 A RAM module of level i $(i > 0)$.

8.5.2.1   *Organization of the RAM*

We organize the RAM in a hierarchical fashion. The elements of level 0 are the bits themselves, each bit consisting of two crossing wires: a select wire and a data wire. When the select wire is driven, the element puts its contents on the data wire. We group α^2 bits into an $\alpha \times \alpha$ square to form a module of level 1. If the width of an element (a bit) is b_0 the elements have to drive wires of length αb_0. A module on level 1 consists of an array of horizontal select and vertical data wires, constituting the α^2 bits of level 0, and some additional logic and wires at the side. We group again α^2 of these modules into a square to form a module of level 2, etc. Figure 8.31 shows three levels of the hierarchy for $\alpha = 4$.

To study the memory in more detail we look at a module of level i (Fig. 8.32). We describe how to extract one of its α^{2i} bits. In order to select one bit of storage, $2i\log\alpha$ address wires are required. We run $i\log\alpha$ of them, called the row address wires, vertically along the side of the module, and the other $i\log\alpha$, the column address wires, horizontally. Its α^2 submodules are organized into α rows of α submodules each. When the select wire of the module is driven, $\log\alpha$ of the row address wires are used (by the decoder) to select one of the α rows of submodules; the select wire running through that row is driven. The other $(i - 1)\log\alpha$ row address wires are run horizontally into each of the α rows of submodules, where they serve as column address wires for the submodules. Of the $i\log\alpha$ column address wires, $(i - 1)\log\alpha$ are run vertically into each of the α columns of submodules, where they serve as row addresses. The other $\log\alpha$ address wires are used by the multiplexer to select one of the α data wires coming out of the columns of submodules. The signal on the selected data wire is driven onto the data wire of the module itself.

If we wish to have a memory of S words with $N + 1$ levels (level 0 through N) we choose $N = \log S/2\log\alpha$, or $S = \alpha^{2N}$. This gives a hierarchical structure with S bits from which we can extract one bit at a time. If we want the word length to be $\log S$, we employ $\log S$ of these structures in parallel: to select one word we select one bit in each of the $\log S$ hierarchies.

8.5.2.2 Area of the RAM

Figure 8.33 allows us to compute the size of a RAM. Let L_i denote the width of a module of level i; then we have the following recurrence relation:

$$L_0 = b_0$$
$$L_i = i\log\alpha + 1 + \log\alpha + \alpha L_{i-1}.$$

The solution to the above relation is

$$L_i = \alpha^i b_0 + \frac{\alpha^i - 1}{\alpha - 1} + \left(\frac{2\alpha^{i+1} - \alpha^i - \alpha}{(\alpha - 1)^2} - \frac{i + 1}{\alpha - 1} \right) \log\alpha.$$

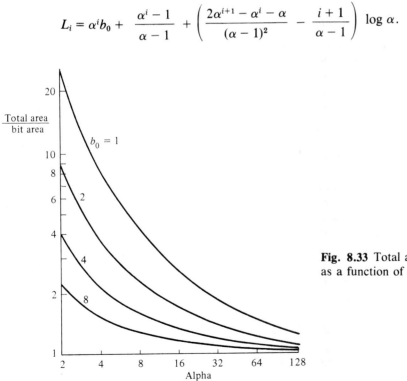

Fig. 8.33 Total area per bit of a RAM as a function of alpha.

We are interested in the width per bit, rather than the width itself. In one direction, horizontal or vertical, module i has α^i bits; we therefore compute L_i/α^i:

$$\frac{L_i}{\alpha^i} = b_0 + \frac{1}{\alpha - 1} + \frac{2\alpha - 1}{(\alpha - 1)^2}\log\alpha - \frac{1}{(\alpha - 1)\alpha^i}\left[\left(\frac{\alpha}{\alpha - 1} + 1 + i\right)\log\alpha + 1\right]. \quad (8\text{–}3)$$

An interesting property of the width per bit, as expressed by Eq. (8–3), is that its limit for $i \rightarrow \infty$ is finite:

$$\lim_{i \rightarrow \infty} \frac{L_i}{\alpha^i} = b_0 + \frac{1}{\alpha - 1} + \frac{2\alpha - 1}{(\alpha - 1)^2}\log\alpha. \quad (8\text{–}4)$$

This means that the width per bit L_i/α^i is bounded from above by Eq. (8–4) independent of the number of levels of a RAM. Expression (8–3) converges in an exponential fashion toward its limit: for small values of i, (8–3) is already very close to (8–4). We therefore use (8–4) as the width per bit for a RAM; its square is then the area per bit. By dividing the area per bit by the bit area b_0^2 we obtain the total area per bit area for a RAM. Figure 8.33 shows this quotient as a function of α for four different values of b_0. It gives the overhead factor in the area that is due to the wires. For a memory of 64K bits with $N = 2$, α should be 16. Expression (8–4) is then roughly equal to $b_0 + 0.6$. This shows that in 3-level 64K dynamic MOS memories, for which b_0 lies between 1 and 2, roughly half of the area will be occupied by wires.

One may wonder why we have not discussed the area that is consumed by the wires for power and ground. The reason for this is that these wires can be thought of as increasing only the width b_0 of each bit; they do this by an amount that is roughly independent of α, as is shown in the following analysis.

For simplicity we assume that the wires for power and ground run in orthogonal directions, say parallel to the data and select wires. We compute how much one of them contributes to the width of a module i. As discussed in Chapter 2, the current density that can reliably be carried by a metal conductor is limited by metal migration phenomena. Assuming that all wires are of the same thickness, the width of a power or ground wire must be proportional to the current it carries and hence to the number of bits served by it. Let the width at the highest level be u; given S and the design of the lowest level memory cell this parameter is easy to compute. The width of the wire in a module on level i is proportional to the current it must supply and is hence $u(\alpha^{2i}/\alpha^{2N})$. In one direction, horizontal or vertical, there are α^N/α^i such modules. The total contribution of all modules on level i is thus $u\,(\alpha^i/\alpha^N)$. Taking the sum of this expression for $i = 0, 1, \ldots, N$ yields

$$\frac{u}{\alpha^N}\frac{\alpha^{N+1} - 1}{\alpha - 1} \approx u\frac{\alpha}{\alpha - 1}.$$

There are $\sqrt{S}$ bits in one direction. The increase of the bit width, due to power and ground, is therefore

$$\frac{u}{\sqrt{S}} \frac{\alpha}{\alpha - 1},$$

which is roughly equal to

$$\frac{u}{\sqrt{S}}.$$

We are interested in the optimal choice of α, but to make that choice we will have to look at the access time as well, which also depends on α.

8.5.2.3 Access Time of the RAM

Each element of level 0 drives a wire of length αb_0 to reach the periphery of its module on level 1; this takes time αb_0. Each module on level 1 drives in the same amount of time a wire that is α times longer to reach the periphery of its module on level 2, etc. With N being the level of the highest module, the time required to extract one bit of storage adds up to $\alpha b_0 N$. We use this figure as the access time. For a RAM of S words, the access time in units of τ is then $\alpha b_0 (\log S/2\log \alpha)$.

8.5.2.4 The Cost of the RAM

We take the product of the area and the access time as the cost function of the RAM. A RAM of S words of $\log S$ bits each has the following area-time product:

$$\left(b_0 + \frac{1}{\alpha - 1} + \frac{2\alpha - 1}{(\alpha - 1)^2} \log \alpha \right)^2 \frac{\alpha b_0}{2 \log \alpha} S \log^2 S . \tag{8-5}$$

Figure 8.34 shows (8–5), normalized with respect to $S\log^2 S$, as a function of α for different values of b_0. One notices that for increasing bit sizes the branching ratio of the hierarchy should decrease. Static memories should therefore have a smaller α than dynamic ones. For dynamic MOS memories the optimal choice for α lies between 8 and 16; for static MOS memories ($b_0 \approx 4$), between 4 and 8. One may speculate that "smart memories" (structures in which part of the processing task is distributed over the memory cells) will have small branching ratios and hence relatively deep hierarchies.

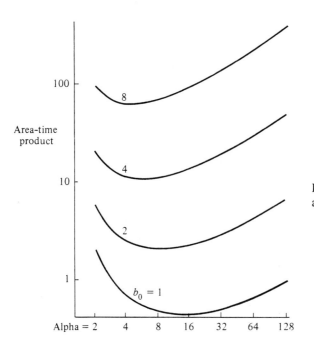

Area-time product

Fig. 8.34 Area-time product of a RAM as a function of alpha.

8.5.2.5 *Energy per Access*

In real systems, the cost of power, cooling, and electrical bypassing often exceeds the cost of the chips themselves. Hence any discussion of the cost of computation must include the energy cost of individual steps of the computation process. In a RAM, each access costs an energy proportional to the length of the wires that must be charged or discharged during a given cycle. Consider a RAM such as that shown in Fig. 8.32. At the highest level (level N) such a device has $S = \alpha^{2N}$ bits. In each cycle, $\log S$ address wires of length L_N will in general change state. In addition one horizontal select line, α vertical data lines, and one multiplexer output line (all of length L_N) will change state. Thus at level N, the energy expended per access will be

$$E_N \approx L_N [\log S + \alpha + 2].$$

At level $N - 1$, $2 \log \alpha$ fewer address wires will be needed. Since only one select line will be active, only α of the α^2 submodules will be active. Each submodule contains wires approximately $1/\alpha$ as large as those at level N.

Thus the total energy per access is

$$E_T \approx L_N [\log S (1 + 1 - 2 \log \alpha/\log S + 1 - 4 \log \alpha/\log S + \cdots) + \alpha + 2].$$

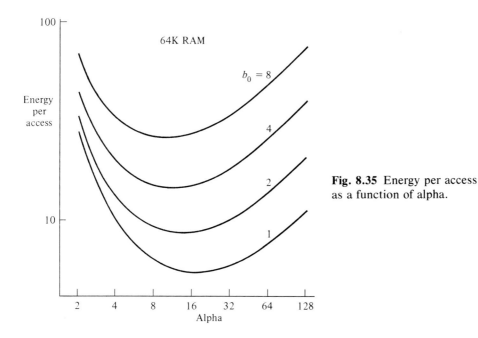

Fig. 8.35 Energy per access as a function of alpha.

This expression evaluates to

$$E_T \approx L_N \log S / \log \alpha [\log S/4 + (\alpha + 2)/2].$$

Using the values from Eq. (8–4), the energy per access of any given size RAM may be evaluated. The results of such an evaluation for a 64K bit RAM are shown in Fig. 8.35. These curves suggest that considerably less power would be required if memory chips, even of current size, were built with smaller submodules and smaller α.

8.5.3 Content Addressable Memory

The basic elements of the RAM are bits. The content addressable memory (CAM) is an example of a word organized memory. We consider a "pure" CAM. It consists of words of w bits each. We access a word by applying w bits of data to the system. We assume that there is only one word in the memory with that content, and the address of that word is produced by the memory.

8.5.3.1 *Organization of the CAM*

The basic elements are the bits of length and width b_1. For reasons that become clear later, the bits do not constitute the modules of level 0: the modules on level 0

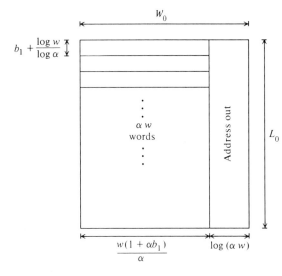

L_0 **Fig. 8.36** A CAM module of level zero.

of the hierarchy consist of αw words of w bits each (see Fig. 8.36). The w data bits are run via parallel wires vertically through the module. Out of each word comes one horizontal match wire going to the right. A word drives its match wire if each data bit received is equal to the corresponding bit stored. When there are αw words in a module of level 0, the address of the matching word leaves the module via the $\log \alpha w$ address wires.

The above organization of a module of level 0 has one defect: it would require the individual bits of storage to drive wires of length $w b_1$, which might be greater than the desired αb_1, to reach the address wires. In Section 8.5.1.2 we concluded that this type of communication should be achieved by hierarchy. We therefore organize the driving of the match wire by the w bits in a word in the same manner as shown in Fig. 8.29(b).

Each word is chopped up into w/α subwords of α bits each (Fig. 8.37). Each of the w/α subwords sends a signal to a "match tree" that has a branching ratio of α and delivers, via $\log_\alpha w$ levels, the logical product of its inputs. The top node of the match tree can drive a wire of length $b_1 \alpha \log_\alpha w = b_1 w$, the length of a word in the memory. Therefore, the word itself can drive a wire of length $b_1 \alpha w$ and we may

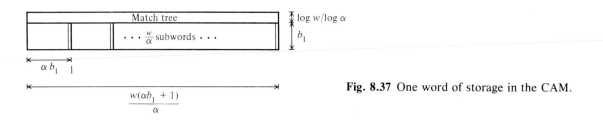

Fig. 8.37 One word of storage in the CAM.

group together αw words into module 0 (Fig. 8.36). Notice that the module's length is roughly equal to α times its width. This will be true for modules on higher levels as well.

We now describe a module of level i (Fig. 8.38). It contains $w\alpha^{4i+1}$ words and consists of α^4 submodules of level $i-1$, grouped into α^2 rows of α^2 submodules each. Each such row contains, besides the α^2 submodules, w data wires to transport the data to each of the submodules and $\log w \; \alpha^{4i-1}$ outcoming address wires to transport to the right the address of the matching word. Each submodule has $w\alpha^{4i-3}$ words and hence one row contains $w\alpha^{4i-1}$ words, thus explaining the number of address wires. A module on level i has α^2 of these rows and thus requires $\log w \; \alpha^{4i+1}$ outcoming address wires; they are placed to the right of the rows.

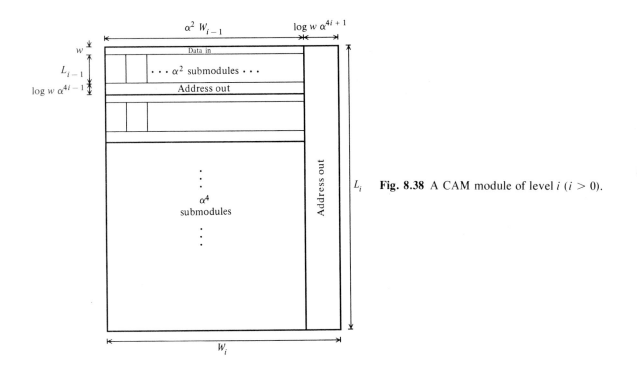

Fig. 8.38 A CAM module of level i ($i > 0$).

In the CAM we have α^4 submodules per module, in the RAM only α^2. This is only an apparent difference: in the CAM we have, for simplicity, combined two steps in the hierarchy; we have however maintained our multiplication factor α for the wire lengths. The length of a module of level $i - 1$, L_{i-1}, is roughly equal to α times the width of a module of level $i - 1$, W_{i-1}. Therefore, module $i - 1$ can already drive wires of length αW_{i-1}. As a consequence, we can put α^2 submodules into one row as this would require the driving of wires only of length $\alpha^2 W_{i-1}$ in each row.

But then we can, and this is the second step, combine α^2 rows as this would require the driving of wires of length about $\alpha^2 L_{i-1}$, which is roughly equal to $\alpha^3 W_{i-1}$.

8.5.3.2   Area of the CAM

We compute the length and the width separately. For the length of a module on level i, L_i, we have the relation (cf. Figs. 8–36 and 8–38)

$$L_o = \alpha w \left(b_1 + \frac{\log w}{\log \alpha} \right),$$

$$L_i = \alpha^2 (w + L_{i-1} + \log w \alpha^{4i-1}).$$

The solution to this recurrence relation is

$$L_i = \alpha^{2i+1} w \left(b_1 + \frac{\log w}{\log \alpha} \right) + (w + \log w) \frac{\alpha^{2i+2} - \alpha^2}{\alpha^2 - 1}$$

$$+ \left(\frac{4\alpha^{2i+2} - 4\alpha^2}{(\alpha^2 - 1)^2} + \frac{3\alpha^{2i+2} - 4i\alpha^2 - 3\alpha^2}{\alpha^2 - 1} \right) \log \alpha .$$

A module on level i has $w\alpha^{2i+1}$ bits in the vertical direction. The length per bit is therefore $L_i/w\alpha^{2i+1}$. This has the following limit for $i \rightarrow \infty$:

$$b_1 + \frac{\log w}{\log \alpha} + \frac{\alpha(w + \log w + 3\log \alpha)}{w(\alpha^2 - 1)} + \frac{4\alpha \log \alpha}{w(\alpha^2 - 1)^2} . \tag{8-6}$$

As in the case of the RAM, $L_i/w\alpha^{2i+1}$ is already very close to the limit for small values of i; the rate of convergence is again exponential. We use expression (8–6) as the length per bit of a CAM.

We find for the width of a module on level i, W_i, the following recurrence relation (cf. Figs. 8–36 and 8–38):

$$W_0 = \frac{w}{\alpha} (\alpha b_1 + 1) + \log \alpha w ,$$

$$W_i = \alpha^2 W_{i-1} + \log w \alpha^{4i+1} .$$

Its solution is

$$W_i = \alpha^{2i}w \left(b_1 + \frac{1}{\alpha} \right) + \frac{\alpha^{2i+2} - 1}{\alpha^2 - 1} \log w$$

$$+ \left(\frac{4\alpha^{2i+2} - 4\alpha^2}{(\alpha^2 - 1)^2} + \frac{\alpha^{2i+2} - 4i - 1}{\alpha^2 - 1} \right) \log \alpha.$$

In the horizontal direction there are $w\alpha^{2i}$ bits. The width per bit, $W_i/w\alpha^{2i}$, has the following limit for $i \to \infty$:

$$b_1 + \frac{1}{\alpha} + \frac{\alpha^2 \log \alpha w}{w(\alpha^2 - 1)} + \frac{4\alpha^2 \log \alpha}{w(\alpha^2 - 1)^2}. \qquad (8\text{-}7)$$

We take the product of expressions (8–6) and (8–7) as the area per bit.

By dividing the area per bit by the bit area b_1^2 we obtain the total area per bit area for a CAM. Figure 8.39 shows this quotient for $w = 32$ as a function of α for different values of b_1.

If we compare Figs. 8.33 and 8.39 we notice that for small values of α the wires in the CAM cause less overhead in area than those in the RAM. For large values of α, it is the RAM that enjoys a smaller overhead in area. For equal bit

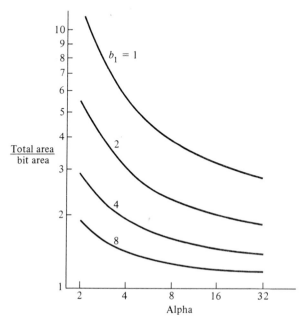

Total area
bit area

Alpha

Fig. 8.39 Total area per bit as a function of alpha for a CAM with word length 32.

sizes, that is, with $b_0 = b_1$, the area overhead factor for the RAM and the CAM are about equal at $\alpha = 8$.

As in the RAM we can compute by how much we should increase the bit size b_1 if we wish to take power and ground into account. We leave it as an exercise for the readers to convince themselves that both power and ground give an increase of $u(\alpha^2/(\alpha^2 - 1))$ to the length and the width of the CAM. This is even closer to u than in the case of the RAM. If we wish to amortize this amount over the bits, the bit width b_1 should be incremented by $(2u/\sqrt{Sw})(\alpha^2/(\alpha^2 - 1))$ for a CAM of S words of w bits each.

8.5.3.3 Access Time of the CAM

For the access time we take the time required to extract the address of the matching word of data from a memory of S words. With the highest level being level N we have $S = w\alpha^{4N+1}$, or

$$N = \frac{\log S - \log w}{4 \log \alpha} - \frac{1}{4}.$$

A word of storage has a response time of $(\log w/\log \alpha)\alpha b_1$; for a module of level 0 this becomes $((\log w/\log \alpha) + 1)\,\alpha b_1$. Each new level of the hierarchy multiplies the wire lengths by a factor α^2 and hence requires an additional time of $2\alpha b_1$. For N levels we find

$$\text{access time} = \left(2N + \frac{\log w}{\log \alpha} + 1\right)\alpha b_1 = \left(\frac{\log S + \log w}{2\log \alpha} + \frac{1}{2}\right)\alpha b_1. \qquad (8\text{--}8)$$

8.5.3.4 The Cost of the CAM

We take again the product of the area and the access time as the cost function. For a CAM of S words of w bits each, formulas (8–6), (8–7), and (8–8) yield the cost function

$$\left(b_1 + \frac{\log w}{\log \alpha} + \frac{\alpha(w + \log w + 3\log \alpha)}{w(\alpha^2 - 1)} + \frac{4\alpha \log \alpha}{w(\alpha^2 - 1)^2}\right)$$

$$\times \left(b_1 + \frac{1}{\alpha} + \frac{\alpha^2 \log \alpha w}{w(\alpha^2 - 1)} + \frac{4\alpha^2 \log \alpha}{w(\alpha^2 - 1)^2}\right)\left(\frac{\log S + \log w}{2\log \alpha} + \frac{1}{2}\right)\alpha b_1 w S.$$

Figure 8.40 shows the cost function as a function of α for a CAM of 64K words of 32 bits each. The curves are fairly independent of the choice of w provided we

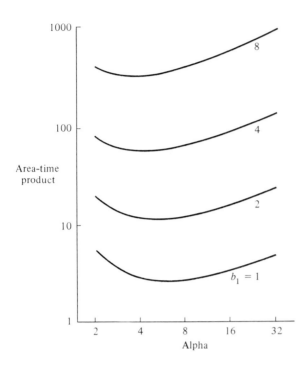

Fig. 8.40 Area–time product for a CAM of 64K 32-bit words.

choose w great enough, say $w \geq 16$. A change in S will basically move the curves only up and down, it will not affect the positions of their minima.

We notice again that increasing the bit size will decrease the optimal choice of α. Comparing Figs. 8.34 and 8.40 we see that content addressable memories should have smaller branching ratios than random access memories. For $b_1 = 4$, which seems a reasonable figure, the optimal choice of α is 4.

8.5.3.5 General Method of Analysis

We have presented a general method for analyzing the cost and performance of recursively defined VLSI structures. Parameters of any such structure may be optimized with respect to some combination of access time, area, and energy.

The results of this study indicate that as more processing is available in each module at level zero, b_0 will be larger and the optimal value of α will decrease. A system with $\alpha = 4$ would seem to be appropriate for structures in which substantial processing is commingled with storage.

Very general arguments were used to generate the basic recursive structure. For that reason it appears that a very large fraction of VLSI computing structures will be designed in this way. The way in which the area, time, and energy measures were established should make it clear how to apply these techniques to other recursively defined computing structures.

8.6 CHALLENGES FOR THE FUTURE

We have seen that it is possible to construct general-purpose computing engines that exploit tremendous concurrency, if computations are properly matched to the machine. The vast quantity of concurrency available in such machines can be an enormous help with the computing tasks we face. However, to date we have no formal way of making the possible concurrency in any given calculation apparent or finding if we have come close to the possible concurrency inherent in the computation.

Our present use of concurrent processing is limited in part by our inability to escape the strong hold that the conventional sequential machine exerts on our thinking. We must approach problems with concurrency in mind, recognizing that communication is expensive and that computation is not. Progress in these endeavors will surely increase when some VLSI computers of the sort we have illustrated in this chapter begin to appear. When the effort of casting the problem as a structure of concurrent processes is rewarded by a tangible increase in performance, the incentive to design concurrent algorithms and machines will surely increase.

The tools that we use to design and implement concurrent processes are primitive. We are badly in need of notation or language that expresses the power and constraints of highly concurrent machines. Whether such machines are general- or special-purpose, a natural way is needed to map problems onto them. Only in this way will it be possible for applications to rapidly find their way into execution in this new computing environment. In addition we need a method of formally proving the correctness of algorithms mapped onto such machines; it is not possible for human programmers to keep track of the exact relationship of the enormous number of tasks executing on such a machine. An ideal notation would allow expression of only those operations that are free of obvious fatal errors such as deadlock.

Perhaps the greatest challenge that VLSI presents to computer science is that of developing a theory of computation that accommodates a more general model of the costs involved in computing. Can we find a way to express computations that is independent of the relative costs of processing and communication, and then use the cost properties of a piece of hardware to derive the proper program or programs? The current VLSI revolution has revealed the weaknesses of a theory too solidly attached to the cost properties of the single sequential machine.

REFERENCES

1. I. E. Sutherland and C. A. Mead, "Microelectronics and Computer Science," *Scientific American*, vol. 237, no. 9, September 1977, pp. 210–228.
2. H. S. Stone, ed., *Introduction to Computer Architecture*, Science Research Associates, Chicago, 1975.
3. C. V. Ramamoorthy and H. F. Li, "Pipeline Architecture," *Computing Surveys*, vol. 9, no. 1, March 1977, pp. 61–102.

4. C. G. Bell and A. Newell, *Computer Structures: Readings and Examples*, New York: McGraw-Hill, 1971. This book gives a general overview of computer structures, describes the PMS and ISP notations for describing the structures, and collects a number of papers concerning alternative computer structures.

5. P. H. Enslow, Jr., "Multiprocessor Organization—A Survey," *Computing Surveys*, vol. 9, no. 1, March 1977, pp. 103–129. This entire issue of *Computing Surveys* is devoted to surveys of parallel processors and processing.

6. D. J. Kuck, "A Survey of Parallel Machine Organization and Programming," *Computing Surveys*, vol. 9, no. 1, March 1977, pp. 29–59.

7. K. J. Thurber and L. D. Wald, "Associative and Parallel Processors," *Computing Surveys*, vol. 7, no. 4, December 1975, pp. 215–255.

8. B. A. Trakhtenbrot, *Algorithms and Automatic Computing Machines*, Boston: Heath, 1963. A delightfully readable introduction to algorithms, theory of computation, and unsolvability.

9. A. V. Aho; J. E. Hopcroft; and J. D. Ullman, *The Design and Analysis of Computer Algorithms*, Reading, Massachusetts: Addison-Wesley, 1974. A survey of algorithms designed using conventional models of computation. Includes discussion of divide-and-conquer, NP-complete problems.

10. R. E. Tarjan, "Complexity of Combinatorial Algorithms," *SIAM Review*, vol. 20, no. 3, July 1978. A most readable discussion; highly recommended.

11. C. D. Thompson and H. T. Kung, "Sorting on a Mesh-Connected Parallel Computer," *Communications of the Association for Computing Machinery*, vol. 20, 1977, pp. 263–271.

12. D. E. Knuth, *The Art of Computer Programming*, Second Edition, vol. 3, "Sorting and Searching," Reading, Massachusetts: Addison-Wesley, 1973.

13. Thomas R. Blakeslee, *Digital Design with Standard MSI and LSI*, New York: John Wiley & Sons, 1975.

14. G. H. Barnes; R. M. Brown; K. Maso; D. J. Kuck; D. L. Slotnick; and R. A. Stokes, "The ILLIAC IV Computer," *IEEE Transactions on Computers*, C-17, 1968, pp. 746–757. (Also appears in Bell and Newell.[4])

15. S. Hammering, "A Note on Modifications to the Givens' Plane Rotation," *J. Inst. Math. Appl.*, vol. 13, 1974, pp. 215–218.

16. A. H. Sameh and D. J. Kuck, "On Stable Parallel Linear System Solvers," *Journal of the Association for Computing Machinery*, vol. 25, 1978, pp. 81–91.

17. D. J. Kuck, "ILLIAC IV Software and Application Programming," *IEEE Transactions on Computers*, C-17, 1968, pp. 758–770.

18. H. S. Stone, "Parallel Computations," *Introduction to Computer Architecture*, edited by H. S. Stone, Science Research Associates, Chicago 1975, pp. 318–374.

19. R. M. Kant and T. Kimura, "Decentralized Parallel Algorithms for Matrix Computation," *Proceedings of the Fifth Annual Symposium on Computer Architecture*, Palo Alto, California, April 1978, pp. 96–100.

20. C. A. R. Hoare, "Communicating Sequential Processes," *Communications of the Association for Computing Machinery*, vol. 21, August 1977, pp. 666–677.

21. O-J Dahl, and C. A. R. Hoare, "Hierarchical Program Structures," in O-J Dahl; E. W. Dijkstra; and C. A. R. Hoare, *Structured Programming*, New York: Academic Press, 1972.

22. G. M. Birtwistle; O-J Dahl; B. Myhrhaug; and K. Nygaard, *Simula Begin*, New York: Petrocelli, 1973.

23. Wm. A. Wulf, ed., *An Informal Definition of Alphard*, Carnegie-Mellon University Technical Report, 1978.

24. B. H. Liskov and S. Zilles, "Specification Techniques for Data Abstractions," *Proceedings, International Conference on Reliable Software, 1975*, pp. 72–87.

25. A. Goldberg and A. Kay, eds., *Smalltalk-72 Instruction Manual*, Xerox Palo Alto Research Center Technical Report, 1976.

26. E. W. Dijkstra, *A Discipline of Programming*, Englewood Cliffs, New Jersey: Prentice-Hall, 1976.

27. H. B. Demuth, "Electronic Data Sorting," PhD. thesis, Stanford University, October 1956.

28. Phillip N. Armstrong, "An Investigation of Sorting and Self-Sorting Memory," *Final Technical Report on Smart Memory Structures*, California Institute of Technology, 1977.

29. S. A. Cook, "The Complexity of Theorem-proving Procedures," *Proceedings, 3rd Annual ACM Symposium on Theory of Computation*, 1971, pp. 151–158.

30. M. Rem and C. A. Mead, *IEEE Journal of Solid State Circuits*, vol. SC-14, April 1979, pp. 455–462.

9
PHYSICS OF COMPUTATIONAL SYSTEMS

Computation is a *physical process*. Data elements must be represented by some physical quantity in a physical structure, for example, as the charge on a capacitor or the magnetic flux in a superconducting ring. These physical quantities must be stored, sensed, and logically combined by the elementary devices of any technology out of which we build computing machinery. At any given point in the evolution of a technology, the smallest logic devices have a definite *physical extent*, require certain minimum *time* to perform their function, and dissipate a *switching energy* when switching from one logical state to another. From the system viewpoint, these quantities are the units of cost for a computation. They set the scale factor on the size, speed, and power requirements of a computing system.

The basic truth that data elements are physical objects, rather than merely mathematical ones, expands the ways in which we view information and computation. Over the years, many eminent scientists have studied the interaction of physics and information, and of physics and computation, in certain isolated areas where these fields overlap. Out of this background is now emerging a more unified discipline that combines computer science, information theory, thermodynamics, and quantum mechanics. We have chosen to call this discipline the *physics of computational systems*.

This physical view of information processing leads to quantitative metrics of computational complexity and enables comparisons to be made over a wide range of algorithms, computing structures, and technologies. But this is not all: It also leads to the identification of fundamental constraints on the properties of computational elements and to fundamental lower bounds on the area, time, and energy required for computation. In what follows we have made connections with the body of existing theory where possible. In several areas where no credible theory exists, we have outlined what one might look like. We hope that our efforts will stimulate further work on these important fundamental problems.

9.1   DIGITAL SYSTEMS

How do digital electronic circuits differ from other electronic circuits? The answer lies not so much in how the circuits are constructed but rather in how they are used. In analog electronic circuits, a small signal, such as that originating from a phonograph pickup or magnetic tape head, is amplified until it has enough power for the intended output device, such as a loudspeaker. There is a class of computational systems that use analog electronic information to represent quantities upon which computations are performed, and analog integrators, adders, and so forth, to perform the computations. Why use an electronic signal to represent only one bit of information, when we could use the same signal as an analog quantity to represent 12 or 16 bits of information? Unfortunately, as we pass an analog signal through a number of computational elements, each element unavoidably introduces noise and distortion. These deviations from ideality accumulate as errors in the output signal. After passing through a very large number of analog computational elements, a signal becomes so degraded that the final output signal no longer represents the outcome of the computation.

Useful computations often require millions or even billions of steps before the desired result is produced. Analog representations of information cannot undergo such a large number of computational transformations without a significant accumulation of errors and loss of information. A digital representation of information is often used because such representations may pass through an indefinite sequence of digital electronic circuits without information loss or accumulation of errors. This indefinite extensibility is the distinguishing property of digital systems constructed of such circuits. As we will see, this property places important physical constraints on the design of the devices and digital logic circuits in digital systems, and on the energy required to perform digital computations.

Logic circuits that can be cascaded in indefinitely extensible numbers must *restore* the logic level at each stage. Depending on the details of the logic family, the logic levels and switching thresholds are usually associated with logic circuit transfer functions involving voltages, or currents. However, independent of the logic family, the actual *signal* that represents *information* is stored as an *energy* associated with the control element of the input switching device of each logic circuit. Within a logic family, the stored energy per bit is scaled with the physical size of the input device: large input devices require a large energy stored on their input in order to carry out their function, while small devices require a small energy. Any given logic family has a well-defined way of apportioning the output signal from one device to its recipients so that they will all function properly.

We will now study the propagation of signal energy through indefinitely long cascades of restoring logic circuits, using transfer functions involving signal energy level. Although we use a particular logic family for illustration, the results are applicable to all restoring logic. In order to avoid the complexities resulting from scaling the amount of signal energy at each stage with device size, fan-out, etc., we use cascades of identical inverters. This study leads to a conceptually correct picture of the fundamental requirements that restoring logic circuits must satisfy.

9.1.1 Restoring Logic

In order for a logic signal to propagate through an indefinite number of such restoring logic circuits, in spite of the unavoidable errors and lack of ideality in the circuits themselves, it is necessary that such errors not propagate from one circuit to the next. Consider the chain of *identical* inverters shown in Fig. 9.1. As long as the node 1 signal energy level stays within the range of valid logic-1, then any change in this energy must result in a smaller change in the energy level at node 2. This condition must be true over the entire range of inputs recognizable as a valid logic-1 and likewise for the range of inputs representing a valid logic-0. The transfer functions of the restoring logic device must therefore have the form shown in Fig. 9.1. Since the slope of the curve must be less than unity over the entire range of valid logic-1 and also over the entire range of valid logic-0, it follows that the slope of the curve must be steeper than unity in the region between valid logic-1 and valid logic-0. Since the point at which the input energy level equals the output energy level is the dividing point between logic-1 and logic-0, this requirement is often stated in the following form: The magnitude of

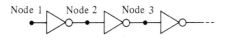

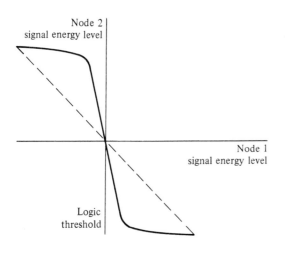

Fig. 9.1 Inverter signal energy transfer characteristics, in a cascade of similar inverters.

the slope of the transfer function must be greater than one at the point where the input level is equal to the output level.

Each restoring logic circuit must be able to drive the inputs of at least one circuit like itself. Thus the energy required to change the output node from a logic-0 to a logic-1 is at least as large as that required to change the input node from a logic-1 to a logic-0. Since this switching must be accomplished within a finite time, it follows that in the region between a valid logic-1 and a valid logic-0, the restoring logic circuit must exhibit *power gain*. The circuit must be able to supply more energy to the input node of the next stage than was supplied to its input node. Therefore, it is not possible to derive all the energy for the switching event from the input signal. We conclude that restoring logic circuits must draw power from some *power supply* separate from the actual signal path. This requirement has a number of important implications.

Perhaps the simplest implementation of a restoring logic circuit is the CMOS inverter, shown in Fig. 9.2. Input information is represented by the energy stored on the gate capacitances of the two transistors. The output node capacitance, C, represents the input gate capacitances of a similar successor device. A given charge Q on the output node results in a certain voltage V_{out} at that node. The positive and negative supply voltages for the CMOS inverter are $+V/2$ and $-V/2$, respectively. At the logic-0 level, the charge $-CV/2$ results in a voltage $-V/2$. Similarly, a charge $+CV/2$ is necessary to charge the node to a voltage $+V/2$. The relationship between V_{out} and Q is shown in Fig. 9.3. The energy stored on the capacitor is indicated for a logic-1 and for a logic-0 by the shaded areas.

Over the range of input voltage representing a logic-0, the upper (p-channel) transistor is turned on and the lower (n-channel) transistor is turned off. The output is thus "connected" to the positive supply and hence "restored" to this level. Similarly, the output is "connected" to the negative supply by the turned on n-channel transistor if the input is a logic-1. The basic scheme is to provide two power supply wires that are at voltages equal to the desired signal levels. Transistors are used as "switches" to connect the output to the appropriate one of these levels. The result of the logic operation is an energy stored on the capacitance C:

$$\text{Energy stored} = \int V_{out}\, I dt = \int V_{out}(dQ/dt)\, dt = \int V_{out}\, dQ.$$

Energy relationships are easily visualized by plotting the charge on the capacitor versus the voltage across it, as shown in Fig. 9.3. The diagonal line represents the constraint imposed by the capacitor:

$$Q = CV_{out}.$$

Let us follow a switching event that takes the system from the lower left corner of the plot (logic-0) to the upper right corner (logic-1) in Fig. 9.4. The vertical

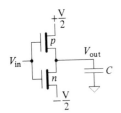

Fig. 9.2 CMOS inverter.

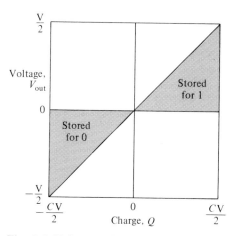

Fig. 9.3 Voltage and charge relationship for the simple inverter shown in Fig. 9.2.

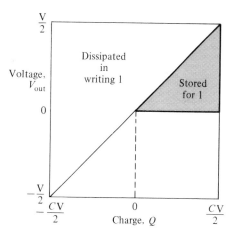

Fig. 9.4 Relationship between stored energy and energy lost during charging the capacitor, for the circuit shown in Fig. 9.2.

distance from the diagonal line to the upper line labeled V/2 is equal to the voltage across the *p*-channel transistor. As charge flows through the resistance of the *p*-channel transistor, a total energy equal to the area above the diagonal line is dissipated. The energy stored on the capacitor is shown by the area labeled "stored for 1." Similarly, an energy equal to the area under the diagonal line will be dissipated in the resistance of the *n*-channel transistor as the system is switched from the upper right corner of the plot to the lower left corner. We define the energy required to switch from one state to the other *and back* as the switching energy, E_{sw}, of the inverter; it is the total area of the box in Fig. 9.3:

$$E_{sw} = CV^2.$$

Notice that this energy is independent of the shape of the Q versus V_{out} relationship for the capacitor, of the characteristics of the transistors, and of the time dependence of the switching event.

It is convenient to think of the "signal" as positive for a logic-1, negative for a logic-0, and zero at the inverter logic threshold. We therefore define a *signal energy level* that has this property:

$$\text{Signal energy level} = \int (V - V_{inv}) d|Q|.$$

This quantity represents the potential of the node relative to the inverter logic threshold, weighted by the magnitude of the charge representing the signal. It is used in all signal energy level plots in this chapter.

We might attempt to reduce the energy dissipation by connecting a third transistor from the node to ground. A logic transition would then consist of activating this transistor prior to activating either the upper or lower transistors. The energy dissipation would thus be cut in half, since the act of discharging would require only $C(V/2)^2$. However, the energy used to charge and discharge the gate capacitance of the third transistor is at least $(C/2)V^2$. The total energy dissipated in switching a minimum node (2 gate capacitances) is not changed.

9.1.2   Steering Logic

Throughout the development of this text, we have shown many examples of logic functions being performed by steering signals through pass transistors to appropriate destinations. In this way a given logic function can often be implemented with a smaller number of stages than would be the case if only restoring logic were used. In such *steering logic*, data flows on signal paths through a network of pass transistors. The route taken by the signals is determined by the settings of "switches" operated by control drivers. Changing the setting of the steering "switches" requires charging and discharging the gates of the pass transistors. This energy is supplied by the control drivers. Changing the logical state of signal path inputs to the steering network requires charging and discharging portions of the signal path. In both cases, an energy of $\approx E_{sw}$ must be dissipated for each pass transistor charged or discharged, and the energy involved must come from restoring logic. This energy is dissipated either by the drivers of the signal paths, by the drivers of the gates of the pass transistors, or both. After passing through a number of steering logic stages, signals of necessity degrade both in terms of logic level and of time performance. Restoring stages must be placed at intervals throughout an array of steering logic to regenerate the signals.

9.1.3   LC Logic

Restoring logic circuits appear to be very inefficient. More energy is dissipated during the charging of the capacitor within such a logic circuit than is finally stored on the capacitor. The origin of this loss is clear: The power supply is at a fixed voltage, and whenever the capacitor voltage is less than the supply voltage, any current through the charging transistor will dissipate energy. How might we construct a logic family that does not dissipate energy? The construction of a lossless logic family would be very desirable, but is it in fact possible? If it is not possible, can we at least determine the minimum amount of energy that must be dissipated per switching event?

We might try to circumvent the "lossy" properties of the restoring logic family by storing energy in one of two alternate forms, and then using a transistor to switch between the two. An example of an attempt at such a scheme is shown in Fig. 9.5. Here the logic signal is stored as either a current through the inductor L or a charge on the capacitor C. A transition of the circuit from one state to the

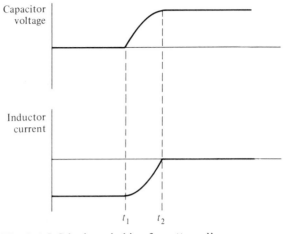

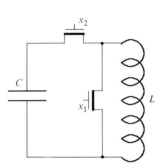

Fig. 9.5 Scheme for attempting to construct dissipationless logic.

Fig. 9.6 LC logic switching from "zero" state to "one" state.

other is illustrated in Fig. 9.6. Waveforms are shown for the voltage across the capacitor and the current through the inductor. Initially the logic signal is stored as a current in the inductor while the voltage across the capacitor is zero. At time t_1, the series transistor x_2 is turned on while the parallel transistor x_1 is turned off. At t_2, the voltage on the capacitor reaches its maximum value, the series transistor x_2 is turned off, and the charge representing the signal is stored on the capacitor C. The current in the inductor is now zero. This quiescent state can be maintained on the capacitor for an indefinite period.

In order to restore the circuit to its original configuration we need only reverse the procedure as shown in Fig. 9.7. At time t_3, the transistor x_2 is turned on, thereby initiating a flow of current from the capacitor into the inductor. The current will increase, reach a peak, and then oscillate back until t_4, when it reaches its maximum negative value once again. At this point the parallel transistor x_1 is turned on and the series transistor x_2 is turned off. The circuit has executed one complete cycle. If the transistors x_1 and x_2 were perfect switches, this form of logic would be dissipationless. We would be able to run an indefinite number of switching events without losing the signal energy. However, in order for the logic to function, the gate voltages on the transistors x_1 and x_2 must come from a signal such as the capacitor voltage V_C.

Figure 9.8 shows the details of the two waveforms in the neighborhood of time t_4. The current is nearing its maximum negative value and the voltage is approaching zero. When the voltage reaches some small value (shown as $-V_0$ in the diagram) the transistor x_2 begins to open. Thus, instead of traveling a straight

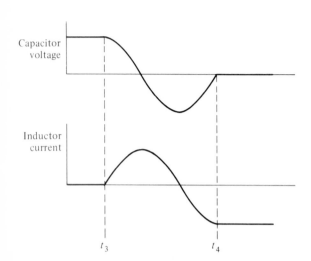

Fig. 9.7 LC logic switching from ''one'' state to ''zero'' state.

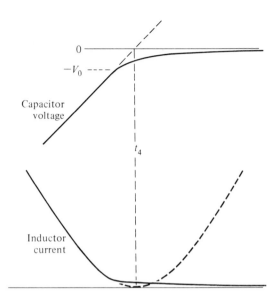

Fig. 9.8 Detail of Fig. 9.6 showing the waveform in the neighborhood of t_4.

trajectory as it would if the transistor were not activated, it will follow a shallower curve, eventually leveling out at zero voltage. No transistor is able to change from a completely on condition to a completely off condition without having a finite voltage change applied to its control electrode. In this case we have assumed the transistor to have zero resistance for gate voltages greater than V_0. It changes to an infinite resistance for gate voltages less than zero. Intermediate in the switching process, the transistor will not act as either a perfect short circuit or a perfect open circuit. It will have a finite voltage across it together with a finite current through it. The energy dissipated in an elementary switching event will be proportional to the current times the voltage, integrated through the switching transient:

$$E_{\mathrm{sw}} = \int V I \, dt = \int V C \, (dV/dt) \, dt = \int C V \, dV = (\tfrac{1}{2}) \, CV_0^2.$$

Thus we see that the total switching energy is equal to that stored on the capacitor with an applied voltage V_0. This irreducible dissipation is required by any switch that cannot sense an infinitesimal voltage difference. *This energy is the same as the switching energy of an ordinary restoring logic circuit powered by a power supply of voltage V_0.*

The conclusion from this example is clear. Storing a large quantity of energy in a switching circuit may have the effect that only a small fraction of that energy will be lost in any given switching event. The necessary irreversible energy loss in any switching event does not, however, depend upon the total energy stored but upon the properties of the switching elements that must sense another logic signal similar to the one they are generating. We will generalize from this and other examples like it, to derive a conclusion for all operable logic families. This conclusion is here presented as a postulate, and like all postulates this one derives its validity from a long, unsuccessful search for counterexamples:

A certain energy, E_0, is required to change the state of a switching device. No logic element constructed using that switching device can dissipate less energy than E_0 per switching event.

9.2 VOLTAGE LIMIT

We have seen that a restoring logic circuit must dissipate at least as much energy as that required to change the state of one of its switching devices. It is thus desirable to use switching devices that can sense as small an energy as possible. The contribution of VLSI to reducing the cost of computation can be traced directly to the reduction in this quantity of energy. There are two independent ways to make this energy small. (1) Make the geometry of the elementary transistors as small as possible. As discussed in Chapter 1, FET gate lengths must not be made smaller than about ¼ micron. (2) Make the operating voltage small. A lower limit on the operating voltage of FET circuits is established in the following discussion. The minimum operating voltage is set by the requirement that the gain of an elementary circuit exceed unity.

When an MOS device is operated near its threshold, the channel resistance R_{ch} is exponentially dependent upon the gate voltage V_g:

$$R_{ch} \propto e^{-qV_g/nkT},$$

where q is the magnitude of the electronic charge. The factor n is due to the substrate effect and is between 1 and 2 for most processes.

A model of a complementary circuit (such as a CMOS inverter) is shown in Fig. 9.9. The resistances R_1 of the lower n-channel device and R_2 of the upper p-channel device are exponentially dependent on the input voltage V_{in} as follows:

$$R_1 = R\,e^{-qV_{in}/nkT}, \qquad R_2 = R\,e^{qV_{in}/nkT}.$$

The output voltage is

$$V_{out} = \frac{V(R_1 - R_2)}{2(R_1 + R_2)}.$$

We are interested in the gain near the switching threshold, which because of the supply voltage convention is at $V_{in} = 0$. Expanding the exponentials as a

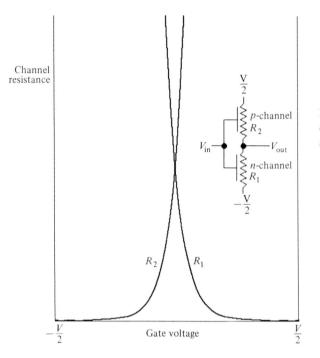

Fig. 9.9 Model of a complementary device such as a CMOS inverter.

power series, and ignoring all but the first order terms in V_{in}, we find

$$V_{out} \approx -(qV/2nkT)V_{in}.$$

The gain of the circuit is thus equal to $qV/2nkT$. Hence realistic supply voltages for complementary circuits should be greater than a few kT/q. At room temperature $(kT/q) \approx 25$ mV. Ratio logic families, such as n MOS, can be analyzed by the same technique. Since they have only one nonlinear device rather than two, their gain is approximately half that given above. They will therefore require twice the minimum supply voltage required by complementary devices. A detailed analysis of the low voltage operation of CMOS circuits has been published by Swanson and Meindl.[1]

9.3   DISCRETENESS OF CHARGE

The analysis of the last section indicates that single inverters can be made to operate with greater than unity gain on power supply voltages of a few kT/q. There is, unfortunately, more to the voltage limit story. Over a large ensemble of transistors, some will have higher or lower threshold voltage than desired due to statistical fluctuations in the number of impurity ions under the gate. It is possible that the pull-up transistor of an inverter could have a particularly high threshold voltage, while its companion would have a particularly low threshold. If the sum of the two variations in threshold voltage exceeds the supply voltage 2V, the device output will always remain in one state. For a Gaussian distribution in threshold voltage with variance ΔV_{th}, the probability of such an occurrence is

$$P = e^{-4V/\Delta V_{th}}.$$

Keyes[2] has estimated the variance $\Delta V_{th} \approx .08V$. We might require that, in a VLSI system containing 10^7 inverters, the probability of all the system's transistors being within threshold limits be greater than 0.9. Such a criterion would require a supply voltage of ≈ 0.7 volt. Unless special attention is directed toward reducing threshold variations, systems with ¼ micron device geometries will be forced to operate with higher supply voltages than the straightforward scaling would indicate.

The minimum operating voltage together with the minimum size of an individual transistor determines the minimum energy dissipated in an elementary logic operation. This energy is then the unit in which all energies in computation are measured. We have defined the total energy required for a complete cycle from logic-1 to logic-0 and back to logic-1 as the *switching energy* E_{sw}. It is equivalent to the area of the entire square in Fig. 9.3.

9.4 TRANSITIONS IN QUANTUM MECHANICAL SYSTEMS

Try as we might, we are unable to devise, even conceptually, a lossless logic element. In such situations, it is wise to ask whether the obstacle we have encountered is a practical or fundamental one. For example, many of the principles of thermodynamics were developed because the efficiency of heat engines could not be improved past a certain point. Where do we look for the physical laws underlying all logic forms and computational systems? Classical physics does not provide a sufficient basis for representing the fundamental behavior of information storage devices. We must instead examine the properties of quantum mechanical systems, as they apply to devices with two stable states, and develop the basic physical principles underlying the representation and manipulation of information. These principles will also help us in a later section visualize the fundamental nature of irreversibility in computational systems.

The quantum theory forces us to think about nature in an entirely new way. In the classical view, matter was considered to be made up of particles such as electrons and protons. These objects were visualized as hard, shiny spheres whose motion followed Newton's laws. On the other hand, light was visualized as an electromagnetic wave, composed of electric and magnetic fields, and could be described by Maxwell's equations. However, early experiments with electrons, and later with other particles, showed diffraction behavior indistinguishable from that observed with light. A vast collection of experimental evidence gradually made it clear that light and matter are special cases of the same kind of stuff. The best way to develop one's intuition here is to view all matter as waves of various types. When we are asked for the position of a particle, we must determine how the particle is localized, compute the shape of a wave packet representing the particle, take the center of the packet as the "position" of the particle, and the average "size" of the wave packet as the "uncertainty" in the position. All our intuition about waves in general can be used directly. The "particle" nature of matter only becomes apparent when waves interact.

If all matter is really made of waves, how are the classical particle attributes such as energy and momentum related to the properties of waves (frequency, wavelength, etc)? *The quantum theory identifies the energy of a particle with the frequency of its corresponding wave, and the momentum of the particle with the inverse of the wavelength.*

The units used to describe the attributes of particles are not the same as those used for waves. Planck's constant, $h = 6.62 \times 10^{-34}$ joule-sec/cycle, is the conversion constant introduced to unify the disparate views. Energy E is related to frequency v by

$$E = hv.$$

Momentum p is related to wavelength λ by

$$p = h/\lambda.$$

Since wave equations are usually written in terms of the angular frequency $\omega = 2\pi v$ and the wave vector $\kappa = 2\pi/\lambda$, it is common practice to use the constant $\hbar = h/2\pi$ in such expressions ($\hbar$ is pronounced "h-bar"). The quantum conversion relations then become

$$E = \hbar\omega \quad \text{and} \quad p = \hbar\kappa,$$

where $\hbar = 1.05 \times 10^{-34}$ joule-sec/radian $= 6.58 \times 10^{-16}$ eV-sec/radian.

One electron-volt (eV) is the energy gained by one electronic charge after being accelerated through a potential difference of one volt. (The charge on the electron is 1.6×10^{-19} ampere-second, and thus one eV $= 1.6 \times 10^{-19}$ joule).

A light wave is made up of a magnetic field and an electric field. The change in the magnetic field with time acts as a source for the electric field. The changing electric field acts as a source for the magnetic field. If we do not wish to bother with the details of the internal workings of the wave, we often refer to its *amplitude* (the magnitude of either the electric or magnetic field; it doesn't matter which one, as long as we stick with it consistently) or to its *intensity* (the absolute square of the amplitude). In the case of matter, the internal workings are more complicated and we refer to a generalized amplitude Ψ called the *wave function*.

Since we are dealing with waves that have both amplitude and phase, it is convenient to represent the wave function as a complex number represented by a point in the complex plane. The real component of a complex number is its x-coordinate and the imaginary component is its y-coordinate. The magnitude of a complex number is the amplitude of the wave and is represented as the length of the line from the origin to the point. The phase of the wave is the angle of the line.

The absolute square of the wave function is the *density* of matter. It can be conveniently calculated by multiplying the wave function Ψ by its own complex conjugate $\Psi*$:

$$\text{Density} = \Psi*\Psi.$$

The density is usually normalized so that when it is integrated over the region of space being considered, the total comes out to be the number of particles in that space. The charge density can then be computed by multiplying the density by the charge per particle. Similarly, the mass density is the density times the mass per particle, etc. To find the center of mass of a wave along, for example, the x-axis, we merely multiply the density at every point by the x-coordinate at that point and integrate over the entire space:

$$\langle x \rangle = \int \Psi*\Psi x \, d(\text{volume}) = \int \Psi*x\Psi \, d(\text{volume}).$$

The computation is exactly the same as finding the first moment of the density function.

Let us examine how the wave nature of matter works in a simple special case. Classically, an atom consists of a positively charged nucleus surrounded by one or more electrons. The electrons are held in "orbits" around the nucleus since their negative charge is attracted to it, much as the planets are attracted to the sun. This classical view has a fatal flaw. If an electron is circling the nucleus, its coordinates are oscillating with time. Such an oscillating charge is known to create an electromagnetic wave, which radiates energy away from the source. It is, in effect, a minute radio transmitter. Why don't the electrons radiate away all their kinetic energy and fall into the nucleus? If the electron is a wave, the only stable orbits will be those where the wave function comes back around the nucleus *in phase* with itself. In other words, there are a number of very special orbits that have the property that the wave function changes by an integral multiple of 2π radians as it encircles the nucleus. These remarkable solutions to the wave equations correspond to *standing waves*. The density function of an electron in a state represented by such a solution does not change with time. Such states are called *eigenstates* of the system. Since the density function of a system in an eigenstate does not change with time, the system does not radiate energy, and thus the orbits of electrons around a nucleus can be stable.

The eigenstates of a physically bounded system form a discrete set. The phase of the wave function can rotate $0, 2\pi, 4\pi$, etc., as it makes one entire circuit in the space. In each eigenstate, the system will have a definite energy. A transition between eigenstates will be accompanied by the radiation or absorption of a definite amount of energy. This quantization of the energy of all physical systems is the origin of the name quantum theory.

Suppose we have a system with two eigenstates, characterized by wave functions Ψ_1 and Ψ_2 with corresponding energies E_1 and E_2. The time dependence of the wave functions can be written explicitly as follows:

$$\Psi_1 = \varphi_1 e^{iE_1 t/\hbar}, \qquad \Psi_2 = \varphi_2 e^{iE_2 t/\hbar}.$$

Here φ_1 and φ_2 are functions only of the coordinates and not of time. The time dependence of the expected value $\langle x \rangle$ of any coordinate x of a system in one of its eigenstates can be computed from the wave function;

$$\langle x \rangle = \int \Psi^* x \Psi.$$

The integral is taken over all space where the amplitude of the wave function is nonzero. It can be seen from these expressions that $\langle x \rangle$ is not a function of time. A similar computation can be performed for any other measurable parameters of the system. *In all cases, the expected values of measurable parameters of a system in an eigenstate do not vary with time.*

Consider a system whose wave function is made up of a combination of the two wave functions:

$$\Psi_s = A\Psi_1 + B\Psi_2 = A\varphi_1 e^{iE_1 t/\hbar} + B\varphi_2 e^{iE_2 t/\hbar},$$

where $A^2 + B^2 = 1$. The total energy of the system is $E_{tot} = A^2 E_1 + B^2 E_2$. The expected value of x for this wave function is

$$\langle x \rangle = \int \Psi^* x \Psi = 4AB \int \varphi_1^* x \varphi_2 \, \cos(E_1 - E_2)t/\hbar + \text{terms independent of time.}$$

The expected values of measurable parameters of a state composed of two eigenstates oscillate with a frequency corresponding to the difference in the energies of the eigenstates.

In particular, the center of charge of the system oscillates with time. This oscillating charge distribution radiates electromagnetic energy. Suppose the system starts in the higher energy eigenstate Ψ_1, corresponding to $A = 1$ and $B = 0$. A slight perturbation might cause a very small oscillation in the charge distribution. This oscillation, no matter how small, will result in energy being radiated away. Any energy lost will lower the total energy of the system to a value less than E_1. Its state will no longer be an eigenstate: B is no longer equal to zero, and A is no longer equal to unity. A transition between Ψ_1 and Ψ_2 is underway. A larger oscillation amplitude results in more radiation, and hence faster energy loss. Thus, in its early stages, such a transition grows exponentially. As the transition proceeds, B grows and A decreases. The state of the system will evolve to a point where $A = B$. This is the point of maximum oscillation amplitude and hence maximum rate of radiated energy loss. Finally, the oscillation will die exponentially, as A tends to zero and B tends to unity. The system has radiated away exactly $E_1 - E_2$ of energy and is in its new eigenstate Ψ_2.

Such a quantum system can absorb energy from the radiation field by the inverse of the process we have described. We can confine two such elementary systems into a single larger combination. If we start with one system in state 1 and the other in state 2, the first will radiate its energy as described previously. If the combination is isolated from the rest of the universe, the radiation cannot escape, and will cause the second system to make a transition from state 2 to state 1. The second will then radiate its energy back to the first. The combination will oscillate forever, transferring energy from system 1 to system 2 and back.

The situation is completely different if we have a large number of physically dispersed absorber systems. The outgoing radiation will excite different absorbers at different times. By the time the absorbers radiate their energy back to the original system, each contribution will arrive with a random phase. The contributions cannot act coherently to drive our original system back into state 1. By introducing a large number of absorbers we have been able to assure that our system remains in state 2 after making its transition. Information concerning the final state of the system has been gained through the loss of the phase information of the wave function. In this way, time is given a preferred direction. The result costs us the transition energy, which we cannot recover. This is the physical basis of the second law of thermodynamics, which we will discuss in Sections 9.5 and 9.11.

9.4.1 Transition Times

The physical picture of quantum mechanical transitions presented above allows us to make quantitative estimates of the time required for transitions between eigenstates. If the energies of the eigenstates of a system are the same, none of the expected values of its measurable parameters can change with time. If the eigenstates are at different energies, it is possible for the system to oscillate with time, and thus to make a transition from one state to another. The rate of buildup of the oscillation is determined by how tightly the system is coupled to the electromagnetic radiation field. The magnitude of this coupling is given by the term $\int \varphi_1^* x \varphi_2$ in the above example. Even for very large coupling terms, the system requires of the order of one cycle to radiate away its energy, thus setting a lower limit to the transition time τ between two states separated by an energy ΔE:

$$\tau > \approx \hbar/\Delta E.$$

Suppose we use a quantum system having two eigenstates as a storage device to represent a bit of information. We will choose a symmetrical system, i.e., one having two eigenstates of the same energy, since it will remain in a given state unless externally perturbed. Starting with such a system in one of its eigenstates, a transition from one state to another can be caused as follows:

1. Perturb the system such that the desired state has a lower energy than the initial state.
2. Excite the system in order to mix the wave function of the initial state with that of the final state.
3. Wait for the resulting oscillating charge distribution to radiate energy until the system has come to equilibrium in the final state.
4. Remove the perturbation so that the two states once again have the same energy.

We might ask if metastable behavior is possible during transitions between the discrete eigenstates of a quantum mechanical system. If we interrupt a quantum mechanical transition midway, we are left with a system with equal probability of being in either state. Since the states now have equal energies, there is no preferred direction to the transition. This situation is the quantum equivalent of the "hung" flip-flop. Even after an arbitrarily long time, there is a finite probability that the system will not have recovered from its mixed state. Quantum mechanics offers no escape from the possibility of incomplete transitions. Designers must cope with the phenomenon of metastability at a system level, as discussed in Chapter 7.

9.5 IRREVERSIBILITY

Except for one, the laws of physics are symmetrical with respect to time. A moving picture of a physical event can be played backwards and still represent a possible physical event. This notion flies in the face of our common perception of the relentless onward march of time. The preferred direction comes from the loss of information contained in the phase of the wave function. Order and coherence are easy to lose but hard to recover. Thermodynamics introduces the term *entropy* as a measure of disorder in a system. A more detailed discussion is given in Section 9.11.

The *second law of thermodynamics* is a concise statement of the asymmetric nature of disorder. It states that the total entropy in the universe always increases. In any closed system, the total entropy can at best remain constant and can never decrease with time. An increase in order or organization in one part of the universe can occur only at the expense of a still larger increase in disorder in another part.

We have attempted to construct a logic element that can be placed in a definite state. This definiteness is a state of low entropy. Consider a flip-flop balanced near its metastable point. In that condition, our uncertainty as to the final state of the flip-flop is very high, and therefore it is in a high entropy condition. Once it has switched to one or the other of its stable states, we are very sure of the outcome. The device has made a decision. It has removed entropy from the information system. This decision has been accomplished by turning electrical energy into heat.

In order to introduce a direction into a transition between states, energy must be lost irreversibly. A system that conserves energy cannot make a transition to a definite state and thus cannot make a decision.

9.5.1 Logically Reversible Systems

In our discussion of the inherent dissipative nature of restoring logic we have not considered the reversibility (or lack thereof) of the computing process itself. There have been a number of discussions on this topic.[3,4] It appears possible to construct universal computing machines that are reversible in the sense that once they have reached a final step of computation they can be played backward and will retrace the steps of computation to their initial state. The point here is not whether such reversible machines exist but that if one were to construct such a machine it would need a control signal that would tell it whether to compute in the forward or reverse direction. Whichever direction its computation was proceeding, a dissipation of energy would be necessary to keep the computation proceeding in that direction. Thus, far from recovering the energy spent in the original computation, one would spend at least as much energy retracing the steps of the computation. If the control signal were not present, and each individual logic element were capable of computing in either direction reversibly, the system would simply have no direction whatsoever. The situation is made clearer with reference to Fig. 9.10.

The elementary logic device in the box accepts inputs from the left and produces outputs on the right that are some logic function of the inputs. It uses a certain energy from the power supply, and this energy is converted into heat. The

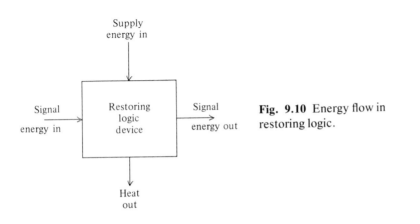

Fig. 9.10 Energy flow in restoring logic.

device is arranged with a control signal such that it can be run backward. With the control signal in the opposite state, inputs are taken from the right and outputs are presented to the left, again with an input of electrical energy from the power supply and the output of heat energy.

Suppose we were to construct a truly reversible logic device. The signals at either side of the box can be either inputs or outputs and energy can flow either into or out of the power supply depending upon the direction of signal flow. (In some schemes the connection to the power supply is not required). It is clear that this device has no preferred direction of computation; it has no way of deciding whether computation should proceed from left to right or from right to left. This is simply a macroscopic example of the necessity of energy dissipation as a mechanism for providing a direction for the computation process. This dissipation is a necessity stemming from the second law of thermodynamics. *The only physical law that gives time a preferred direction is the second law of thermodynamics.*

Conceptual heat engines that are claimed to be reversible are the topic of any elementary treatment of thermodynamics. Such reversible engines are fitted with two heat reservoirs. If mechanical work is done on the heat engine by an external system, heat is pumped from the low-temperature reservoir to the high-temperature one. If the engine is allowed to do mechanical work on an external system, heat flows from high to low temperature. Such discussions omit the mechanism that controls the energy flow into or out of the heat engine. Such a mechanism must dissipate energy and therefore destroys the reversibility of the entire system. The difference in viewpoint is exactly the same as that we encountered with LC logic. The energy processed by the heat engine can be made arbitrarily large if a large (and therefore slow) enough engine is constructed. The necessary dissipation in the control device that operates the valves, etc., can be made negligible compared with the energy scale of the entire engine. However, the dissipation in these devices can never be eliminated. Schemes employing sensing and control devices that dissipate no energy can be used to construct perpetual motion machines. Perhaps the best known example is the *Maxwell Demon.*[5]

Logical and energetic reversibility are separate and distinct ideas. It is possible to construct logically reversible computing systems, but not energetically reversible ones. Logically reversible operations may be a convenient formalism for viewing certain logic families such as magnetic bubbles. Creating or destroying a bubble is extremely awkward, but bubbles can be steered on different paths with ease. Schemes for implementing combinational functions in a logically reversible way have been proposed that do not create or destroy ones or zeroes.[3,4] Such schemes would seem to be ideally matched to the properties of bubble devices. However, such logically reversible schemes are not energetically reversible. Magnetic bubble devices require a static *bias* field perpendicular to the plane of the substrate. In addition, the bubbles are caused to move by a *clock* magnetic field vector in the plane of the substrate. The clock field vector rotates 360 degrees every clock cycle. In such a logic family, the direction of the computation is determined by the direction of rotation of the clock magnetic field. Energy stored in the domain walls of a bubble at any location must be dissipated as the bubble is moved to another location. The dissipated energy is supplied by the clock magnetic field.

9.6 MEMORY

The requirements described above for restoring logic circuits allow an essentially indefinite spatial extension of combinational logic operations. However, modern computation is normally not done in a strictly combinational manner. One step of a computation produces an intermediate result. This result then forms the input to a second step and so forth until the entire computation is complete. Such a sequential mode of computation, no matter how concurrent in nature, requires that intermediate results be stored in some form of memory device. Binary information implies elementary memory elements of a bistable nature—one state denoting a logic-0, the other a logic-1. A mechanical system that behaves in this way is the inverted pendulum shown in Fig. 9.11(a). The force of gravity holds the pendulum stably in either the rightmost or the leftmost position. Switching from one state to the other can be accomplished by pushing the weight up to its maximum position and letting it fall onto the opposite stop.

Physicists view bistable systems of this sort in terms of a diagram such as that shown in Fig. 9.11(b). In such diagrams the potential energy of the physical system is plotted as a function of its spatial or electrical coordinate. If the pendulum is left in one of its stable states, given by the minima in the potential diagram, it will stay there indefinitely until enough external energy is provided to surmount the potential maximum and to allow the system to re-equilibrate in the other potential minimum. Note that the energy provided by the external switching source is lost in the impact when the pendulum falls to its stop (and perhaps bounces a bit until the energy is dissipated). Many writers have considered particles in potential wells of this shape to derive minimum switching energies for computation.[2,3]

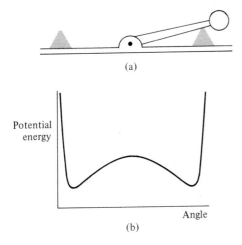

(a)

Potential
energy

Angle

(b)

Fig. 9.11 (a) Inverted pendulum.
(b) Potential energy of pendulum.

The slope of the potential energy curve, i.e., the derivative of the potential energy with respect to the angle of the pendulum, has the units of a torque. This torque is being supplied by gravity and pushes the pendulum toward one of its stable positions.

The energy required to switch from one state to the other can be supplied deliberately, or by some random occurrence. Suppose our pendulum were mounted on a railroad car. While the train is stationary, we expect the device to remain in its initial state. However, when the train passes over a very rough stretch of track, the pendulum may bounce into the other state. The potential maximum must be high enough to prevent such random events.

We must ask if the requirements we have established for restoring logic allow us to construct memory devices of the required reliability. Suppose a number of identical inverters, each having the energy transfer characteristic previously shown in Fig. 9.1, are connected in tandem as shown in Fig. 9.12. The resulting signal energy transfer characteristic between nodes 1 and 3 contains a point where the signal energy level on node 3 is equal to the signal energy level on node 1, as shown in Fig. 9.12. With this particular signal energy on node 1, we could disconnect the output of the second inverter from the input to the third, connecting it instead to the input of the first. Neither node energy in the resulting device would show any inclination to change with time toward either a logic-1 or a logic-0. This situation is precisely that of the "hung" flip-flop described in Chapters 1 and 7. If, however, the signal energy at one node were displaced slightly from this point of metastable equilibrium, it would evolve exponentially with time as shown in Fig. 9.13.

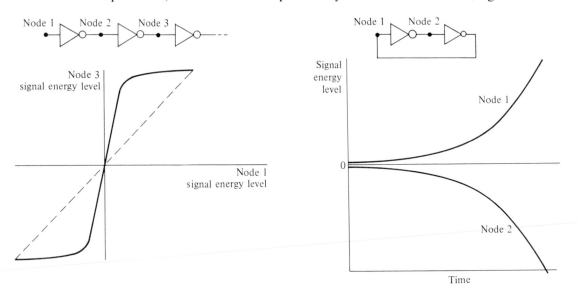

Fig. 9.12 Signal energy transfer characteristic of inverter pair.

Fig. 9.13 Time evolution of the two nodes of a flip-flop.

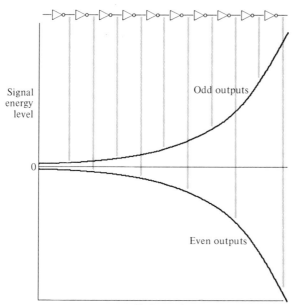

Fig. 9.14 Growth of node signal energy in a string of inverters.

We will analyze the time evolution of this system in detail later in this section. However, it is clear that for the signals to move away from the logic threshold, each inverter must exhibit power gain. A small deviation from threshold at one input results in a larger deviation in the opposite direction at the second input, resulting in a still larger deviation in the first input, etc. Hence, qualitatively, we can see that the requirements for restoring logic that allow it to perform indefinitely extensible combinational functions are precisely the same as those that allow us to implement stable, reliable, flip-flop style memory. In fact, the flip-flop can be viewed as an implementation of a recursive definition of restoring logic. It simply maps the spatial behavior of an indefinitely long string of restoring devices into the time domain. This mapping is illustrated in Fig. 9.14. Here, a string of inverters has a signal very close to the logic threshold applied at their input. The voltages at the outputs of the even stages and the odd stages diverge in opposite directions from the logic threshold as shown in the figure. Since each deviation from the logic threshold is larger than that of the stage before by the gain of the inverter, this curve is seen to be similar in form to that of Fig. 9.13 where the time behavior of the even and odd output of the flip-flop is shown as a function of time. The details of this mapping from the spatial to the time domain for any logic family will provide us with a fundamental measure of the time scale introduced by devices of that family.

9.6.1 Time Behavior of the Flip-Flop

A simple small-signal model of a flip-flop composed of CMOS inverters with both amplifying stages in their gain region is shown in Fig. 9.15. The inverters have been represented by an input capacitance and a current source supplying a current equal to the gate charge Q divided by the transit time τ. Both stages are treated as identical. The differential equations describing the time evolution of the two node voltages are

$$dQ_1/dt = -Q_2/\tau \quad \text{and} \quad dQ_2/dt = -Q_1/\tau.$$

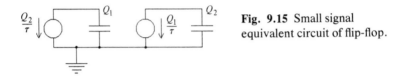

Fig. 9.15 Small signal equivalent circuit of flip-flop.

It can be verified by direct substitution that the following simple exponential form (shown in Fig. 9.13) satisfies the two differential equations:

$$Q_2 = -Q_1 = Q_0 e^{t/\tau} = CV_0 e^{t/\tau}. \tag{9.1}$$

Let us compare this solution with that for the cascaded inverters of Fig. 9.14. With a voltage V_0 at the input of the cascade, the value of the voltage at the nth node will be equal to

$$V_n = V_0 A^n = V_0 e^{n \ln A}. \tag{9.2}$$

Comparing Eqs. 9.1 and 9.2, we see that the growth of the node voltage in the inverter spatial cascade corresponds in the following manner to the growth of the node voltage in the flip-flop with time:

$$n \ln(A) \quad \longleftrightarrow \quad t/\tau$$

[in the spatial domain]   [in the time domain]

Hence the factor $\ln(A)$ per stage plays the same role in the growth in signal per stage in combinational logic as does the factor $1/\tau$ per unit time in the time evolution of a nearly balanced flip-flop. This is a fundamental result that maps the spatial domain of combinational logic into the time domain of flip-flop-like devices.

Thus, we can model much of the behavior of combinational restoring logic by studying the behavior of simple cross-coupled flip-flops in the time domain. For this reason, the flip-flop occupies a central place in the theory of the physics of computational systems. Since the flip-flop is inherently a self-contained device and represents all the degrees of freedom available in a logic family, we can use it to construct fundamental characterizations of logic families.

9.6.2   Energetics of the Flip-Flop

Our inverted pendulum is not an accurate physical model for the flip-flop, since the pendulum uses a conservative field, (i.e., gravity) and therefore cannot be arranged in such a way to provide necessary power gain. Another way of stating the problem is that there is no mechanism possible using inverted pendula coupled in any way to perform restoring logic operations on other pendula. We supply power to the pendulum only when we are increasing the elevation of its mass in the gravitational field. Suppose we attempt to change the state of a flip-flop by supplying a current into the side of the flip-flop that is at the lower potential. If the current we supply is just large enough, the potential on that side of the flip-flop will be raised, turning on the transistor on the opposite side and changing the state of the flip-flop. However, we could supply a lower current (and therefore lower power) for an indefinite period of time without changing the state of the flip-flop. In this way, a large quantity of energy can be supplied to the flip-flop without changing its state, provided that the energy is supplied slowly enough.

Stored energy represents information. In order to change the energy level representing one bit of information, there is a minimum amount of energy we must dissipate, set by the properties of the switching devices used. For that reason, the rate at which a system can process information is related to the system power dissipation. The factor that relates the two is E_{sw}, the energy dissipated per device per switching event.

What "force" holds the flip-flop of Fig. 9.13 in one of its two stable states? Suppose we attach a voltage source from ground to one of the nodes, and adjust its value somewhere between $-V/2$ and $+V/2$. The circuit will "fight back" by drawing a current from the voltage source. The magnitude of that current versus the value of the node voltage is shown as I in Fig. 9.16. If we disconnect the voltage source when the node is at some intermediate voltage, the current being absorbed by the transistors of the circuit can no longer be supplied by the source and will act to discharge the capacitance of the node and therefore return the node voltage to its stable state. The current thus acts like a "restoring force," similar to the gravitational restoring force on the inverted pendulum. The curve labeled F in Fig. 9.16 is the integral of the current from $-V/2$ to the voltage in question. Since the current is just the derivative of F with respect to the node voltage, F acts as some kind of "potential" analogous to the gravitational potential energy in the pendulum. It is called the *dissipative function,*[6] or *thermodynamic potential* of the flip-flop. The circuit is stable only at a local minimum in F.

F has the units of power: it is precisely half of the total power P dissipated in any circuit whose only dissipative elements are linear resistors. From Fig. 9.16

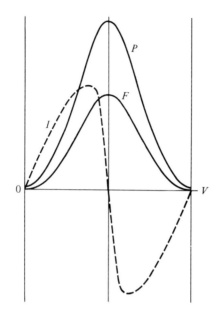

Fig. 9.16 Power dissipation (P), node current (I), and dissipative function (F) of a CMOS flip-flop versus voltage of driven node.

we can see that F is still very nearly half of P for the flip-flop, even though the resistance of the transistors is exponentially dependent on their gate voltage. What are the general relations between F and P? The field of nonequilibrium thermodynamics, of which this question is a part, has not as yet provided the answer. Since F represents useful work done to keep the flip-flop in its stable state, it seems that the total power dissipated by the flip-flop must be greater than F. We do not have a proof of this property, even though it seems intuitively obvious. It is interesting that the conditions of Sections 9.2 and 9.8 that are required for logic to operate are precisely the ones that make the problem untreatable by ordinary nonequilibrium theory. A thermodynamics of systems of this type would be a profound contribution. We discuss the matter further in Section 9.11.

Active memory circuits like the flip-flop (also called "static" in the popular literature) are held in one of two stable states by a local minimum in the dissipative function. In order to switch between the two states, not only must the energy E_{sw} be supplied to the circuit but it must be supplied within a certain time. An amount of power must be supplied that exceeds the maximum in the dissipative function. Passive memory circuits (also called "dynamic") can be switched by supplying the switching energy as slowly as desired. Notice that the minimum energy required to switch a flip-flop is indeed E_{sw}, since one node makes a positive transition while the other node makes a negative transition.

For efficient logic families like CMOS, the dissipative function is similar in form to the actual power dissipation of the circuit. However, for less efficient families, the dissipative function may represent only a small fraction of the total energy consumed by the circuit. Many logic families dissipate much more power than that actually required to change the stored energy. It is always possible to waste power. The simplest example is to simply connect a resistor from VDD to ground. The power that is wasted will have no effect whatsoever upon the performance of the device. It is even possible to waste power in such a way that it depends upon one of the voltages in the circuit. For example, a transistor can be connected with its drain to VDD and its source to ground. The gate can be connected to one of the nodes of a flip-flop. The total power dissipated by the flip-flop will then be the actual power dissipated by the internal circuit plus the power wasted in this parasitic transistor.

The ratio of the actual power to the dissipative function is a measure of ideality of these restoring logic circuits. Power must be dissipated in order to construct a controllable potential function and thus allow logic decisions to be made and information stored. However, it is possible to construct circuits that dissipate much more power than that required to restore the bistable element into one of the two stable states. The smaller the power dissipated over and above the dissipative function itself, the more ideal the logic circuit.

Ideally we would apply the results of nonequilibrium thermodynamics to the flip-flop as a system. However, no general results are known for systems in which more than kT of energy is lost in a single event. This is an area of opportunity for fundamental work.

9.7 THERMAL LIMIT

The energy E_{sw}, representing the difference between a logic-1 and a logic-0, must be very large compared with the thermal energy kT, where $k = 8.62 \times 10^{-5}$ eV/°K is Boltzmann's constant. At room temperature, where $T \approx 293$°K, the thermal energy $kT \approx 0.025$ eV $= 4.0 \times 10^{-21}$ joule. The probability P of a thermal fluctuation of energy E_{sw} within the response time τ of the circuit is given by the Boltzmann relation:

$$P = \exp[-E_{sw}/kT],$$

leading to a failure rate of $(1/\tau)\exp[-E_{sw}/kT]$ failures per second, per device.

It is thus possible to reduce the probability of error due to thermal fluctuations, to any degree required, by increasing the signal energy used to represent the information. This then is the definition of our "indefinite" number of computational steps. It is possible to reduce the ultimate error rate to any desired extent but not eliminate it altogether.

Many authors have concluded that the minimum value of E_{sw} is equal to kT. No workable system can be built with such low stored energies. From the above discussion, it is clear that the minimum switching energy is given by

$$E_{sw} > kT \ln(\text{mean time between errors}/\tau).$$

Even for very large systems computing for very long periods of time, a switching energy of $E_{sw} = 100kT$ provides a very low probability of failure due to thermal fluctuation. There does remain a finite chance that some storage element will switch spontaneously due to thermal noise, but in today's systems and even foreseeable VLSI systems, the probability of such a random switching event due to thermal noise is much less than that of a failure due to electrical noise, cosmic rays, or mundane device failure mechanisms. For example, in systems with poorly designed timing constraints, synchronizer failures occur many orders of magnitude more frequently than thermal failures (this observation is the origin of the *Seitz* Criterion given in Chapter 7).

9.8 QUANTUM LIMITS

The low voltage limit on the operation of amplifying electronic devices discussed earlier is a fundamental one, arising from the quantization of the electronic medium (the charge on the electron) and the energy uncertainty due to thermal fluctuations kT. It is therefore a result of both quantum and thermodynamic principles. While we have derived the limit by a classical circuit theory approach, the quantum nature of the electron appears in the channel resistance's exponential dependence on the gate voltage. This dependence also involves the ratio of the thermal energy kT to the charge q on the electron, due to the thermal distribution of individual electrons in the source of the FET and the fact that each electron must independently surmount a potential barrier in the semiconductor. Hence, we may think of the set of electrons that surmount the potential barrier as a collection of individual events whose statistics are governed by thermal fluctuations and the magnitude of the electronic charge. Information is represented by a quantity that is quantized. Each quantum of that quantity must carry an energy greater than kT. Hence, even in logic in which a very large number of electrons are used to represent a logic signal, the voltage must be larger than kT/q. A similar limit applies to logic in which the quantum of information storage is the magnetic flux unit Φ_0. In this case, the current supplied into the control inductance of any device must be at least kT/Φ_0.

A lower bound on the switching energy of logic devices was found in the previous section to depend on thermal considerations. The lower bound on the *size* of properly operational FETs is determined not by thermal considerations but by the *discreteness of electrical charge* mentioned earlier and by the *wave nature of the electron*.

The wave function of a free electron can be written as $\Psi = e^{i\kappa x}$. The dependence of the wave vector κ on the energy E is given by

$$\kappa = (2mE/\hbar^2)^{1/2}.$$

If the energy E is positive, then κ is real and Ψ represents a traveling wave. If the energy is negative, then κ is imaginary and Ψ is a damped exponential. The simple fluid model of the FET, given in the Section 1.15 (A Fluid Model for Visualizing MOS Transistor Behavior), may help us to visualize how electrons are contained in a transistor. The "outline of the dam" in the figures in that section (in other words the line representing the barrier over which "fluid" must flow) is called the *conduction band edge*. The conduction band edge is the zero of energy for electrons in the semiconductor. Electrons with higher energies can move freely about. Electrons with lower energies cannot. Consider an electron in the source region in "case 1" of Fig. 1.37(b) that is moving toward the drain region. When the electron encounters the barrier, it does not have the energy to surmount it. Therefore, it is reflected and ends up traveling in the opposite direction. However, the reflection is not abrupt. The wave function is exponentially damped to the right of the conduction band edge. If the barrier is thin enough, the damped exponential may have a substantial amplitude on the opposite side. In this way the electron can pass through a classically "forbidden" region. This process is called *tunneling*.

For a transistor to operate properly, the current due to electron tunneling must be smaller than other currents in the circuit. Suppose the barrier (usually the depletion region around the drain) has a thickness Δx. The ratio of the wave function on the two sides of the barrier is $e^{-|\kappa|\Delta x}$. Since the electron density is proportional to the square of the wave function, the density ratio on the two sides of the barrier will be $e^{-2|\kappa|\Delta x}$. In order to keep tunneling from dominating the device behavior, this ratio must be small, which implies

$$2|\kappa|\Delta x \gg 1,$$

and thus

$$\Delta x \gg (0.5)(\hbar^2/2mE_b)^{1/2}.$$

For a value of the barrier energy, E_b, of one eV, we find that Δx is about 1.0×10^{-3} micron. Gate oxides and junction depletion layers must be many times this thickness. In 1978, gate oxide is already less than 0.1 micron thick. We are thus within sight of a fundamental size limitation due to quantum phenomena.

When making the preceding calculation, we must use for the mass m the effective mass of the electron in the particular semiconducting material from which the device is constructed. The preceding calculation assumes m to equal one *free electron mass*, $m_e = 0.91 \times 10^{-27}$ g, which is approximately the effective mass in silicon. However, effective masses in high-mobility materials are much less than one m_e. For this reason it is dangerous to compare performance figures for devices made of different materials. A high-mobility material may give a very short transit time for today's dimensions. However, it is usually not possible to construct as small a device with this material as with a lower mobility, higher effective mass semiconductor. The smallest operational devices of high-mobility materials will thus be larger, and generally no faster, than those made of silicon.

9.9   TWO TECHNOLOGIES—AN EXAMPLE

In this section we will apply physical considerations to the comparison of two very different technologies for constructing computational systems. The technologies selected for this example are based on (1) semiconductor FET devices, and (2) superconducting logic devices. The material presented in this example demonstrates the importance of considering not only device physics and device design but also system physics and system architecture, when making such comparisons.

Several types of limits on the performance of semiconductor FET logic families have been noted in the foregoing discussion: those dealing with the temperature of operation, those arising from quantum phenomena, those associated with the granularity of charge in the semiconductor substrate, and voltage constraints arising from gain considerations. Of these, the limit due to quantum phenomena appears the least restrictive.

Therefore any quantum-limited process operating at very low temperature merits serious study. For this reason, superconducting logic families have attracted much attention. In such logic families, information is stored as a magnetic flux trapped in a superconducting ring and is switched by means of a Josephson[7] (or similar) junction. Devices have been demonstrated that exhibit very fast switching times and low operating power. It is important to understand the relative merits of such a radically different technology from the point of view of overall system design. We should therefore find some way to compare it directly with semiconductor technology over the range of sizes implied by scaling into submicron dimensions.

Over a wide range of technologies, complexity levels, and functionality, one clear metric of system cost has emerged: the maximum power dissipated by a system. In modern technology the basic electronic components account for a very small fraction of the cost of a complete system. The cost of the "system power overhead," including power supplies, power distribution, filtering, decoupling, cabling, connections, lights, switches, packaging, thermal management, and mechanical artifacts, usually accounts for the lion's share of the total system cost. The electronic devices affect the system power overhead through the energy E_{sw} required to perform an elementary logic operation, while system architecture affects it through the communications overhead superimposed on the basic logic functions.

In comparing radically different technologies, it is important to compare total system cost and therefore total system power. Total system power scales as the *switching energy* of the basic technology. Propagation delay can often be traded off against power dissipation over a wide range in any given technology, but their product cannot be reduced below the switching energy. In a charge controlled semiconductor device such as the MOSFET, the irreducible switching energy is $E_{sw} = C_g V^2$, for gate capacitance C_g and total supply voltage V.

In a superconducting device $E_{sw} = LI^2$, where L is the inductance of the superconducting loop plus the associated junction, and I is the supply current. In both technologies, parasitics will increase E_{sw} to several times the values computed for minimum devices. However, for purposes of comparison, we will consider only the minimum devices themselves.

Since all energies in both types of logic are multiples of kT, it might appear that operating a computer at very low temperatures would reduce the total power required. That this is not the case is easily demonstrated. Suppose that to perform a computation a machine dissipates energy $E_L = nkT_L$ as heat at some low temperature T_L. To maintain the low temperature, this heat energy must be transported to and released at room temperature, T_H, by some refrigerator. The total energy to run the system is equal to E_L plus the work required to run the refrigerator. Thermodynamics[8] shows us that a refrigerator operating on the Carnot cycle requires the least amount of work input per unit of heat transported from the low-temperature environment to the high-temperature environment. On input of work W, a Carnot refrigerator can transport, from the T_L to T_H environments, a quantity of heat energy Q given by

$$Q/W = T_L/(T_H - T_L).$$

Thus the work W required to transport E_L from T_L to T_H is, in general,

$$W \geq E_L(T_H - T_L)/T_L.$$

The total energy, E_{tot}, required for the computation is therefore

$$E_{tot} \geq nkT_L + nkT_L[(T_H - T_L)/T_L] = nkT_H.$$

As T_L is lowered, the switching energy is lowered, but the work input to the refrigerator must be increased by at least an equal amount. The total energy cost, including that necessary to run the refrigerator, is thus independent of the temperature of the computer's switches. This energy cost is, at minimum, identically equal to nkT at the temperature of the ultimate heat sink. In some space applications a heat sink at very low temperatures is available. However, for terrestrial computers, refrigerating electronic devices in order to reduce the energy of computation is logically equivalent to constructing a perpetual motion machine. For this reason, we will use kT at the heat sink temperature in system energy calculations, independent of the actual temperature at which switching devices operate.

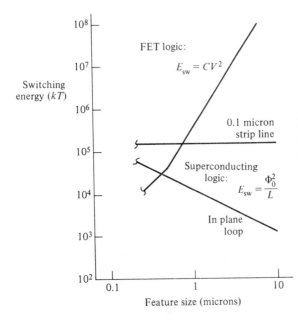

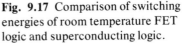

Fig. 9.17 Comparison of switching energies of room temperature FET logic and superconducting logic.

Now we turn to the details of the technology comparison. The switching energy of MOSFET logic is $E_{sw} = C_g V^2$. The most straightforward MOSFET scaling results from reducing all dimensions by the same scaling factor. If this type of scaling is applied to the MOS family, the gate capacitance decreases linearly with the scaling factor. In order to keep the electric fields constant, the supply voltage is scaled by the same scaling factor. The switching energy is thus reduced by the *third power* of the scaling factor, as illustrated in the top curve in Fig. 9.17. The lower size limit shown is a conservative estimate set by device physics factors previously discussed.

Were it possible to build FET devices that operate with one electronic charge on their gate, their performance would not benefit from scaling to smaller dimensions. In such a device, the switching energy can be expressed in terms of q, the charge of the electron:

$$E_{sw} = CV^2 = q^2/C.$$

Since C decreases as the device dimensions are scaled down, the switching energy actually increases. This relationship illustrates a general principle:

A logic device working at its quantum limit requires a higher switching energy as the dimensions of the device are made smaller.

Even at present dimensions, superconducting logic operates at or near its quantum limit. The flux in a superconducting ring must be an integral multiple of the flux quantum $\Phi_0 \simeq 2 \times 10^{-15}$ Webers. The switching energy for a device operating with one flux quantum can be written as

$$E_{sw} = LI^2 = \Phi_0^2/L.$$

Note that the inductance $L = \Phi_0/I$ is directly proportional to the size of the loop. The above dependence for superconducting logic is illustrated in the bottom curve in Fig. 9.17. The lower size limit shown is set by the penetration depth λ of the superconductor.[9] Magnetic field strength decreases with distance, x, into the superconductor as $e^{-x/\lambda}$. If the thickness of the superconducting ring is less than a few λ, the ring cannot localize the flux within it. A typical value of λ is 0.1 micron.

In many superconducting circuits, the inductance is formed by a strip-line parallel to the substrate ground plane and spaced from it by a thin (≈ 0.1 micron) insulating layer. Scaling of such devices without decreasing the insulator thickness results in the middle curve shown in Fig. 9.17.

Comparing the curves, it is clear that, when an accounting is made of the total system energy, and when the effects of scaling to submicron dimensions are taken into account, room temperature FET logic is a remarkable technology. At achieveable submicron dimensions, it can actually outperform its superconducting counterpart. Lower switching energies in the superconductor technology can be achieved only by sacrificing density. This trade-off may be desirable under some circumstances. It seems more likely, however, that maximum computation per unit cost will be achieved by jointly minimizing switching energy and maximizing circuit density.

The absolute speed attainable with the superconducting logic is, however, considerably better than that of its FET counterpart. For a critically damped Josephson junction, the time response τ is either

$$\tau_{LC} \simeq (LC)^{-1/2} \text{ (limited by the resonant circuit)}$$

or

$$\tau_Q = \hbar/E_{sw} \text{ (limited by the quantum transition time).}$$

In these equations, L is the loop inductance used above and C is the junction capacitance. The current carrying capability of the Josephson junction varies exponentially with dielectric thickness. Since the current level of the Josephson circuits increases only gradually as the devices are scaled down, the dielectric thickness can be assumed approximately constant as the devices are scaled. Hence τ_{LC} will scale down as the 3/2 power of the device's linear dimensions. The quantum transition time is set entirely by the switching energy. For the FET, the oxide thickness must be scaled, and the delay time varies linearly with the scaling factor.

At 1 micron feature size, for example, the switching time of a minimum dissipation superconducting device is $\approx 2 \times 10^{-11}$ sec, limited by the quantum transition time. By using higher currents *and* increased switching energy, the switching time can be made as short as $\tau_{LC} \approx 2 \times 10^{-13}$ sec. For an FET with the same feature size, the transit time is $\approx 5 \times 10^{-11}$ sec. The trade-off between power and delay time extends to shorter times for Josephson devices than it does for FETs. Although the minimum switching energy (referred to at room temperature) will be about the same as that of future minimum feature size FET's, one will have the option of switching Josephson devices on the order of one hundred times faster than the fastest FETs, but at the cost of some hundreds of times the power. Extremely fast switching times put even more premium on management of communication paths (in 10^{-13} sec, a signal in a superconducting transmission line travels ≈ 10 microns).

One basic problem with low-temperature logic is that the lower switching energy levels result in poor noise immunity. They therefore require better shielding to reduce the effect of external electromagnetic occurrences to a level well below the switching energy.

Another problem is that the low switching energy creates a mismatch to the outside world for which a penalty in additional power consumption has to be paid, since the drivers to the outside world consume a large amount of power and introduce extra delays. As long as information is not required to exit the low-temperature environment, chip-to-chip communication can be done at high bandwidth. Note that, in this respect, superconducting logic is superior since it is much better matched to the impedance of transmission lines than is FET logic. In any event, exponentially staged drivers are required when driving from the low-energy environment to the outside world, as discussed in Chapter 1. These drivers introduce a minimum delay τ_{dr}:

$$\tau_{dr} \gtrsim \tau e \ln(Y),$$

where Y is the ratio of energy required at the destination to that of the elementary logic device. If the switching energy of a logic element is a factor of 100 smaller due to operation at low temperature, a factor of at least 10 in driver delay is introduced. Furthermore, the dissipation of the last stage of the driver is determined by the energy level necessary *in the outside world*, not in the low-temperature environment. The cost of this driving energy is at least 100 times higher than that for a room-temperature driver of the same capability, due to the constraints imposed by the laws of thermodynamics.

Architects comparing alternative technologies for building computing systems take into account many costs other than just total switching energy. The weights assigned to the various factors usually depend upon their proximity to absolute constraints imposed by physical law or by system performance and cost considerations. In certain situations, we may be perfectly willing to pay the price for large increments in energy, energy conversion equipment, mass, volume, and structural and operational complexity, to achieve an increment of system performance.

Suppose, for example, we now had to specify a very high performance, general purpose computer for the late 80s or early 90s. Since switching speed translates directly into time performance in the classical stored program computer, we might see no other alternative for high performance than a machine based on superconducting devices. Such a decision recognizes that no present alternatives exist for trading off processing speed against concurrency in multiple processors for general purpose computation. That such alternatives must ultimately exist is of course evident from observations of the information processing capability of living organisms.

Superconducting devices meet the requirement for high speed in the classical computer, and a number of machines based on that technology will likely be built before viable high-concurrency alternatives appear. However, in the longer term, in applications where mass, volume, structural complexity, and cost are real constraints, semiconductor devices operated at heat sink temperature will generally have the advantage. Thus, the switching technology likely to dominate the terrestrial environment, used for personal computing and personal communications on a vast scale in an enormous number of different applications, is semiconductor technology. For other applications, the semiconductor technology itself may benefit in a variety of ways from low-temperature operation (Ref. 9 of Ch. 1), as for example in the reduction of subthreshold current in submicron FET's.

9.10 COMPLEXITY OF COMPUTATION

Classical complexity theory counts the number of steps required by a particular algorithm to perform a given computation. That number is called the *complexity* of the algorithm. The number of elementary logic events is closely related to the number of computational steps. We have seen that each elementary logic event required a minimum energy, E_{sw}. Therefore, the minimum energy required to perform a computation is equal to the number of elementary logic events multiplied by the switching energy per logic event.

The complexity of a physical computation is most appropriately measured in energy units. It must include the elementary switching energies mentioned above, plus the energy cost of communicating information throughout the system. Thus, the computation energy provides us a unified measure of the efficiency of the algorithm, the computational structure, and the mapping of one onto the other. Any change that results in a lower energy cost of computation for a given technology implies an improvement in one or more of these three basic ingredients of the design of computing systems.

It is useful to define complexity in such a way that it does not depend on the particular state of the technological evolution. This can be accomplished by normalizing the energy to the switching energy of the elementary logic devices:

$$\text{Complexity of computation} = \frac{\text{Energy required for computation}}{E_{sw}} .$$

This measure of the complexity of a computation is equivalent to the total entropy of the computation as defined in the next section.

9.11 ENTROPIC VIEW OF COMPUTATION

The term *entropy* occurs in two widely separated fields: thermodynamics[8] and information theory.[10,11,12] The underlying ideas are similar in both fields: entropy is used as a measure of the amount of disorder in a system.

In thermodynamics, entropy is defined as proportional to the logarithm of the number of ways of arranging the internal particles of a system while holding the external aggregate properties constant. Suppose a system consists of two containers holding a total of 10 blue molecules and 10 red molecules. There is only one way in which the molecules can be arranged so that all 10 blue ones are in one container, and all 10 red ones are in the other. On the other hand, there are a very large number of ways we can arrange to have 5 of each color in each container. The second arrangement has much more disorder and therefore much more entropy than the first.

In information theory, entropy is defined as the logarithm of the number of possible messages that can be represented by a bit string of given length. A simple example serves to clarify the idea. A string containing n bits can represent at most 2^n different messages. A string containing m bits can represent 2^m different messages. If certain bit combinations were not allowed, fewer messages could be represented. If no constraints are placed on the way the bits are used (if no combinations are excluded, for example), then the entropy is equal to the number of bits in the string. The entropy of the two strings taken together is just $\log(2^{n+m}) = n + m$ bits, exactly the total number of bits in the combined message. Entropy as a measure of information content thus conforms to our intuitive notion that twice as long a message should be able to carry twice as much information.

How can the basic ideas of information theory be extended to the notion of computation? We can view the methods of classical complexity theory in the following way. A computation poses a question about a certain collection of data. The operation of the computational algorithm is similar to a giant game of "twenty questions." Each logical decision cuts down the solution space to some fraction of its former size. The number of decisions is the logarithm of the ratio of the size of the total possible solution space to the size of the resolution element of the final answer.

The relation between this metric of classical computational complexity and the entropy of information theory is thus clear. Both give the number of decisions required to specify one correct answer element in some total space. In the information theory case, the space is that of all messages representable with a given length and format. In the computation case, the space is that of all solutions possible with a given algorithm and quantity of starting data.

We will call the number of logic operations required to perform a given computation on a given assemblage of data the *logical entropy*, S_L, of the data relative to that computation. The purpose of the computation is to produce just one correct outcome, i.e., the computation must reduce the logical entropy to zero. Information theory concludes that with sufficiently complex coding and decoding schemes, information can be communicated reliably with only $\approx kT$ of energy per bit[11]; however, for a simple one-bit code much more energy is required to obtain an equally low error probability. In order that the reliability of the computation be maintained as discussed in Section 9.7, many kT of energy must be used to process each bit of information. If the actual energy used is E_{sw}, then the energy efficiency, relative to the information theoretic limit, is $\approx kT/E_{sw}$.

Each bit of logical entropy requires at least E_{sw} of energy dissipation for its removal:

$$-\Delta S_L/\Delta E_{DL} \leq 1/E_{sw} \text{ bits processed/joule.}$$

Stated another way, the increment of energy dissipated to reduce the logical entropy is greater than or equal to the switching energy multiplied by the actual reduction of the logical entropy in bits processed:

$$\Delta E_{DL} \geq -E_{sw}\Delta S_L \text{ joule.}$$

In any physical computing system, the logical entropy treated by classical complexity theory is only part of the story. There is also a *spatial entropy* associated with a computation. Spatial entropy may be thought of as a measure of data being in the wrong place, just as logical entropy is a measure of data being in the wrong form. Data communications are used to remove spatial entropy, just as logical operations are used to remove logical entropy. Communication is thus as fundamental a part of the computation process as are logical operations.

Consider a spatially distributed structure consisting of information storage sites and communication pathways between the sites. An elementary communication event in an assemblage of data in this structure is the transmission of one bit of information from one storage site to another. By analogy with information theory, an abstract kind of spatial entropy might be defined as the logarithm of the number of possible combinations of communications that can be performed in a given spatial assemblage of data. However, such a definition does not characterize the energy requirements for communications for a real computation in a real physical structure. By analogy with our definition of logical entropy, we define the spatial entropy, S_S, of a given assemblage of data in a given structure relative to a given computation as the number of communication events required for that computation, with each event weighted by its energy dissipation relative to E_{sw}. At least E_{sw} of energy dissipation is necessary for each communication event, since at least one switching event must occur to cause a communication event. The transmission of a bit of information may require some additional dissipation of energy, the amount depending upon the length and physical characteristics of the transmission pathway. Therefore, the spatial entropy, S_S, is just the total required communication energy divided by E_{sw}. Each bit of spatial entropy requires at least E_{sw} of energy dissipation for its removal:

$$-\Delta S_S/\Delta E_{DS} \leq 1/E_{sw} \text{ bits communicated/joule.}$$

In other words, the increment of energy dissipated to reduce the spatial entropy is greater than or equal to the switching energy multiplied by the actual reduction of the spatial entropy in bits communicated:

$$\Delta E_{DS} \geq - E_{sw} \Delta S_{S} \text{ joule.}$$

We now develop a constraint relationship between reductions in logical and spatial entropy during a computation, and necessary associated increases in thermodynamic entropy in the overall system. In order to best do this, we first take a closer look at the concept of thermodynamic entropy, S. The original definition of entropy, and the idea of changes in entropy being a useful measure of changes in the state of a system, emerged during early studies of heat engines, before the underlying nature of entropy as a measure of disorder had been visualized. In these studies, the incremental change in entropy of a system during reversible or quasi-static processes is defined as the ratio of heat energy input into a system divided by the absolute temperature at which this input occurs. The difference in entropy between two states of a system is then calculated by a process of integration using any incrementally reversible or "quasi-static" path between the states. The calculated difference is the same for any such special paths. This state variable was found to have very interesting properties. For example, it could be used to predict whether or not a postulated process could actually occur. All actual processes are irreversible, and it turns out that during an irreversible process between two states of a system a larger increase in thermodynamic entropy occurs than would be predicted by the integration for a reversible or quasi-static path between the states. Thus, no postulated process can occur that would result in a decrease in thermodynamic entropy in an isolated system.

The realization that entropy is actually a measure of disorder or uncertainty emerged during the later work of Boltzmann and Gibbs during the development of statistical mechanics. Unfortunately, thermodynamic entropy, S, has retained its historical units of an energy per degree absolute temperature. These units tend to obscure and make mysterious what otherwise might be a more straightforward, dimensionless concept of disorder, such as those introduced for logical and spatial entropy. For this reason, we define a *dimensionless form of thermodynamic entropy*, $S_{T} = S/k$. By choosing a proportionality constant, k, having the correct value and dimensions, we find that S_{T} actually *equals* the logarithm of the number of ways of arranging the quantum states of the particles of a system while keeping the aggregate properties constant. (S_{T} as defined here is in many treatments of statistical mechanics referred to as the logarithm of the "thermodynamic probability," and written as $S/k = \ln W$, or as $S/k = \ln \Omega^*$). Thus S_{T} is an extensive but dimensionless quantity, having the units of "particles," that measures the relative disorderliness of a system of particles. The correct constant, k, is none other than Boltzmann's constant. It may be thought of as the gas constant per particle, or per atom, and is also the conversion factor from the scale of energy to the scale of temperature. (The gaseous state is a state of maximum entropy. In such a state, adding heat energy just increases temperature. In lower entropy states, more interesting things may happen upon the input or removal of heat energy, sometimes vividly reflecting major changes in the relative disorder of a system of particles, as for example during the melting of a block of ice by adding heat energy at 273 °K.)

The total energy dissipated during a computation necessarily produces an increase in thermodynamic entropy in the overall system:

$$\Delta S \geq (\Delta E_{DL} + \Delta E_{DS})/T,$$

and thus

$$\Delta S_T kT \geq - E_{sw}\Delta S_L - E_{sw}\Delta S_S.$$

This leads us to the "bottom line." In order to reduce logical and spatial entropy in any real physical computational process, energy must be dissipated, causing a corresponding increase in the thermodynamic entropy of the system, as follows:

$$\Delta S_T + (E_{sw}/kT) \, \Delta S_L + (E_{sw}/kT)\Delta S_S \geq 0.$$

The relationship between logical and thermodynamic entropy can be clarified by considering a CMOS inverter, such as that in Fig 9.2. The charge on the output node can be viewed as made up of a number of particles; electrons with charge $-q$ and holes with charge $+q$. The total number of particles is fixed at $N_0 = CV/2q$. Our resources consist of N_0 electrons and N_0 holes. The logarithm of the number of ways to achieve any given voltage on the node is the dimensionless thermodynamic entropy, S_T, of the node at that voltage (relative to an additive constant). When the node contains N_0 electrons, its voltage is $-V/2$. When the node contains N_0 holes, its voltage is $+V/2$. These are both states of zero entropy (relative to an additive constant), since there is only one way we can deploy our resources to achieve each of them. All other states are mixtures of some electrons and some holes and may be achieved by deploying our resources in a number of different ways. When the output node is at zero voltage, halfway between the positive and negative supply rails, it contains $N_0/2$ electrons and $N_0/2$ holes. This is the point of the maximum thermodynamic entropy. It is likewise the point of maximum uncertainty in the logic state of the node, and hence of logical entropy. Energy is the only common coinage in the world of entropy. The amount of energy required to change the thermodynamic entropy by one unit is kT. The amount of energy required to change the logical entropy by one bit is E_{sw}. The term E_{sw}/kT appearing above is just the scale factor between the amounts of energy required to produce unit changes in the two levels of entropy.

We may now view the voltage limit developed in Section 9.2 in a more fundamental way. Consider the flip-flop of Fig. 9.13 as a thermodynamic system. It can interact with the external environment by absorbing electrons and holes from the two rails of the power supply. In the storage nodes of the circuit, excess electrons and holes annihilate each other in such a way that the constraints described above are satisfied. Each rail can be viewed as containing a gas of the appropriate particles, each with an average energy $qV/2$. The environment can also absorb heat from the flip-flop. Under what conditions can this device spontaneously evolve from a "hung" state of high internal entropy to one of lower entropy? Each particle that can surmount the barrier of a transistor and enter the flip-flop decreases the entropy of the environment by one unit. The energy $qV/2$ carried into the flip-flop by each such particle is, on the average, dissipated inside and causes an equal quantity of heat to flow back into the environment. This heat increases the entropy of the environment by $qV/2kT$ units. The thermodynamic entropy change in the overall system, that of the environment plus that of the flip-flop, must be greater than zero:

$$\Delta S_{T(total)} = qV/2kT - 1 + (\Delta S_T \text{ of flip-flop}) > 0.$$

In order that the entropy of the flip-flop decrease, the first two terms must be greater than zero, leading to a result identical to that of Section 9.2:

$$qV/2kT - 1 > 0,$$

and thus,

$$V > 2kT/q.$$

Computation in a macroscopic system can be viewed in the same terms. With respect to a given computation, an assemblage of data in an information processing structure possesses a certain logical and spatial entropy. Under proper conditions, we may be able to remove all of the logical and spatial entropy from a system leaving it in the state of zero logical and spatial entropy; we call this state "the right answer." A sufficient dissipation of energy and outflow of heat must occur during the computation to increase the entropy of the environment by at least an equal amount. When viewed this way, information processing engines are really seen as just "heat engines," and information as just another physical commodity, being churned about by the electronic gears and wheels of our intricate computing machinery.

9.12   CONCLUSION

We opened this book with a discussion of the physical properties of elementary switching devices. We have now closed with a discussion of physical principles that influence the higher level properties of computational systems.

The communication of information over space and time, the storage and logical manipulation of information by change of state at energy storage sites, and the
transport of energy into and heat out of systems, depend not only on abstract
mathematical principles but also on physical laws. The synthesis and functioning
of very large scale systems, whether artifical or natural, proceed under and indeed
are directed by the constraints imposed by the laws of physics.

We look forward to the further development of the physics of computational
systems. We hope others will uncover new insights and examples in this important
area of investigation, for reporting in future editions of this text.

REFERENCES

1. R. M. Swanson and J. D. Meindl, "Ion-Implanted Complementary MOS Transistors in Low Voltage Circuits," *IEEE Journal of Solid-State Circuits*, vol. SC-7, April 1972, pp. 146–153.
2. R. W. Keyes, "Physical Limits in Digital Electronics," *Proc. IEEE*, vol. 63, May 1975, pp. 740–767.
3. C. H. Bennett, "Logical Reversibility of Computation," *IBM Journal of Research and Development*, vol. 17, November 1973, pp. 525–532.
4. T. Toffoli, "Computation and Construction Universality of Reversible Cellular Automata," Technical Report No. 192, The Logic of Computers Group, University of Michigan, June 1976.
5. L. Brillouin, *Science and Information Theory*, New York: Academic Press, 1956.
6. L. D. Landau and E. M. Lifshitz, *Mechanics*, Reading, Massachusetts: Addison-Wesley, 1960.
7. M. Tinkham, *Introduction to Superconductivity*, New York: McGraw-Hill, 1975.
8. F. W. Sears, *Thermodynamics, the Kinetic Theory of Gases, and Statistical Mechanics*, Reading Massachusetts: Addison-Wesley, 1953.
9. E. A. Lynton, *Superconductivity*, London: Science Paperbacks, Chapman and Hall Ltd., 1971.
10. C. E. Shannon, "Communication in the Presence of Noise," *Proc. IRE*, vol. 37, January 1949, pp. 10–21.
11. C. E. Shannon and W. Weaver, *The Mathematical Theory of Communication*, University of Illinois Press, 1949.
12. R. G. Gallager, *Information Theory and Reliable Communication*, New York: Wiley, 1968.

FURTHER SUGGESTED READING

CHAPTER 1

Gibbons, J. F., *Semiconductor Electronics*, New York: McGraw-Hill, 1966. A classic text containing an excellent introduction to basic semiconductor theory, electronic devices, and electronic circuits.

Penney, W. M., and L. Lau, eds., *MOS Integrated Circuits*, New York: Van Nostrand Reinhold, 1972. A good early reference on MOS integrated circuit design.

Keyes, R. W., "Physical Limits in Digital Electronics," *Proceedings of the IEEE*, vol. 63, no. 5, May 1975, pp. 740–767. A comprehensive survey paper on this topic.

Richman, P., *MOS Field-Effect Transistors and Integrated Circuits*, New York: Wiley-Interscience, 1973. Provides a thorough discussion of the physics, fabrication, and function of MOS devices and circuits.

Seitz, C. L., ed., *Proceedings of the Caltech Conference on VLSI*, Caltech, January 1979. Provides an overview of a wide range of topics relevant to VLSI systems and technology.

CHAPTER 2

Grove, A. S., *Physics and Technology of Semiconductor Devices*, New York: Wiley, 1967. The comprehensive text on process technology and device physics.

Oldham, W. G., "The Fabrication of Microelectronic Circuits," *Scientific American*, September, 1977. Provides an excellent overview of the fabrication process. (This issue of *Scientific American* is devoted to all aspects of microelectronics).

Muller, R. S., and T. I. Kamins, *Device Electronics for Integrated Circuits*, New

York: Wiley, 1977. Provides insight into the device physics relevant to integrated circuit technology.

Mead, C. A., "Shottky Barrier Gate Field Effect Transistor," *Proc. IEEE*, vol. 54, February 1966, pp. 307–308. This paper describes the first operating MESFET.

Cobbold, R. S. C., *Theory and Applications of Field-Effect Transistors*, New York: Wiley-Interscience, 1970. A comprehensive reference on all types of field-effect transistors.

Seitz, C. L., "Self-timed VLSI Systems," *Proceedings of the Caltech Conference on VLSI*, January 1979. Describes the effects of scaling on system timing, and contrasts the scaling of bipolar and MOS technologies.

d'Heurle, F. M., and P. S. Ho, "Electromigration in Thin Films," Chapter 8 in *Thin Films—Interdiffusion and Reactions*, J. M. Poate; K. N. Tu; J. W. Mayer; eds., New York: Wiley-Interscience, 1978. A thorough discussion of the metal migration phenomenon with a complete set of references to the original work.

CHAPTER 3

Bell, C. G., and A. Newell, *Computer Structures: Readings and Examples*, New York: McGraw-Hill, 1971. Contains an excellent discussion of the levels in the hierarchy of computer architecture, and many specific examples of computer structures.

Soucek, B., *Microprocessors and Microcomputers*, New York: Wiley, 1976. An introductory reference containing sections on basic digital design, and on the interfacing and programming of a number of present day microprocessors.

Dietmeyer, D. L., *Logic Design of Digital Systems*, Boston, Mass.: Allyn and Bacon, 1971. A comprehensive text on switching theory and logic design.

Kohavi, Z., *Switching and Finite Automata Theory*, New York: McGraw-Hill, 1970. Another good text on switching theory.

Caldwell, S. H., *Switching Circuits and Logical Design*, New York: Wiley, 1958. An early text containing interesting material on relay contact networks.

Unger, S. H., "A Computer Oriented Towards Spatial Problems," *Proc. of IRE*, vol. 46, no. 10, October 1958, pp. 1744–1750. An early paper describing a spatially distributed processor, anticipating present strategies for commingling processing and memory.

Shannon, C. E., "Symbolic Analysis of Relay and Switching Circuits," *Trans. of AIEE*, vol. 57, 1938, pp. 713–723. The classic paper first describing a method for the mathematical treatment of switching circuits.

Boole, G., *An Investigation of the Laws of Thought*, London, 1854. Reprinted by Dover Publications, 1953. Contains a presentation of the algebra of logic on which Shannon based his switching algebra.

CHAPTER 4

Newman, W. M., and R. F. Sproull, *Principles of Interactive Graphics*, 2d ed., New York: McGraw-Hill, 1979. The authoritative reference on all aspects of computer graphics.

Donovan, J. J., *Systems Programming*, New York: McGraw-Hill, 1972. An introductory text that provides practical information on the implementation and use of assemblers, macroprocessors, compilers, loaders, operating systems, etc.

Freeman, P., *Software Systems Principles*, Science Research Associates, Inc., 1975. Presents the basics of a broad range of topics important in the building of software systems.

Gries, D., *Compiler Construction for Digital Computers*, New York: Wiley, 1971. Provides an excellent introduction to techniques for the construction of compilers and interpreters.

Aho, A. V., and J. D. Ullman, *Principles of Compiler Design*, Reading, Mass.: Addison-Wesley, 1977. Another excellent text on compiler construction.

Johannsen, D. L., "A Design Technique for VLSI Chips," *Proceedings of the Caltech Conference on VLSI*, January 1979. Describes the first "silicon compiler," which "compiles" a chip layout from a "high-level" description of its floor plan and instruction set.

Lyman, J., "Lithography chases the incredible shrinking line," *Electronics*, vol. 52, April 12, 1979, pp. 105–116. Provides a good overview of the rapidly evolving art of fine-line lithography.

Brooks, F. P., Jr., *The Mythical Man-Month, Essays on Software Engineering*, Reading, Mass.: Addison-Wesley, 1975. An interesting set of essays on experiences in the engineering of large software systems. Before mobilizing a large group of people to construct a complex system, one would be well advised to read this fascinating book.

CHAPTER 5

Bell, C. G.; J. C. Mudge; and J. E. McNamara, *Computer Engineering*, Bedford, Mass.: Digital Press, Digital Equipment Corp., 1978. An excellent treatment of real-world computer engineering and the evolution of commercial computer systems, providing valuable perspective for the practicing computer architect.

Luecke, G.; J. P. Mize; and W. N. Carr, *Semiconductor Memory Design and Application*, New York: McGraw-Hill, 1973. The design of memory chips is an important topic in its own right; this entire reference is devoted to that subject. It describes many cells and subsystems commonly used in commercial memory parts. Many of these fit naturally into the design methodology discussed here and can be used for on-chip memory subsystems.

CHAPTER 6

Kleene, S. C., *Introduction to Metamathematics*, Groningen, Netherlands: Wolters-Noordhoff Publishing, 1952; Amsterdam, Netherlands: North-Holland Publishing Company, 1952. This classic text provides a comprehensive treatment of the foundations of mathematics and computation: mathematical logic, recursive function theory, computability and decidability, metamathematics and proof theory.

Minsky, M. L., *Computation: Finite & Infinite Machines*, Englewood Cliffs, N.J.: Prentice-Hall, 1967. An introduction to computability; contains several examples of exceedingly simple bases for universal computation.

Steele, G. L., Jr., and G. J. Sussman, "Design of LISP-Based Processors or, SCHEME: A Dielectric LISP or, Finite Memories Considered Harmful or, LAMBDA: The Ultimate Opcode," Memo No. 514, M.I.T. Artificial Intelligence Laboratory, March 1979. This paper discusses some novel, open areas for architectural exploration still available using von Neumann style machines. It also describes the prototype LISP microprocessor that Steele designed as his project for the 1978 M.I.T. VLSI system design course.

Winston, P.H., *Artificial Intelligence*, Reading, Mass.: Addison-Wesley, 1977. An introduction to AI; describes a rich world of interesting programs one can build using classical machines given an appropriate set of heuristics and a powerfully expressive language. The AI field may provide clues to exciting applications as ultra-concurrent VLSI systems provide real computing power.

Weizenbaum, J., *Computer Power and Human Reason*, San Francisco: W. H. Freeman, 1976. Indicates certain limits to our understanding of human information processing, and cautions the reader to not take too literally the word *Intelligence* in Artificial Intelligence.

CHAPTER 7

Koestler, A., *The Ghost in the Machine*, Chicago: Henry Regnery Co., 1967. A broad view of hierarchies.

Simon, H. J., "The Architecture of Complexity," *Proc. of the American Philosophical Society*, vol. 106, no. 6, December 1962. Another advocacy of hierarchies.

CHAPTER 8

Rem, M., "Associons and the Closure Statement," MC Tract 76, Mathematical Centre, Amsterdam, 1976. An elegant notation, with formally defined semantics, for global concurrent systems.

Backus, J., "Can Programming be Liberated from the Von Neumann Style? A Functional Style and Its Algebra of Programs." *Communications of the ACM*, vol. 21, no. 8, August 1978, pp. 613–641. An insightful view of the limitations of von Neumann style machines and traditional computer languages.

Berkling, K., "A Computing Machine Based on Tree Structures," *IEEE Transactions on Computers*, vol. C-20, no. 4, April 1971. Describes the mapping of Algol- or LISP-like programming notations onto a binary tree for execution.

Seitz, C. L., ed., *Proceedings of the Caltech Conference on VLSI*, Caltech, January 1979. Contains papers covering a wide range of topics relevant to ultra-concurrent VLSI systems.

CHAPTER 9

Feynman, R. P.; R. B. Leighton; and M. Sands, *The Feynman Lectures on Physics*, vols. 1–3, Reading, Mass.: Addison-Wesley, 1963–65. Highly readable and understandable, these are an excellent series of tutorial textbooks.

Leighton, R. B., *Principles of Modern Physics*, New York: McGraw-Hill, 1959. An excellent reference text.

Jaynes, E. T., "Information Theory and Statistical Mechanics," *Physical Review*, vol. 106, May 1957, pp. 620–630. This and the following paper comprise the first treatment to successfully unify the two disciplines. The result is a great simplification and unification of the underlying principles of thermodynamics.

Jaynes, E. T., "Information Theory and Statistical Mechanics II," *Physical Review*, vol. 108, October 1957, pp. 171–190.

Penfield, P. L., Jr.; R. Spence; and S. Duinker, *Tellegen's Theorem & Electrical Networks*, Cambridge, Mass.: M.I.T. Press, 1970.

Landauer, R., "Irreversibility and Heat Generation in the Computing Process," *IBM Journal of Research and Development*, vol. 5, July 1961, pp. 183–191.

Landauer, R., and J. A. Swanson, "Frequency Factors in the Thermally Activated Process," *Physical Review*, vol. 121, March 1961, pp. 1668–1674.

Swanson, J. A., "Physical versus Logical Coupling in Memory Systems," *IBM Journal of Research and Development*, vol. 4, July 1960, pp. 305–310.

Keyes, R. W., and R. Landauer, "Minimum Energy Dissipation in Logic," *IBM Journal of Research and Development*, vol. 14, March 1970, pp. 152–157.

Keyes, R. W., "Physical Problems and Limits in Computer Logic," *IEEE Spectrum*, vol. 6, May 1969, pp. 36–45.

Landauer, R., "Fundamental Limitations in the Computational Process," *Ber. Bun. Gessell. Phys. Chem.*, vol. 80, no. 11, 1976, pp. 1048–1059.

Landauer, R., and J. W. F. Woo, "Minimal Energy Dissipation and Maximal Error for the Computational Process," *Jour. Appl. Phys.*, vol. 42, May 1971, pp. 2301–2308.

Millar, W., "Some General Theorems for Non-Linear Systems Possessing Resistance," *Philosophical Magazine*, vol. 42, October 1951, pp. 1150–1160.

Landauer, R., "Wanted: A Physically Possible Theory of Physics," *IEEE Spectrum*, vol. 4, September 1967, pp. 105–109.

Anacker, W., "Computing at 4 degrees Kelvin," *IEEE Spectrum*, May 1979, pp. 26–37. A detailed overview of a proposed system to be constructed out of superconducting logic. The reader will find it instructive to compare this technology with the scaled FET technology described in a number of papers in the Friedrich, Kosonocky, Sugano entry below.

Friedrich, H.; W. F. Kosonocky; and T. Sugano, eds. Joint Special Issue on Very Large-Scale Integration, *IEEE Transactions of Solid State Circuits*, vol. SC-14, no. 2, April 1979; also published in *IEEE Transactions on Electron Devices*, vol. ED-26, no. 4, April 1979.

APPROXIMATE VALUES
OF PHYSICAL CONSTANTS

The following values are given in the "rationalized mks" system of units, in which
meter, kilogram, and *second* are the basic units for mechanics, and are
supplemented with the *coulomb,* a unit of charge. The units of charge, current,
voltage, power, energy, resistivity, capacitance, and inductance are the same as
those in common use. The ampere is a coulomb per second. The energy required
to lift one coulomb of charge through a potential difference of one volt is one joule.
A joule per second is a watt. An ohm is defined as the resistance required to
develop a volt across it when an ampere flows through it. A farad is the capacitance that can store one coulomb at one volt. One henry is the inductance across
which one volt is developed when the rate of change of current through it is one
ampere per second.

The term *rationalized* is applied to this system of units because the choice of
value for the permeability of vacuum, μ_0, as $4\pi \times 10^{-7}$ allows Maxwell's
equations to be written without the appearance of factors of 4π. The velocity of
light, which can be measured, then determines the value of ϵ_0, the only other
physical constant that must appear in Maxwell's equations.

Certain *energies* below are expressed in terms of the electron-volt, or eV, an
informal unit equal to the energy required to lift an electron through a potential
difference of one volt.

Charge on the electron	$q \simeq 1.6 \times 10^{-19}$ coulomb
Electron volt	$eV \simeq 1.6 \times 10^{-19}$ joule
Planck's constant	$h \simeq 6.6 \times 10^{-34}$ joule-second/cycle
	$\simeq 4.1 \times 10^{-15}$ eV-second/cycle
h-bar	$\hbar = h/2\pi \simeq 1.05 \times 10^{-34}$ joule-second/radian
	$\simeq 6.6 \times 10^{-16}$ eV-second/radian

1 cycle/second	$= 2\pi$ radians/second $= 1$ hertz
Flux quantum	$\Phi_0 = h/2q \simeq 2.1 \times 10^{-15}$ volt-second [or weber]
Boltzmann's constant	$k \simeq 1.4 \times 10^{-23}$ joule/°K
	$\simeq 8.6 \times 10^{-5}$ eV/°K
Permeability of vacuum	$\mu_0 = 4\pi \times 10^{-7}$ henry/meter
Permittivity of vacuum	$\epsilon_0 \simeq 8.85 \times 10^{-12} \simeq 10^{-9}/36\pi$ farad/meter
Velocity of light in vacuum	$c = (\epsilon_0\mu_0)^{-1/2} \simeq 3.0 \times 10^8$ meters/second
Wavelength of visible light in vacuum	0.4 to 0.7 micron (4000 to 7000 Å)

The following are approximate values for properties of materials used in integrated system fabrication:

Dielectric constant, ϵ/ϵ_0, for silicon $\simeq$ 11.7; for $SiO_2 \simeq$ 3.9 .

Volume resistivity, ρ, for aluminum $\simeq 2.6 \times 10^{-8}$ ohm-meter (the resistance per unit length of a conductor of cross section A is ρ/A).

Coefficient of thermal expansion, for Si $\simeq 2.5 \times 10^{-6}$ /°C; for $SiO_2 \simeq 0.5 \times 10^{-6}$ /°C.

Silicon's atomic number is 14, atomic weight is 28, melting point is 1415°C, and specific gravity is 2.3.

An excellent treatment of the physical properties of common semiconductors is given in A. S. Grove, *Physics and Technology of Semiconductor Devices*, New York: Wiley, 1967. Standard references on physical constants and the properties of materials are *American Institute of Physics Handbook*, McGraw-Hill, and *Handbook of Chemistry and Physics*, Chemical Rubber Publishing Company.

INDEX

References to color plates appear in boldface type.